CW01214220

Herefordshire Maps

Herefordshire Maps
1577 to 1800

by

Brian S. Smith

Logaston Press

LOGASTON PRESS
Little Logaston Woonton Almeley
Herefordshire HR3 6QH

First published by Logaston Press 2004
Copyright © Brian S. Smith

All rights reserved. No part of this publication
may be reproduced, stored in a retrieval system,
or transmitted, in any form or by any means,
electronic, mechanical, photocopying, recording
or otherwise, without the prior permission,
in writing of the publisher

ISBN 1 904396 24 0 (paperback)
1 904396 25 9 (hardback)

Set in Times New Roman by Logaston Press
and printed in Great Britain by
Cromwell Press, Trowbridge

Front cover: Adapted from John Owen and Emanuel Bowen's
Britannia Depicta or Ogilby Improv'd of 1720, plate 143.

For Alison

Contents

Preface and acknowledgements	ix
Plates	xi
Bibliography and abbreviations	xiii
Measures and scales	xv
Parishes (map)	xvi

Part One

1	Introduction	3
2	The landscape of Herefordshire	5
3	The printed county maps	17
4	Estate maps	31
5	The surveyors	41
	References (Part One)	55

Part Two

Carto-bibliography of the printed maps of Herefordshire before 1800	61
Catalogue of the manuscript maps of Herefordshire before 1800	91
Herefordshire map-makers before 1800	185

Index   199

Published with the generous support of grants

from the Marc Fitch Fund and from

the Geoffrey Walter Smith Fund of

the Woolhope Naturalists' Field Club.

Preface and Acknowledgements

Globes, atlases and jigsaws of the world, Europe and the British Isles, produced experimentally in the 1930s by George Philip & Son, the London map publishers, filled my grandfather's house. My own first tentative attempts to draw maps, one of the Isle of Wight and the other of Anglesey, look very similar in my childish hand. Later, I learnt to use maps and as an archivist I had custody of printed and manuscript maps like those described here, opening opportunities for research on individual historical maps and their map-makers.

This more ambitious project sets out to describe the surviving maps of Herefordshire from the earliest in 1577 to the end of the eighteenth century, a watershed in surveying and draughtsmanship in Britain. The initial research was presented in a paper to the Woolhope Naturalists' Field Club, the county's society for the study of archaeology, local and natural history. It soon became clear to me that the subject deserved fuller exploration and it has taken a further three years to extend the research. Even now, the task cannot be considered complete, for manuscript maps not seen by me will inevitably continue to come to light in private archives and among uncatalogued collections in public repositories. Moreover, the large quantity of nineteenth-century maps, both printed and manuscript, demand similar examination.

This study can, therefore, claim to be no more than a start. It is, however, not only the first book about Herefordshire maps, it is also the first to bring together a listing of both the printed and manuscript maps of an English county. It is offered, in appreciation of a lifetime spent among manuscripts and maps, to historians as a tool, to collectors of printed maps as a guide and to Herefordshire readers in general as a new source of information about their county.

Acknowledgements

The underlying debt that I owe to four great bibliographers, Thomas Chubb, R.A. Skelton and Donald Hodson for printed maps and Susan Bendall for manuscript maps, cannot be exaggerated. As John Speed so disarmingly excused himself, 'I have put my sickle into other mens corne'. My appreciation of their bibliographies is expressed more specifically later and may be measured by the frequent citation of their works. My research would have been impossible without the advice and assistance of many archivists, map custodians, librarians and, not least, their production staff, in the repositories named in the references and footnotes. They have steered me through their catalogues and laboured to present me with unwieldy maps for examination. I cannot name them all but some specific acknowledgements are included in the footnotes. Inevitably, the heaviest burden has fallen upon Sue Hubbard and Elizabeth Semper O'Keefe (Herefordshire Record Office), Robin Hill (Hereford City Library) and Joan Williams and Rosalind Caird (Hereford Cathedral Library).

In addition, the librarians of the Society of Antiquaries and Hay-on-Wye (Powys) have supplied me with reference books. Peter Barber, Keeper of Maps at the British Library and Sarah Bendall, Emmanuel College, Cambridge, have been unstinting in their encouragement and advice. I am particularly grateful to Peter Barber for reading the draft typescript of the introductory chapters and to Sarah Miller of Ross Books and Sally Forwood of Forwood's, Hay-on-Wye, map and print-sellers, for reading the carto-bibliography of the printed maps.

I have been particularly fortunate in my publisher. Andy Johnson and Ron Shoesmith at Logaston Press have been unreservedly enthusiastic, accessible and supportive, as well as quite remarkably tolerant and patient throughout the whole course of publication.

Privately owned estate maps

I wish especially to acknowledge the generosity and interest of the many owners who have allowed me to examine their estate maps, either by making them available to the public in libraries and record offices or to me in the privacy of their home. The inclusion of entries here does not imply that the maps are therefore generally accessible for research. Although there is normally no restriction on access to maps in public repositories, because of their size special arrangements may be needed for their consultation. It is, therefore, always advisable to make an appointment before visiting either a public repository or private archive.

Private archives containing Herefordshire estate maps in the care of an archivist include those belonging to the Dean and Canons of Windsor, the Dean and Chapter of Hereford, Oxford colleges, the Marquess of Bath and the Marquess of Salisbury. An appointment within the limited opening hours of private archives is essential and a fee may be charged towards the costs of providing access. The catalogue entries indicate to whom enquiries should be addressed.

The location of other privately owned maps that I have examined is given here as 'Private'. All enquiries about these should be addressed in the first place to the County Archivist, Herefordshire Record Office, The Old Barracks, Harold Street, Hereford HR1 2QX, who will know the current wishes of their owners regarding access. Entries for the maps of the Stoke Edith estate, which are not accessible for research, have not been included at the owner's wish.

Finally, having revisited some maps two or three times and noticed additional points on each occasion, I am only too aware of the omissions, misunderstandings and imperfections that may remain. I could have happily spent hours more in re-examining and elaborating the descriptions of individual maps or in pursuing every byway of evidence. And this pleasurable work would never have been concluded. I echo the plea of John Smith of Nibley, the steward of the Berkeley estates and author of *Lives of the Berkeleys* in 1618:

> I end, with this request to you … that if any of you shall observe any slip of pen, number, marginall, or other small mistake, … that yee would reforme them favourably … No man can depart from the condition of erring … Figures or notes in my pocket paper-books cursarily taken in my searches, may have been mistaken, both in paper and parchment; And to have reviewed my paynes in both would have exceeded the paynes I reviewed.

Brian Smith
Vowchurch
August 2004

Plates

Between pages 30-31
1	Christopher Saxton 1577, proof copy. British Library, Royal MS.18.d.iii
2	Christopher Saxton 1577. British Library, MAPS.118.e.1
3	Alltyrynys, Walterstone, c. 1600. The Marquess of Salisbury
4	Joannes Jansson 1646. HRO, K38/D4
5	John Ogilby 1675. HRO, K38/D6
6	John Seller 1694. British Library, MAPS. C.24.a.9
7	Emmanuel Bowen c. 1762-64. British Library, MAPS. C.26.c.29
8	John Cary 1787. HRO, K38/D14
9	E. Noble 1789. HRO, K38/D16

Between Pages 62-63
10	John Speed 1610. HRO, K38/D2
11	John Speed, draft for Hereford city 1606. Merton College, Oxford, D3/30, no. 7
12	Thomas Jenner's 1643 issue of Jacob van Langeren 1635. British Library, Maps.C.7.a.31
13	Joannes Blaeu 1645. DCA, B5/300
14	Unidentified French map, c. 1700. Hereford City Library
15	Richard Blome 1673. HRO, K38/D5
16	Robert Morden 1695. HRO, K38/D8
17	Thomas Jefferys and Thomas Kitchin 1748. British Library, MAPS. C.24.aa.6
18	Alexander Hogg 1784. HRO, K38/D9

Between Pages 78-79
19	Isaac Taylor, Herefordshire 1754: cartouche. HRO, C99/III/232
20	Isaac Taylor, Herefordshire 1754: conventional signs. HRO, C99/III/232
21	Isaac Taylor, Herefordshire 1754: Central Herefordshire. HRO, C99/III/232
22	Isaac Taylor, Herefordshire 1754: Colwall area. HRO, C99/III/232
23	Isaac Taylor, Herefordshire 1754: Kington area. HRO, C99/III/232
24	Isaac Taylor, Herefordshire 1754: Ross area. HRO, C99/III/232
25	Isaac Taylor, Herefordshire 1754: Leominster area. HRO, C99/III/232
26	Isaac Taylor's house, Ross. B.S. Smith
27	Isaac Taylor, Gloucestershire 1777, with his pair of dividers. B.S. Smith
28	Isaac Taylor, Hereford city 1757. HRO, AP25/2
29	Isaac Taylor, Hereford city 1757: cartouche. HRO, AP25/2

Between Pages 110-111
30	Mocktree and Bringewood, Downton 1662. HRO, T87/1
31	Lower Lye, Aymestrey 1708. British Library, Egerton MS.2873.C
32	Ewyas Harold 1718. British Library, Add. MS. 60746
33	The Hazle, Ledbury 1720. HRO, J95/1
34	The Homme, Weobley 1733. HRO, W89/1

35	Richards Castle 1743. HRO, F76/III/22	
36	Stretton Sugwas 1757. HRO, C99/III/216	
37	Mocktree and Bringewood, Downton 1662: cartouche. HRO, T87/1	
38	Eardisland 1718: signature. Private	
39	The Homme, Weobley 1733: scale bar. HRO, W89/1	
40	Longtown castle 1718. British Library, Add. MS. 60746.	
41	Moccas 1772. Woolhope Naturalists' Field Club	
42	Bredwardine 1772. Woolhope Naturalists' Field Club	

Between Pages 126-127

43	Welsh Newton 1686. HRO, C82/1
44	Garway, c. 1700. HRO, AL40/7687
45	Goodrich castle 1717. HRO, AW87
46	'Haversham' 1731. B.S. Smith
47	Grendon Warren 1732. HRO, AA59/8245
48	Bridstow 1755. HRO, C59/6
49	Aconbury 1757. HRO, C99/III/217
50	Great Doward, Whitchurch 1758. HRO, AS40
51	Much Birch 1768. HRO, J78/1

Between Pages 158-159

52	Downton 1780. HRO, BL35
53	Bringewood Furnace, Downton 1780. HRO, BL35
54	Leintwardine 1780. HRO, BL35
55	Pipe and Lyde 1784: cartouche. DCA, 4696/27
56	Pipe and Lyde 1784: drawing instruments. DCA, 4696/27
57	Pipe and Lyde 1784: reference table. DCA, 4696/27
58	Pipe and Lyde 1784: north and east points. DCA, 4696/27
59	Bridstow 1788. HRO, A97/1
60	Ledbury 1788. HRO, G2/III/55
61	Bosbury 1791. HRO, AS80/5
62	Longtown, late 18th century. HRO, R57/4
63	Stretton Sugwas 1794. HRO, C99/III/218
64	Longtown 1794. HRO, N44/39

Permission to reproduce their maps has been generously given by the Duchy of Cornwall, the Marquess of Salisbury, the Lord Biddulph, Mr. R. Marshall, Mr. J. Scudamore, Mr. W.D. Turton, Farrer & Co, Haynes & Storr, the Church Commissioners, the Dean and Chapter of Hereford Cathedral, the Governing Body of Merton College, Oxford, Hereford City Library, Herefordshire Record Office, Woolhope Naturalists' Field Club. Plates 1, 2, 6, 7, 12, 17, 31, 32, 40 have been reproduced by permission of the British Library.

Geoff G. Gwatkin of Ross kindly permitted me to base the map of the Parishes and Townships of Herefordshire 1840 (p. xvi) on his published research. The photographs of maps in the Dean and Chapter's archives (DCA) were taken by Gordon Taylor and of maps in the Herefordshire Record Office (HRO) and Woolhope Naturalists' Field Club collections either by Derek Foxton or myself, aided in the Record Office by Mr. S. Rexworthy.

Bibliography and Abbreviations

The titles of publications and repositories listed below are those that have been used constantly in the preparation of this book. The many other sources used for specific matters are noticed in the end-notes to the introductory chapters and the preliminary remarks to the Carto-bibliography of Printed Maps and the Catalogue of Manuscript Maps.

Bendall	S. Bendall, *Dictionary of land surveyors and local map-makers of Great Britain and Ireland 1530-1850*, 2 vols., London 1997.
BL	British Library, London.
Chubb	T. Chubb, *The printed maps in the atlases of Great Britain and Ireland. A bibliography 1579-1870*, 1927, reprinted Folkestone 1970.
DCA	Dean and Chapter Archives, Hereford Cathedral.
Delano-Smith & Kain	C. Delano-Smith and R.J.P. Kain, *English maps: a history,* London 1999.
DNB	*Dictionary of national bioography*, 27 vols., 1882-1901, with later supplements. Its enlarged successor, the *Oxford dictionary of national biography,* 60 vols. and on-line access, is due to be published in September 2004.
HCL	Hereford City Library.
Hodgkiss	A.G. Hodgkiss, *Discovering antique maps*, 5th. edn., Princes Risborough 1996.
Hodson	D. Hodson, *County atlases of the British Isles* vol. 1, *Atlases published 1704 to 1742*; vol. 2 *Atlases published 1743-1763* (Tewin, Herts. 1984, 1989); vol. 3 *Atlases published 1764-1789*, London 1997.
HRO	Herefordshire Record Office.
Moreland &Bannister	C. Moreland and D. Bannister, *Antique maps,* 3rd. edn. London 1989, paperback edn. 1993.
NLW	National Library of Wales, Aberystwyth.
PRO	The National Archives (Public Record Office), Kew.
RO	Record Office.
Robinson	C.J. Robinson, *A history of the mansions and manors of Herefordshire*, London 1872, reprinted Almeley 2001.
Skelton	R.A. Skelton, *County atlases of the British Isles 1579-1850. A bibliography, 1579-1703*, London 1970.
Trans. WNFC	*Transactions of the Woolhope Naturalists' Field Club.*
Whitehead 2001	D. Whitehead, *A survey of historic parks and gardens in Herefordshire*, Hereford and Worcester Gardens Trust 2001.

Note. In Part One references to maps listed in the Carto-bibliography of Printed Maps and the Catalogue of Manuscript Maps are not given in the end-notes, as these are readily identifiable by their date, place or authorship.

Measures and Scales

Statutory English measures

Length
12 inches	=	1 foot
3 feet	=	1 yard
5½ yards	=	1 perch
4 perches	=	1 chain (made of 100 metal links, each of 7.92 inches)
80 chains	=	1 mile (of 1,760 yards).

Area
1 square yard	=	the area equivalent to a square, each side of which measures one yard
30¼ square yards	=	1 perch
40 perches	=	1 rood (1,210 square yards)
4 roods	=	1 acre (4,840 square yards).

Conversion to metric measurements
1 yard	=	0.914 metres
1 chain	=	20.12 metres
1 mile	=	1.609 kilometres
1 acre	=	0.405 hectares.

Herefordshire customary or computed measures
6 computed miles	=	8 statutory miles
3 computed acres of arable	=	2 statutory acres
3 computed acres of woodland	=	5 statutory acres.

Scales
1 inch to 1 chain	=	1:792
1 inch to 2 chains	=	1:1,584
1 inch to 3 chains	=	1: 2,376
1 inch to 4 chains	=	1:3,168
1 inch to 5 chains	=	1:3,960
1 inch to 6 chains	=	1:4,752
1 inch to 8 chains	=	1:6,336.
1 inch to 1 mile	=	1:63,360
6 inches to 1 mile	=	1:10,560
20 inches to 1 mile	=	1:3,168
25 inches to 1 mile	=	1:2,534
26.6 inches to 1 mile	=	1:2,376.

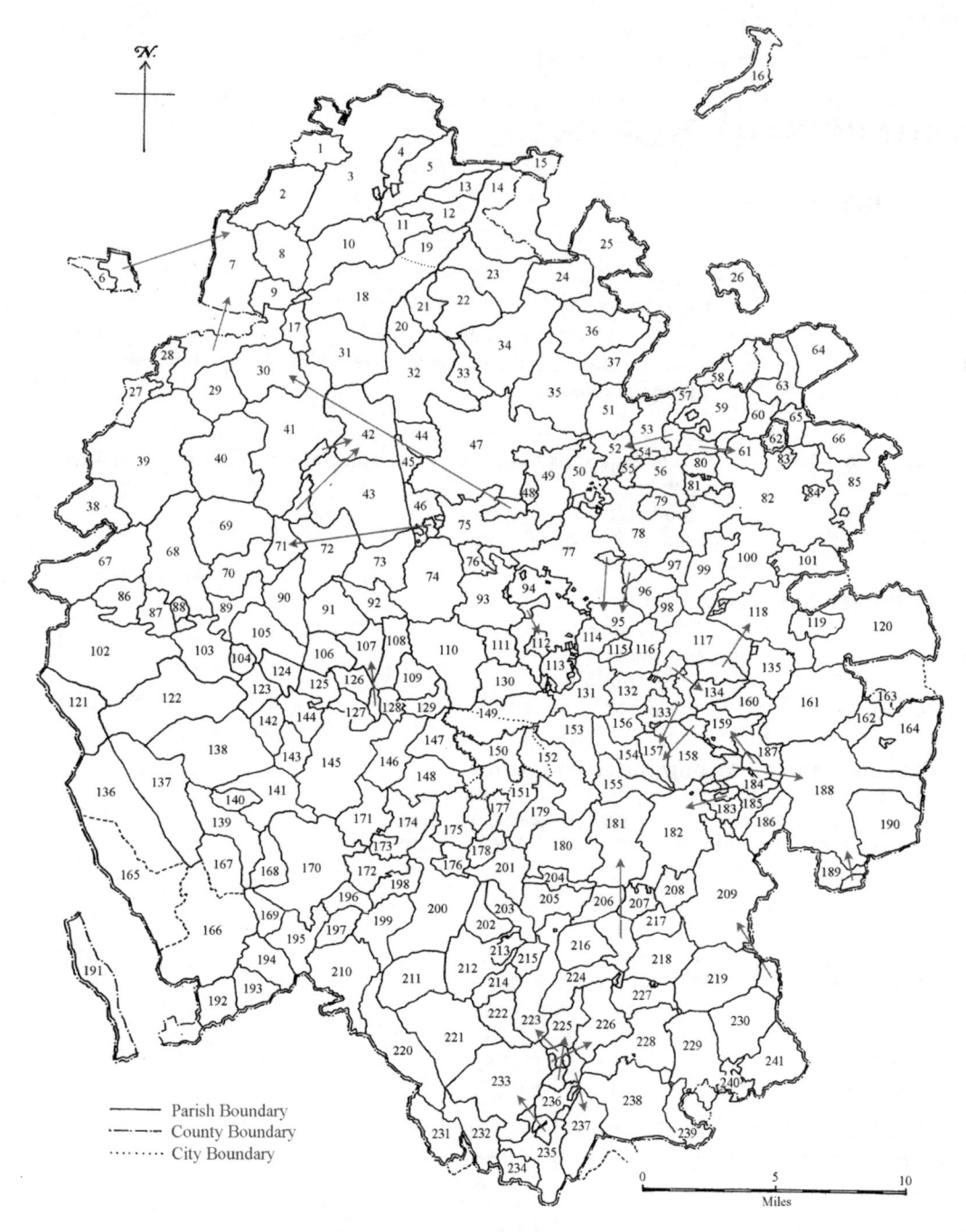

Herefordshire Parishes 1840

xvi

PARISHES AND TOWNSHIPS 1840

Based, with his permission, upon G. Gwatkin, *Gazetteer and administrative maps of Herefordshire*, privately printed, Ross, 2002. For a listing of the dates of the many changes in status and boundaries of Herefordshire parishes, see F.A. Youngs, *Guide to the local administrative units of England, vol.2: Northern England*, Royal Historical Society, Guides and handbooks, No. 17, 1991, pp. 121-137. Ecclesiastical and civil parishes are not necessarily coterminous or spelt in the same way. Names in italics indicate the status in 1840 of civil parishes transferred to Herefordshire or created after that date. The links between detached parts of some parishes have not all been shown. Place-name spellings are those of village names on current Ordnance Survey maps.

Abbey Dore	170	Buckton, township of Bucknell		Eastnor	190
Aconbury	201	Bullingham, with Grafton, chapelry		Eaton Bishop	146
Acton Beauchamp, formerly Worcs.		in Bullinghope		Edvin Loach	62
Adforton, township in Leintwardine		Bullinghope, Lower and Upper	177	Edwyn Ralph	61
Allensmore	174	Burghill	110	*Egleton, township in Bishop's Frome*	
Almeley	69	Burrington	5	Elton	12
Amberley, chapelry in Marden		Byford	125	Evesbatch	119
Ashperton	159	Byton	17	Ewyas Harold	195
Aston	13			Eye	34
Aston Ingham	241	**Callow**	178	Eyton	33
Avenbury	100	Caple, How	217		
Aylton	185	Caple, King's	216	**Felton**	95
Aymestrey	18	Cascob (part)	6	Ford (extra-parochial)	48
		Clehonger	148	Fownhope	181
Bacton	168	Clifford	102	Foy	218
Ballingham	206	Clodock	166	Frome, Bishop's	118
Bartestree, chapelry in Dormington		Coddington	162	Frome, Canon	160
Bicknor, Welsh, formerly Monmouthshire.		Collington	60	Frome, Castle	135
Birch Little	203	Colwall	164		
Birch, Much	202	Combe, township in Presteigne		**Ganarew**	234
Birley	46	Cowarne, Little	97	Garway	220
Bishopstone	126	Cowarne, Much	117	Goodrich	237
Blakemere	142	Cradley	120	*Grafton, township in Hereford, All*	
Bockleton	57	Craswall (township)	136	*Saints, see Bullingham*	
Bodenham	77	Credenhill	109	Grendon Bishop	56
Bolstone	204	Croft	21	Grendon Warren	79
Bosbury	161	Cusop	121		
Brampton Abbots	227	Cwmyoy (part)	191	**Hampton Bishop**	152
Brampton Bryan	2			*Hampton, New, extra-parochial in*	
Bredenbury	81	**Dewchurch, Little**	205	*Leominster*	
Bredwardine	103	Dewchurch, Much	200	*Hampton Charles, hamlet in Bockleton*	
Breinton	147	Dewsall	176	Hampton Wafer (extra-parochial)	55
Bridstow	226	Dilwyn	43	*Hardwicke, in Clifford*	
Brilley	67	Dinedor	179	Harewood	215
Brimfield	24	Dinmore (extra-parochial)	76	Hatfield	53
Brinsop	108	Docklow	52	Haywood (extra-parochial)	175
Brobury	104	Donnington	189	Hentland	223
Brockhampton, former chapelry of		Dormington	154	Hereford (All Saints, St John, St	
Woolhope	207	Dorstone	122	Nicholas, St Owen, St Peter)	150
Brockhampton, with Norton, township		Downton	4	Hereford (St Martin)	151
in Bromyard	84	Dulas	169	Hereford, Little	25
Bromyard	82			Holme Lacy	180
Bucknell (part)	1	**Eardisland**	42	Holmer	149
		Eardisley	68	Hope Mansell	239

xvii

Hope, Sollers	208
Hope under Dinmore	75
Humber	50
Huntington	38
Ivington, in Leominster	
Kenchester	128
Kenderchurch	197
Kentchurch	210
Kilpeck	199
Kimbolton	35
Kingsland	32
Kingstone	171
Kington	39
Kinnersley	70
Kinsham	9
Knill	28
Laysters, see Leysters	
Lea	240
Ledbury	188
Leinthall Earls, former chapelry of Aymestrey	19
Leinthall Starkes	11
Leintwardine	3
Leominster	47
Letton	89
Leysters	37
Lingen	8
Linton	230
Llancillo	193
Llandinabo	213
Llangarron	233
Llangrove, in Llangarron	
Llanrothal	231
Llanveynoe, chapelry in Clodock	
Llanwarne	212
Longtown (chapelry in Clodock)	165
Lucton	20
Ludford (part)	15
Lugwardine	153
Luston, township in Eye	
Lyonshall	40
Madley	145
Mansel Lacy	107
Mansell Gamage	106
Marcle, Little	186
Marcle, Much	209
Marden	94
Marstow	236
Mathon (part). *Mostly Worcs. until 1897*	163
Michaelchurch Escley	137

Middleton-on-the-Hill	36
Moccas	123
Monkland	44
Monnington	124
Mordiford	155
Moreton Jeffries	98
Moreton-on-Lugg	111
Munsley	187
Newhampton (extra-parochial)	54
Newton (township in Clodock)	167
Newton, Welsh	232
Norton, see Brockhampton with Norton	
Norton Canon	90
Ocle, Livers (extra-parochial)	115
Ocle Pychard	116
Orcop	211
Orleton	23
Pembridge	41
Pencombe	78
Pencoyd	214
Peterchurch	138
Peterstow	225
Pipe and Lyde	130
Pipe Aston, see Aston	
Pixley	184
Presteigne (part))	7
Preston on Wye	144
Preston Wynne	114
Pudlestone	52
Putley	183
Pyon, Canon	74
Pyon, King's	73
Radnor, Old (part)	27
Richard's Castle	15
Rochford	26
Ross	228
Rowlstone	194
St Devereux	198
St Margarets	139
St Weonards	221
Saltmarsh (extra-parochial)	83
Sapey, Upper	64
Sarnesfield	71
Sellack	224
Shobdon	31
Sollers, Bridge	127
Stanford Bishop	101
Stapleton, township in Presteigne	
Staunton-on-Arrow	30
Staunton-on-Wye	105

Stoke Bliss	58
Stoke Edith	157
Stoke Lacy	99
Stoke Prior	49
Storridge, in Cradley	
Stottesden (part)	16
Stretford	45
Stretton Grandison	134
Stretton Sugwas	129
Sutton St Michael	112
Sutton St Nicholas	113
Tarrington	158
Tedstone Delamere	66
Tedstone Wafer	65
Thornbury	59
Thruxton	173
Titley	29
Tretire with Michaelchurch	222
Treville (extra-parochial)	172
Tupsley, township in Hampton Bishop	
Turnastone	140
Tyberton	143
Ullingswick	96
Upton Bishop	219
Vowchurch	141
Wacton	80
Walford	238
Walterstone	192
Wellington	93
Wellington Heath, in Ledbury	
Weobley	72
Westhide	132
Weston Beggard	156
Weston under Penyard	229
Whitbourne	85
Whitchurch	235
Whitney	86
Wigmore	10
Willersley	88
Willey, township in Presteigne	
Winforton	87
Wisteston, former chapelry in Marden	
Withington	131
Wolferlow	63
Woolhope	182
Wormbridge	196
Wormsley	92
Yarkhill	133
Yarpole	22
Yatton, chapelry in Much Marcle	
Yazor	91

Part One

1 Introduction

A Delight of Maps

Maps are the most delightful of all sources of historical evidence. Before 1800, whether revealing the broad picture of the whole world or viewing a single plot of ground under a magnifying glass, maps bring together the accuracy of an archive and the artistic skill of the draughtsman. Archives, however interesting their content, look dull. Paintings, however visually appealing, are biased by the artist's interpretation. But in maps, like these of Herefordshire, are seen the hills and woods, orchards and rivers of a landscape that William Camden, the Elizabethan antiquary, described as 'admirably well provided with all necessaries for life. Insomuch, that it may scorn to come behind any County in England for fruitfulness'.[1] Also displayed are the roads marked out by their milestones, the inns, smithies and farmhouses, the open sheepwalks of the western mountains, the crowded streets of Hereford city crammed within its medieval walls.

Hereford is famous for a map. Its cathedral's *Mappa Mundi* is the largest surviving complete medieval map of the world. It is also among the earliest on which Hereford is marked. But although thousands come to see it every year and it has been the subject of much scholarly research, there is not a single guide to the many maps of Herefordshire itself.

The aim of this book, therefore, is to describe the surviving maps relating to the county between 1577 and 1800. It includes both the printed maps of the whole county and the unique manuscript maps of small areas within the county. Elsewhere, carto-bibliographies have concentrated exclusively on either the printed county maps, for example Gloucestershire, Warwickshire and Monmouthshire, or on the manuscript maps, as for Essex, Glamorgan and Devon.[2] The relationship between the two kinds of maps, the historical interest of both and the relatively modest number for the small county of Herefordshire, determined the decision to describe the whole range in one volume. Here, therefore, are the unique large-scale manuscript maps drawn for one specific purpose, to illustrate a legal dispute, describe a newly acquired property, or assist the management of an estate, surveyed for the sole use and benefit of a single client. Here also are the small-scale county maps, descended from maps of the world and its parts, which with the invention of printing were designed commercially for a wide variety of users. Both distantly draw on mediterranean and arabic sources revolutionised by the technical advances of the western European renaissance to flower in Tudor England. Both strands progressed separately but with some cross-fertilisation. Some surveyors, like Christopher Saxton, John Rocque and, most especially in Herefordshire, Isaac Taylor of Ross, made both printed maps and estate surveys. Common to both were surveying techniques, decorative features and conventional signs. Both were similarly influenced by mathematical advances, new instruments and changing artistic fashion.

The evidential value of all the maps binds them together most strongly. Throughout this carto-bibliography attention has been directed towards describing and explaining the content and context of the maps rather than the minutiae of those bibliographical details essential for serious collectors. With the few exceptions acknowledged in the listing, all the manuscript maps have been examined personally, as have all the copies of printed maps available in the public collections in Hereford. The descriptions of the majority of the printed maps have been based upon the unsurpassable bibliographies of Chubb, Skelton and Hodson.[3] In serving the collector these indispensable works cannot be bettered, but for the benefit of the historian more notice has been given here to the significance of the printed county maps as historical evidence.

The Covering Dates

By coincidence the earliest printed and manuscript maps of Herefordshire are both dated 1577. The choice of a terminal date has proved more difficult. Other catalogues of manuscript maps run to 1840, the approximate date of the tithe maps which covered most parishes in England at a large scale and formed the basis of local maps until the publication of the twenty-five-inches-to-one-mile Ordnance Surveys maps later in the nineteenth century, in Herefordshire between 1878 and 1887. Other carto-bibliographies of printed maps continue until 1900 in order to include the expanding range of nineteenth-century Ordnance Survey maps and the many specialist maps published in Victorian Britain.

The turn of the nineteenth century is, however, a significant watershed. The Ordnance Survey was founded in 1791 and its first maps, of Kent, were published in 1801. Its surveyors' drawings of Herefordshire at a scale of two inches to one mile date from 1812 to 1817 and were published privately by Henry Price at the familiar scale of one inch to one mile in 1817. The occupation of surveying became more professionalised in the late eighteenth century, not only with the emergence of the Ordnance Survey but also with the employment of engineers to survey public works like canals, harbours and roads, specialist landscape gardeners and the first business firms of estate managers and advisers. With this professionalism came a growing conformity in training, working methods and draughtsmanship. From the late eighteenth century maps lose their individuality and artistic exuberance. The developments of nineteenth-century mapping deserve a separate study. This is a good time to stop and take stock of the achievements of the previous two centuries.

The existence of all known issues of printed maps is well recorded, and only one additional map, not previously known, has been discovered. The number of manuscript maps, on the other hand, is limitless. Some have undoubtedly been overlooked in unexpected places or in uncatalogued collections in public repositories. Others certainly remain safely but unrecorded in private hands. Any information about maps not included in this listing would be greatly appreciated.[4]

2 The Landscape of Herefordshire

THE ELIZABETHAN LANDSCAPE

When Christopher Saxton drew the first map of Herefordshire in 1577 he entitled it *FRUGIFERI AC AMENI HEREFORDIAE COMITATUS DELINIATIO* – 'A Map of the Fruitful and Pleasant County of Hereford'(Plate 2). Some two hundred years later, in 1789, William Marshall, the agriculturist, summed up his introductory remarks on the county, 'Herefordshire may be said, without flattery, to be altogether beautiful'.[1] Both descriptions of this most rural and least populated English county remain valid today.

The county has been likened to a saucer, circular in shape and some thirty miles in diameter, more or less level in the middle with an outer rim of high ground. The county town and cathedral city of Hereford lies at its centre and the rim of hills stretches from horizon to horizon. George Bickham, who in 1753-4 drew a cartographical vista of the county, chose Ross for his viewpoint, looking northwards over the Wye valley from The Prospect, where in 1705-6 John Kyrle, the town's philanthropist, had created a public garden and fountain. Nearly two centuries earlier, in 1576, Elizabeth I's Privy Council had ordered the magistrates in Wales to provide Saxton with guides who could lead him to towers, high places and hills where he could get good views of the countryside.[2] In England he is thought to have used the newly re-equipped beacon sites for some of his triangulation points and it is likely that the Worcestershire Beacon on the Malvern hills, the eastern boundary of Herefordshire, would have been one of his all-round viewpoints. There was a beacon on its summit from at least 1539 and there is a crude drawing of it on a map of 1633.[3] The view stretches across the low wooded hills and fields of Herefordshire from the Shropshire hills in the north to May Hill in the south and westward to the dark backcloth of the Black Mountains marking the Welsh border. The reverse view from the ridge of Crib y Garth on the Black Mountains is equally spectacular, with the advantage that on a clear day the city of Hereford can be picked out. From there, on New Year's Eve 1999 the darkness of the English countryside as far as the distant glow of Gloucester and Cheltenham suddenly erupted on the stroke of midnight in a starry blaze of fireworks to celebrate the new millennium, brighter than any Tudor beacon.

The name of Crib y Garth, locally called The Cat's Back, is a reminder that Herefordshire lies in the Welsh borderland, where the tide of English settlement before the Norman Conquest ran out of strength in lapping against the foothills of the Welsh mountains. At the end of the eighth century the western limit of Mercia ran northwards along the river Wye from Monmouth as far as Bridge Sollers, seven miles west of Hereford, from where Offa's Dyke struck north-westward to Kington. Later, the Normans, shortly before and after the Conquest, established the present boundary of their military Marcher earldom and the administrative county west of the Wye, but it remained a somewhat flexible border for centuries. Hereford's first royal charter of privileges granted by Richard I in 1189 was addressed 'to our citizens of Hereford in Wales'.[4] Some parishes west of the Golden Valley stayed within the diocese of St. David's until the late nineteenth century and the detached Herefordshire island of Fwthog in the heart of the Black Mountains was not transferred to Monmouthshire until 1891.[5] In Saxton's time Welsh was still widely heard in the county. About one-quarter of the inhabitants of the western hundreds of Webtree and Wormelow had Welsh patronymic names in 1524-25 and four-fifths in Ewyas Lacy hundred; even 16% in Hereford city bore Welsh patronymic names, though that number was declining as they adopted, as the English had in the middle ages, the custom of retaining the same surname in succeeding generations.[6] The Welsh influence persists in the intonation of Herefordshire dialect and in many place and field-names.[7]

The rim of hills is formed of the hard Pre-Cambrian granites and Cambrian conglomerates of the Malverns, infused with igneous dykes and volcanic streaks, and the later softer bands of Devonian Old Red Sandstones and Silurian Limestones which predominate in the Bromyard Downs and the Mortimer Forest area running into south Shropshire. Farther west the Pre-Cambrian gabbros of Stanner Rocks close to Kington form the borderland which stretches south to the impressive crags of the Old Red Sandstone in the Black Mountains. An underlying geological weakness has left surface evidence of landslips on Hatterrall Hill in the Black Mountains, at the broken gully of Black Darren on their eastern flank and the crooked church of Cwnyoy on the western.[8] Saxton was given eye-witness accounts of a three-day landslip on Marcle Hill in 1575, very shortly before his survey of Herefordshire, and on his map marked 'Kinnaston chap[el] W[hi]ch was dreven downe by the removing of the ground'. At intervals since then further landslips and earthquakes have been experienced in the county.[9] Where the Old Red Sandstone lies close to the surface many small quarries provided an easily worked purple-pink or greyish building stone, which also splits into roofing slates. Hereford cathedral, for example was supplied with stone from Capler quarry by the river Wye near Fownhope.

At the end of the last ice age the final entry of glaciers into Herefordshire from mid-Wales some 50,000 years ago straightened river valleys, breaking up the softer marl-rich beds of the Old Red Sandstone and leaving a blanket of moraine across much of the county, especially over its western half. As the ice melted and shrank great lakes formed at the margins, still evident in the flat lands around Wigmore and Letton. The final disappearance of the ice 14,000 years ago left isolated humps of harder Old Red Sandstone, with intervening heaps of glacial moraine, which underlie the distinctive landscape of central Herefordshire with its wooded, flat-topped hills and conical knolls. Rising prominently among these is the Woolhope Dome, smoothed by earlier ice movements and dominated by its hard Silurian limestones, also found elsewhere at nearby Shucknall and around Wigmore. Devonian and Carboniferous limestones, conglomerates and sandstones form the rock towers and craggy outcrops of the lower Wye valley below Ross. Around Ross itself the Ryelands soil is of a lighter sandy nature in contrast to central Herefordshire.[10]

The Wye is the principal river, flowing sedately eastwards through the middle of the county after its hectic passage from mid-Wales until, just downstream of Hereford, it turns south and is joined by the smaller rivers of Lugg and Arrow from the north-west of the county and the Frome from the north-east. Broad riverside meadows absorb most of the periodic flash floods of the smaller rivers and the wider overflowing of the Wye when it is swollen by heavy rainfall or snowmelt in the Welsh mountains. The changing courses of the rivers are reflected in the old parish boundaries still marked on current Ordnance Survey maps and are traceable in the pools and ox-bow of the Wye at Clifford, seen best from the top of Merbach Hill. A flood in 1730 shifted the river at Whitney 650 metres and the church, which found itself on the opposite bank, had to be rebuilt. The first bridge at Whitney, built in 1774, was swept away in the great flood of 1795 and its successor in 1814. Lower downstream Moccas bridge was washed away in 1960 and every few years flooding closes the river-crossing at Hereford to traffic for some hours. Below Hereford the Wye meanders in great loops until forced over the rapids and through the limestone gorge at Symond's Yat to Monmouth, where it is met by the hill streams of west Herefordshire — Dore, Escley, Monnow and Olchon.

Woodland, forests (in the legal sense of royal hunting preserves) and deer parks were spread all over the county. Saxton marked Mocktree Forest and Bringewood Chase in the far north of the county, Deerfold Forest near Wigmore, Haywood Forest south of Hereford as well as the dense woods near Lyonshall, in the Golden Valley and around Bromyard and Woolhope. By then the medieval hunting grounds of the bishops of Hereford on the western side of the Malvern Hills had been cleared, but although the process of felling woodland for agriculture was to continue, Herefordshire remains well wooded. Until brick-making became fashionable and universal in the eighteenth century and better transport allowed the introduction of building materials from other parts of the country in the nineteenth century Herefordshire houses were primarily built with timber from local sources.

Herefordshire's heyday had been in the early middle ages when the castles of Marcher barons like those of Clifford, Ewyas Lacy (Longtown), Pembridge (in Welsh Newton), Weobley and Wigmore, together with the smaller motte and bailey castles of their retainers, dominated the countryside as a bulwark and refuge against the Welsh. They were already in decay when Owain Glyn Dwr's incursions at the beginning of the fifteenth century caused them to be hastily re-commissioned for the defence of the kingdom. With the failure of his bid for Welsh autonomy in 1415 and the extinction of the old baronage at the end of the Wars of the Roses in 1485 Herefordshire lost its strategic importance. Only half a century later the other great land-owning bastion of medieval society was overthrown with the reformation of the church and the dissolution of the monasteries. One-fifth of the landed property of the nation was put on the market by the Crown to the benefit of the rising merchants and gentry of Tudor England. The eight deer parks marked by Saxton indicate the homes of the influential Elizabethan gentry who had succeeded the Marcher barons, families like Baskerville of Eardisley, Hoskyns of Moorhampton in Abbey Dore, Parry of Newcourt in Bacton and Scudamore of Holme Lacy, who served the Tudors at court and in government.

The distribution of the villages on Saxton's map is substantially the same as the present day. Indeed, most of them had been established centuries earlier by either the Romano-British Welsh or the Anglo-Saxons in the first millennium AD. Some of the hamlets and individual manor houses are perhaps less prominent now than in Saxton's day, but in the absence of later industrialisation and large expansion of towns no significant new places have grown up since the sixteenth century. No accurate population figures are available before the first census in 1801, when the population of the whole county was 89,191, living in 17,944 houses, but in 1664 there were, by comparison, 13, 101 houses charged with, or exempted from, paying the Hearth Tax.[11] The sparse settlement of west Herefordshire in 1577, which dated from before the Norman conquest, confirms its relative backwardness. In 1610 Rowland Vaughan of Whitehouse in Vowchurch described, with some exaggeration, the endemic poverty and lawlessness of the Golden Valley.[12]

Saxton's lifetime coincided with a substantial growth in the population of England and an unprecedented increase in poverty, encouraging migration from the countryside to the towns and prompting the first poor laws of 1597 and 1603. In the middle ages the site and status of Hereford as a frontier town had raised it to equal importance and size with Gloucester and Worcester and bigger than Shrewsbury. It was still strongly walled when John Leland visited it about 1536 and estimated that its castle was as large as that at Windsor.[13] Remaining timber-framed houses bear witness to the prosperity of its citizens in the late fifteenth century. The city's continuing prosperity may be judged from the endowment of almshouses for the poor, the report that High Town was jammed with merchandise, the corporation's successful appeal to Elizabeth I for a new and expensively illuminated charter of privileges in 1597, the building of the famously splendid and huge Market Hall in High Town (*c.* 1600, demolished 1862), Old House (1621) and its neighbours, and the continuing existence of fourteen city gilds.[14] When John Speed printed his map of the city in 1610 he showed the houses tightly packed within the encircling city walls and suburban development beginning to stretch along the roads radiating from the five town gates (Plates 10, 11).

Nevertheless, there were also signs of the weakness of Hereford's economy. Its trade had been affected early in the sixteenth century by the heavy tolls levied on traders to meet the corporation's expenditure. The demolition of the dean and chapter's two fulling mills, ordered by Henry VIII before 1527, must have contributed to the decline of the cloth trade, although the subsequent tales of ruin and poverty would have been exaggerated in order to obtain permission for rebuilding the mills in 1555. There were still six fulling mills in Hereford in 1691 but the industry gave way to the leather trade, already important in Elizabeth I's reign, and glove-making.[15] Undoubtedly, the number of the city's poor did increase, as everywhere else, and periodic outbreaks of plague in the county, especially in 1580, 1609-10, 1636-7, combined with other years of high mortality following bad harvests, disproportionally affected the poor in Hereford and other towns.[16] But the underlying intractable problem was Hereford's deficient communications and its distance from London and the coast. The gloving industry and agricultural markets and fairs could only support a local, not a regional, centre.

Hereford was, and is, the only town of any size and importance in the county. Leominster, twelve miles to the north at the crossing of the Lugg, Arrow and subsidiary streams, was the second town of the county, like Hereford returning two members to Parliament and similarly dependent upon its markets and the cloth and agricultural trades. It had a reputation for lawlessness and religious dissent but its support for Queen Mary had earned it a royal charter in 1554. The four smaller market towns of Bromyard, Kington, Ledbury and Ross served their hinterlands within the distance of a convenient day's travel of some ten miles but attracted more distant traders for their sales of wool, leather and corn. Other smaller market towns like Pembridge and the medieval castle boroughs of Wigmore and Weobley were failing, though in 1628 Weobley regained its right to return two members of parliament, largely at the instigation of the Tomkins family of Garnstone.[17]

The whole county, therefore, depended in Elizabethan times upon agriculture and its associated trades. It was good wheat-growing country, with ample woodland for pigs and extensive meadows and hill pasture for rearing cattle and sheep. Its own breed of sheep, the small but highly-praised wool-bearing Ryelands from the Ross area, had long been famous, succeeding the medieval Leominster 'ore' for the highest quality of wool grown in England.[18] Even well after the decline of the cloth industry in Herefordshire towns, wool was being sold to clothiers in Gloucestershire and Yorkshire.[19] Additionally, farmers and dealers could cream off profits from the droves of Welsh black cattle and lambs passing on their way to the English lowlands and London. There were as yet no other significant farming specialities, although Rowland Vaughan's pioneering scheme begun about 1588 for 'floating' water-meadows in the Golden Valley to improve their productivity was imitated elsewhere in the country and has been claimed as a first step in the so-called agricultural revolution.

The printed county maps can only illustrate the surface of the county. Other evidence, documentary and architectural, reveals the vibrant and entrepreneurial activity bubbling beneath the surface of Elizabethan and Jacobean England. Four of the earliest Herefordshire estate maps of lands in Bringewood 1577, Marden 1582, at Alltyrynys in Walterstone and Strangworth in Pembridge a little later, give a hint of this seething activity. Each related to property disputes. It was a period for new ventures, the assertion of rights, but with resort to litigation rather than to arms. It provides the starting point for examining the changes in the landscape recorded in the maps over the following two centuries.

THE CHANGING FACE OF HEREFORDSHIRE

One of the most obvious changes that followed was the continued reduction of the Herefordshire woodland. Between Saxton in 1577 and Henry Price in 1817 the only printed county map that can be relied upon to show the extent of the woodland accurately is that of Isaac Taylor in 1754. Other map-makers strewed symbols of trees haphazardly across the face of their maps. But the general impression is doubtless correct, that woods from Mocktree in the far north of the county to Archenfield in the south-west were much reduced.

In addition to the longstanding needs for timber for building there was a new voracious demand for fuel for industrial purposes. A case was brought in 1595 against Rowland Vaughan for felling 1,200 timber trees in Snodhill Park to provide fuel for the iron furnace which he, his brother Henry Vaughan of Moccas and James Baskerville were operating at Peterchurch.[20] Bringewood was being extensively felled from at least 1575 and the building of an iron furnace nearby on the river Teme at Downton by 1601 accelerated the process.[21] James Scudamore had an iron mill at Linton (Ross) before 1618 and his son John, 1st Viscount Scudamore, built Carey Forge on the Wye at Brockhampton (Ross) in 1628, which was supplied with fuel from his woodland on the Holme Lacy estate and at Abbey Dore.[22] There were other furnaces and forges before the Civil War, for example at Pontrilas, St Weonards and Whitchurch, and from 1684 at New Weir, near Symond's Yat, where there had been a mill from at least 1587-88; and not all of the forges are marked on the county maps.[23] It has been estimated from the accounts of mid seventeenth-century ironmasters that the whole process from smelting to the production of one ton of bar-iron required 24 cords (about 4,000 cubic feet or 113 cubic metres) of wood for making charcoal. The Bringewood forge alone was consuming up to a quarter of a million cubic feet (7,075 cubic metres) of wood a year.[24] Such figures explain why iron-working developed in Herefordshire despite the lack of iron ore locally. It was easier to bring the iron ore to the furnace than transport these huge quantities of timber to the iron mines. Eventually, in the mid eighteenth century the Herefordshire ironworks could not match their competitors near the coalfields and the industry died out. In the meantime, even with good management the woodland was being destroyed.

The timber was also being burnt by the local potteries, the short-lived late sixteenth and early seventeenth-century glassworks at St Weonards and May Hill and in the many small lime kilns and brick works. Tanning thrived on the bark produced in the woodlands. There were tanners in the towns, supplying the glovers and saddlers, and tanhouses are named on eighteenth-century estate maps of Marden and Canon Pyon. Field-names on estate maps also betray the whereabouts of the many fields called Brick Close, Brick Kiln Field and Limepitts. None was of more than local significance, but they contributed to both the destruction of the woodland and the diversity of opportunities for employment.

The loss of woodland was balanced and masked by the massive growth in the number of orchards. John, 1st Viscount Scudamore (1601-71) is commonly credited with popularising the Redstreak cider apple and the cultivation of orchards on his Holme Lacy estate from 1635. But years earlier in 1597 John Gerard, the herbalist, had already commented on the number of fruit trees surrounding country houses and in the hedgerows near Hereford. Cider-making had been a feature of west Gloucestershire since the thirteenth century, the bishop of Hereford planted fruit trees on his manor of Prestbury near Cheltenham in 1247 and a variety of 'best apples called Reddesters' were paid in tithes at Elmore, Gloucestershire in 1530. It seems likely, therefore, that detailed research on medieval archives will eventually produce similar early evidence for Herefordshire. Certainly orchards were sufficiently widespread for John Evelyn to comment in 1664 that 'all Herefordshire is become, in a manner, but one entire orchard' and the Harleys of Brampton Bryan were requiring their tenants to plant orchards of 40 to 50 trees from the 1670s.[25] Although most of the cider produced was retained locally for domestic use, a surplus was already being exported to London when Edmund Gibson revised Willliam Camden's *Britannia* in 1695.

A century later the agricultural writers William Marshall and John Clark gave long descriptions of cider production, which by that period amounted to some 15,000 hogsheads a year (over 3,680,000 litres) for local consumption in the four counties of Gloucestershire, Herefordshire, Monmouthshire and Worcestershire. A further 15,000 were sold to the dealers in Ledbury, Hereford, Upton on Severn and Worcester for the markets of London and Bristol, and thence to the East and West Indies.[26] Marshall considered May Hill to be the hub of the cider-making country, adding that 'in the deep-soiled district about Dymmock and Marcle ... the whole country may be said to be a forest of fruit trees'.[27] The apple trees were grown in the hedgerows, in pasture and in ploughland (as advocated by Marshall) but most closely and effectively in orchards. Despite the low priority given to the crop by farmers and poor yields every three years or so the production of cider was held to be well worth while, particularly as a cash crop for poor cottagers, who from a few trees might expect to sell a good surplus beyond their own needs. A useful by-product was the feed for a growing number of pigs from windfall apples and cider-making waste. Estate maps of all parts of the county show how much land was given over to orchards and the field-name Redstreak Orchard occurs in the late eighteenth century at Almeley, Blakemere, Much Marcle and Stoke Prior.

The second and comparable innovation was the cultivation of hops, introduced from Germany to Kent early in the sixteenth century to flavour ale. A hopyard at Whitbourne is recorded in 1577 and cultivation spread rapidly in the seventeenth century. Early hopyards are named on maps before the 1720s on the estates of some of the principal landowners in the county, at Wigmore (Harley), Holme Lacy (Scudamore), Goodrich (Henry Grey, Duke of Kent), and Marden (Coningsby). Later eighteenth-century estate maps mark hopyards in the west and north of the county at Almeley, Blakemere, Leintwardine, Madley and Mansel Lacy, but the heart of the hop-growing region remained in the Bromyard area, stretching across east Herefordshire and into the neighbouring parts of west Worcestershire.[28] The principal market for hops was Worcester. Until about 1960 the hopyards and oasthouses for drying the hops were a distinctive and familiar feature of the Herefordshire countryside, together with the seasonal arrival of Black Country hop-pickers.

Hops were a favourite with farmers and had the added advantage that between the rows of bines young cider apple trees could be planted and allowed to grow until they were well established. In the course of crop rotation the hops could then be moved on elsewhere and the mature apple trees allowed to flourish as an orchard or the land put down to pasture without the risk of the young fruit trees being damaged by stock.[29] The woodlands also benefited as hops were grown on a network of ropes strung between ten to twelve-foot high poles. Each acre in a hopyard required 700 poles, which before the invention of creosote in the mid-nineteenth century lasted only a couple of years, accounting for the number of alder beds and ash coppices in the hop-growing areas.[30]

ENCLOSURE AND IMPROVEMENT

Apple trees and hops could only be grown effectively in enclosed fields, not subject to the constraints of open-field farming. This partly explains why at first they were regarded as an adjunct to the house and its garden and were grown in those parts of southern England where open fields did not exist. In the mid sixteenth century John Leland, the king's antiquary, had commented on the enclosed fields in Archenfield and all the way between Hereford and Ludlow and it has been reckoned that in 1675 only 8% of the total county was open-field arable.[31] A century later Marshall asserted that 'Herefordshire is an inclosed county. Some few remnants of common fields are seen in what is called the upper part of the county; but in general it appears to have been inclosed from the forest state; crooked fences, and winding narrow lanes'.[32] It is certainly true that there are very few parliamentary inclosure acts for Herefordshire.[33] Piecemeal enclosure is a feature of hilly and wooded country and in this, as in other characteristics of the county, Herefordshire is perhaps more akin to Wales than midland England. Estate maps from the seventeenth century confirm Marshall's observation. Nevertheless, the road maps of Ogilby in 1675, Taylor's first one-inch-to-one-mile map of the county in 1754 and many estate maps all record the presence of open fields all over Herefordshire, especially well illustrated in the maps of Marden *c.*1720, Much Marcle 1741 and 1791, Richards Castle 1743 (Plate 35), Stretton Sugwas 1757 (Plate 36), Mansel Lacy 1770 and Ullingswick 1783. Other eighteenth-century estate maps show small areas of open-field lands, the consolidation of neighbouring strips in an open field in the hands of a single landowner or tenant, or enclosed fields that from their shape or names were patently former strips in open fields.[34] All this suggests that the open-field system may have been more widespread than Marshall and the agricultural historians have allowed, but that the enclosure of medieval open fields had proceeded by unrecorded private agreements or the imposition of the lord of the manor at an early date.

It was a continuing process. Between 1757 and 1794 Guy's Hospital, which Clark praised as an example of a well managed estate, enclosed the open fields at Stretton Sugwas without recourse to an act of Parliament (Plate 63).[35] By contrast, at Much Marcle, where 95 landowners were involved, Edward Wallwyn of Hellens, a newcomer with a large estate in the parish, took the course of persuading his fellow landowners to obtain an inclosure act. It was passed in May 1795 but the award was not completed until mid 1797 and the realignment of the fields and planting of new hedges presumably did not take place until after harvest during the autumn and winter of 1797-8. The legal, surveying and road-making costs, excluding new fences, were considerable at 16*s.* 6*d.* per acre and these, combined with changes in tenancy agreements, hit the smaller landowners and tenants. There was little opposition to the enclosure of Much Marcle, and that came from other landowners rather than the small occupiers, who had little say in the decision-making process as they owned only a minority of the land. Here and elsewhere the interests of the small tenants and commoners were ignored. Enclosure was generally, and correctly, held to be necessary for the improvement of agriculture and increasing the rentals of landowners. This, it was argued less sympathetically, would have the effect of providing more employment for the feckless rural poor. The net result at Much Marcle was a totally different landscape of hedged fields, new minor roads, larger farm holdings and one new homestead.[36] In those areas where land had long been enclosed the face of the landscape changed but slowly. Without the upheaval of an inclosure award the routes of roads ran their ancient course and, although farm boundaries might be altered with a change of tenants or by purchase and sale, field boundaries long remained undisturbed. On maps of the Holme Lacy estate many of the riverside fields were identical from the beginning to the end of the eighteenth century.

The enclosure of waste lands on the hills, especially in the west of the county, and on the fringes of parishes elsewhere, mostly occurred after 1800. Clark estimated in 1794 that there were 20,000 acres of waste land in Herefordshire, half of it on the Welsh border.[37] The printed county maps support his statement and the Moccas estate atlas of 1772 contains large-scale maps of the open pastures on Cusop Hill and on 'King Arthur's Stone Mountain' between Bredwardine and Dorstone. Both areas have since been enclosed into large blocks of straight-fenced fields, with bracken and scrub replaced by good grazing. These maps and others, like those of Goodrich in 1717 and Eastnor in 1726, also show the small plots nibbling at the edges of the common lands or grouped by small greens on the poorest patches of land at the extremities of the parish. Contrary to popular belief these are usually not squatters' cottages, but encroachments licensed by the lord of the manor, who owned the soil of common and waste land. By providing homes and a livelihood for the local poor, many of whom both worked as tradesmen, masons, tailors, lime-burners and labourers and exercised their rights of grazing and fuel-gathering on the common lands, landowners gained a small income from additional rents. Such commoners, who had a reputation for laziness and lawlessness, were those most seriously affected when the open commons and greens were enclosed. A few large commons survive, like Ewyas Harold Common and Vagar Hill in the west and Bromyard Downs in the east, relics of waste lands since prehistoric time.

Enclosure, which brought about the most important change in the landscape before the nineteenth-century spread of factory-based industries and the consequential sprawl of urban development, was driven by powerful landowners eager to improve their landed property, spurred on by the agricultural reformers. Some improvements, though benefiting from enclosure, had no obvious effect on the landscape and so Herefordshire's fourth specialisation, following chronologically after sheep, cider and hops, goes unrecorded on the maps. This was the development of the Hereford breed of red, white-faced cattle. John 1st Viscount Scudamore is said to have bred such cattle in the seventeenth century, but the breeding of Herefords is usually attributed to farmers in the north of the county early in the eighteenth century, the Tomkins and Galliers families of King's Pyon and Leintwardine.[38] Similarly, the replacement of antiquated timber-framed houses and farm buildings, which Clark thought so desirable, is not reflected in the maps. Early maps by draughtsmen who marked buildings by accurate little drawings of them, as in Kent and Essex, do not exist for Herefordshire. The only surveyors in Herefordshire who inspire confidence in the accurate representation of buildings are William Deeley (Stoke Edith 1680) and Joseph Dougharty (Grendon Warren 1732; Plate 47). The drawing of an apparently brick-built small house and dovecote at Richards Castle in 1743 and inset large drawings on maps of Aconbury Court 1757 (Plate 49) and Great House, Court Farm and Brick House at Canon Pyon *c*.1760 are exceptional.

The appearance of the new country houses built by the men who led the way in the improvement of their estates and controlled politics in the county – Brampton Bryan (from 1661, Harley), Holme Lacy (from 1674, Scudamore), Stoke Edith (from 1696, Foley), Hampton Court (from *c*.1700, Coningsby) – is known from their survival or from topographical paintings, engravings and photographs, not from maps. On his county map of 1754 Isaac Taylor distinguished between old and new country seats but purely by the use of conventional signs. Later in the century the landscape gardeners, Lancelot 'Capability' Brown (1715-83), Humphry Repton (1752-1818) and others were to change the landscape of the parks surrounding some of the great Herefordshire houses.[39] Similar extensive alterations were achieved by the promoters of the 'picturesque' movement, Uvedale Price of Foxley (1747-1829) and Richard Payne Knight of Downton (1750-1824). Both inherited their estates and both completely reshaped them. Price engaged Nathaniel Kent to survey the Foxley estate and offer advice on its improvement in 1774, and Knight had James Sherriff survey Downton in 1780 (Plates 52-54).

COMMUNICATIONS

All the writers and reformers from Saxton and Camden onwards were agreed on the idyllic beauties of the Herefordshire landscape. They were all also equally agreed on the appalling condition of its roads. That experienced traveller Celia Fiennes, writing in and after 1698, described the road between Stretton Grandison and Worcestershire as 'deep and difficult' and 'the worst way I ever went in Worcester or Herrifordshire — its allways a deep sand and soe in the winter and with muck is bad way, but this being August it was strange and being so stony made it more difficult to travel'.[40] Nearly two hundred years later William Marshall railed at length that 'Even the entrance into the county, — from the foot of Mayhill to Ross; — the principal thoroughfare from London to Hereford ... would not be deemed a sufficient by-road, in many parts of the kingdom. The narrow forest lanes and hollow ways still remain ...' and so on for a whole page.[41] With heavy sarcasm, a forecast that property values might be increased by up to a quarter and a final dig that Herefordshire was a hundred years behind the rest of the country, he blamed neglect of the public roads on the very same people who were improving their private estates.

In fact, estate accounts and correspondence suggest that travel and traffic was considerable, if slow. Turnpike trusts had been formed to raise funds by tolls to maintain the main roads, beginning with the Ledbury trust in 1721 and the Gloucester to Hereford road, which Marshall later complained of, in 1726. By 1767 ten trusts operated webs of roads radiating from the towns in Herefordshire or just over its border, their routes and sometimes their turnpike gates being marked on both the printed county maps and relevant estate maps. Clark criticised the turnpike trustees for not re-routing the main roads along the contours of the county's many hills rather than going straight over them, echoing Marshall's complaint that it took six or seven horses to drag a load of corn to market. At that time Hereford was five hours from Kington by coach and six hours from Worcester; London was at least thirty-six hours away. Clark also condemned the minor roads, then still maintained by parish labour, although a few were newly set out when open fields were enclosed.[42]

More important for local access was the replacement of fords and ferries by bridges. John Leland had described the few bridges in the county in the mid sixteenth century. When Saxton drew his map in 1577 the only bridge over the Wye in the county was at Hereford. Following acts of Parliament to permit the charging of tolls Wilton bridge was built in 1579, Hay in 1763, Bredwardine in 1769 and Whitney, the only crossing to survive as a toll bridge, in 1774.[43] The map-makers were slow to mark them on the maps and also were inclined to mark ferries as solid bridges. Ogilby in 1675 was more reliable, giving details of the construction of the more numerous bridges on the main roads over minor streams and rivers (Plate 5).

The disadvantage of a bridges and fords against ferries is the obstruction they cause to river traffic. The descriptions of the steep and stony roads, so narrow that two carriages could not pass, demonstrates the difficulty of transporting heavy goods, such as coal, building materials, timber, casks of cider and waggonloads of corn, hops and bark. The Wye in particular and the Lugg to a lesser extent were used for water-borne transport but their passage was limited by the impassable low water level of the Wye in summer, floods in winter and the obstruction by weirs belonging to water mills all the year. An Act of Parliament in 1662 had authorised a scheme for building locks to control adequate water levels in the river.[44] The county magistrates estimated that the river was navigable for 200 days a year in 1675 when they ordered eight new coal barges.[45] Further acts of 1695 and 1727 for the Wye and the Lugg were needed to achieve the desired result of removing the mills and their weirs and opening the river to the traffic of shallow-draught trows.[46] These small sailing vessels had to be drawn upstream by teams of men because the towpath crossed so frequently from bank to bank that horses were a hindrance.[47] Isaac Taylor depicted a trow in the title cartouche of his map of the city in 1757 (Plate 28), and there are drawings of river boats on maps of Wilton in 1755 (Plate 48) and of the New Weir near Symond's Yat in 1758, as well as in James Wathen's sketches and other illustrations.[48] Herefordshire's grain, cider, timber and bark were sent downstream by the Wye to Bristol, where prices were almost as good as in London, or taken overland to Gloucester and down the Severn. In return the trows brought back manufactured goods and, above all, coal. Fresh interest in improving the navigation of the Wye led to an abortive proposal by Isaac Taylor in 1763 and the appointment of the canal engineer Robert Whitworth to draw up plans for improving the lock at New Weir and the whole length of the river between Hereford and Tintern in 1779.

The rivers, however, remained unreliable. A secure waterway to the river Severn, and thence not only to Bristol but also to the Thames and the Midlands, was the object of the two schemes for acts of Parliament in 1791. One was from Stourport-on-Severn to Leominster and Kington, the other from Gloucester to Hereford, with a branch to the newly opened coalfield at Newent. The intended whole route of both canals is marked on Henry Price's map of the county in 1817, but still hopefully, for the engineering problems and costs proved larger than the promoters had anticipated. By 1796 only the length of the canal between Leominster and Neen Sollers was opened and it was 1798 before the canal from Gloucester at last arrived at Ledbury. It had the immediate desired effect of bringing down the price of coal from 24*s*. a ton to 13*s*. 6*d*. and a longer-term consequence in drawing the expansion of Ledbury in the direction of the canal wharf situated in the fields south of the medieval town.[49] But it was 1845 before it reached Hereford.

HEREFORD CITY

There are few maps of Hereford city after Speed's small but useful printed plan of 1610 and the only other town to be mapped before 1800 was Ledbury, where a local man, J. Lidiard, drew a manuscript sketch map about 1785-88 (Plate 60). In Hereford neither the corporation nor the dean and chapter, who were the principal landlords, commissioned surveys of their property before 1800 and the two manuscript maps of the estates of the bishopric in 1721-22 have been so badly damaged by damp as to be almost totally illegible.[50] Otherwise, for the whole period from 1610 to 1800 reliance has to be placed on Isaac Taylor's plan of the city in 1757 (Plate 28). Fortunately, it is a first-class map, which with the evidence of the buildings still standing and drawings, engravings and written descriptions, provides a reasonably good picture of the eighteenth-century townscape.

The most striking comparison between 1610 and 1757 is the almost identical size of the city, which in 1757 contained 1,279 houses and 5,592 inhabitants, of whom two-thirds lived within the walls. The suburbs outside the gates along the radial streets appear to have been no larger than in Speed's time, underlining the static economy of the city. One of the reasons was the destruction of the suburbs during the Civil War, when houses and the two extra-mural churches of St Martin and St Owen were demolished to destroy the cover for attacking forces or damaged in the course of skirmishes. The lead roof of the chapter house was used to repair the already decayed castle. Both were left in ruin by the war, to be finally demolished in the mid eighteenth century. Within the walls Dr William Brewster, who left part of his fine library to All Saints' church, was among the first to build a distinguished brick house in Widemarsh Street in 1697. The rebuilding of town houses cannot have proceeded far enough to impress Daniel Defoe, who described the city as 'mean built' in 1725. The confining city gates survived long after Taylor had drawn his map, being demolished between 1782 and 1798, and in that same period Broad Street was extended northwards to All Saints' church in 1790 and the old timber-framed market hall in High Town was drastically remodelled in 1792. With the other timber-framed buildings packed into High Town the market hall remained standing for half a century longer.[51] Further changes were brought about under the Paving, Licensing and Lighting Act 1774, when the streets were newly paved, 'in the London manner', prompting the city householders to stretch ropes along the outside of the pavements at fair time to keep the cattle away from their windows. As there could be 1,000 noisy and smelly large beasts filling the streets at the October fair it was perhaps a not unreasonable precaution, if adding to the crush of animals in the thoroughfare.[52] By the time of the first census in 1801 the city's size had grown further. There were then 1,715 families living in 1,392 houses in the city, forming a population of 6,828 inhabitants.

The practice of holding the fair in the streets was condemned as deplorable by Marshall in 1789, for by then taste and sensibility were becoming more fastidious, even in a distant market city far away from the influences of fashion in London and the provincial spas. A new class of travellers was emerging. Those Herefordshire promoters of the 'picturesque', Uvedale Price and Richard Payne Knight, changed the landscapes of their own estates but it was their contemporaries, William Gilpin (1724-1804), Samuel Ireland (d.1800) and Joseph Farington (1747-1821) who led the tourists to Herefordshire and especially to the Wye Valley and Ross. Writers of the Romantic movement, antiquaries and leisured residents were opening the county to a wider public, where, to quote Ireland, 'so various and such an interesting picturesque scenery is perhaps nowhere to be found, either in this or any other country'.[53] This more leisured aspect of Hereford is charmingly captured in James Wathen's sketches of the city in 1799-1800.[54]

3   The Printed County Maps of Herefordshire

Hereford and Early Maps of the World and Britain

The county maps rolling off the Elizabethan and Jacobean printing presses appeared in that exciting and expansionist age marked by the literary triumphs of Shakespeare's plays and the Authorised Version of the Bible and the ventures overseas of the East India Company and the English settlements in North America. Their originality as regional maps owed much to the continental precedents of accurate nation-maps and marine charts that had replaced the medieval *mappae mundi* or world-maps.

The largest of these surviving *mappae mundi*, measuring over 5 by 4 feet (1.58 x 1.30 metres) is at Hereford cathedral. Scholars are still unravelling its secrets. The Hereford map is thought to date from about 1300, with its main features copied from a map, now lost, that had been designed and created in Lincoln some thirty years earlier by Richard of Holdingham or Sleaford, a canon of that cathedral. Though probably begun at Lincoln it seems to have been completed at Hereford. It portrays a world with its heart in the Mediterranean and Black Sea, centred on Jerusalem and illustrated by the stories of the Jewish Bible and classical Greece, being adorned with both mythical creatures and genuine geographical features.

The British Isles lay at the very edge of this world, squeezed into the bottom left-hand corner as three elongated blobs representing the 'islands' of England, Scotland and Ireland. Looking more closely, landmarks may be identified in their correct geographical relationship to each other. The river Severn is easy to pick out, as are Snowdon and the mountains of west Wales, the castles of 'Cunway' and 'Carnarvan', the medieval centres of government of London and Winchester, and the prominent hilltop cathedral of Lincoln. To the west of the Severn and the cathedrals of Gloucester and Worcester have been added, rather faintly but in a contemporary hand, the river 'Wie' and the word 'H'ford'.[1]

By 1300 more accurately drawn maps of a practical nature were appearing alongside these encyclopaedic pictorial *mappae mundi*, most notably the itineraries mapped across western Europe to Rome and other Christian shrines. As a strategic Marcher town on the still troubled border with Wales, boasting a shrine to its bishop; St Thomas Cantilupe (d.1282, canonised 1320), Hereford had earlier been marked on a map of England by Matthew Paris in the mid-thirteenth century. On the mid-fourteenth century Gough map it is shown standing on the highway from London to St David's, and thence to Ireland. Bishop Thomas Cantilupe's accounts for his journey to Rome in 1282 and the household accounts of his successor Bishop Richard Swinfield, travelling from Hereford to London in 1291, provide detailed evidence of journeys along these medieval roads.[2]

In Europe the manufacture of paper from the fourteenth century and the invention of printing in the mid fifteenth century allowed multiple copies of maps to be widely disseminated for the first time. The business of map-engraving and publishing spread northwards from Italy into Germany and the Low Countries during the sixteenth century and in the early seventeenth century became concentrated first in Antwerp and then in Amsterdam. The links between Protestant England and the Low Countries were close, with the result that many early county maps were produced by Flemish and Dutch engravers and printers, either in their homeland or as refugees in London.

Maps of the British Isles, portraying Britain in a recognisable shape, were printed on the continent from 1511. Those by Sebastian Münster (Basel 1540) and George Lily (Rome 1546) mark Hereford. Lily's map also marks Wigmore, Ross, Hay and the Forest of Dean. The map of the British Isles by Gerard Mercator in 1564 is a considerable advance. Mercator marks the rivers Wye, Teme, Lugg and Frome as well as the place-names of 'Bremart' (Bromyard), Hereford, Ledbury, Leominster, Ross, Weobley and Wigmore. Between Hereford and Ross is 'Mordwans', perhaps a mistake for Mordiford, an unusually minor place compared with the other market towns, castles and abbeys named elsewhere. Based on Mercator's map, Abraham Ortelius in his *Theatrum Orbis Terrarum* (Antwerp 1570) marks Hereford, Ledbury and Weobley. In the 1573 issue of the *Theatrum* Ortelius published the first printed map of Wales, supplied by Humphrey Lhuyd, the Welsh antiquary (d.1568). Lhuyd took a generous view of Wales in his own lifetime, for the map covers not only Herefordshire but also all lands west of the river Severn. His choice of places to mark was clearly influenced by his antiquarian knowledge and interests. The presence of Leintwardine and Wigmore, Leominster, Ewyas (as a district rather than a place), Dore, Wilton and Ross may all be explained as places of importance or of historical interest as the sites of castles and monasteries, but why Acton (Beauchamp), Cradley and Mathon? The engraver in Antwerp made a few copying errors, perhaps not surprisingly as Lhuyd gave some names threefold, in English, 'British' and Latin. Kington appears as 'Ynton' and the river 'Aur' is the *Dwr* or Dore, shown flowing north to join the Wye east of a place on the right bank of the river near Hay named 'Riedhelig Clifton'. This must refer to Clifford, the Norman castle opposite Rhydspence on the river's left bank. This area, west and south of the Wye, he marked as *Reinuc,* the name of the post-Roman Welsh kingdom of Rhieinwg, which embraced Brecon, Radnor and Herefordshire as far east as Aconbury. The principal rivers are marked, some hills including the Black Mountains, Malvern Hills and the hills north of Skenfrith, and the woods of the Forest of Dean and between Ledbury and Hereford.[3]

Such was the slowly improving extent of maps of the whole kingdom when in 1573 William Cecil, Elizabeth I's first minister, arranged for Christopher Saxton to survey all the English and Welsh counties in greater detail.

Christopher Saxton and the First County Maps

The Sysil or Cecil family had been settled at Alltyrynys in Walterstone in south west Herefordshire since at least the mid fifteenth century. They were connected with other gentry families of the southern Marches, like the Parrys and Vaughans, who followed the Welsh Tudor kings to London to serve them in government and at court. William Cecil's grandfather, David Cecil (d.1541), had, as a younger son, left Alltyrynys for Lincolnshire and his father, Richard Cecil (d.1587), was a Northamptonshire squire who held office in Henry VIII's household. William Cecil (1520-98) was educated at Cambridge and Gray's Inn before also entering the royal service. He survived the hazards of serving successive Tudor sovereigns and rose rapidly from the office of Master of Requests in 1547 to Secretary of State 1550-53 and 1558-72 and in 1572 became Lord High Treasurer until his death. In 1571 Elizabeth raised him to the peerage as Lord Burghley, the title being taken from the estate where he built his great country house near Stamford on the Lincolnshire-Northamptonshire border.[4] Despite his eminence he never lost touch with his Herefordshire cousins, personally drafting the will of Blanche Parry of Bacton (c.1507-1589), the queen's nursemaid and lifelong lady-in-waiting.

As the queen's chief minister at a time of internal revolt from roman catholic sympathisers in the north of England, the fear of plots against the queen, especially after her excommunication by the pope in 1570, and the threat of war with France or Spain, Cecil was acutely aware of the need for better maps to assist the administration and defence of the realm. He was personally deeply interested in maps, sketching his own and annotating others that came into his hands, and he had been considering a county-by-county survey from the 1560s. He always carried a pocket-sized map of the British Isles made by a member of his household, the experienced cartographer Laurence Nowell, but when in 1563 Nowell offered to produce a map of England and Wales he was turned down.[5] Preference had already been given to John Rudd (1498-1579), vicar of Dewsbury in Yorkshire 1554-70 and a prebendary of Durham cathedral with court connections going back to 1540. Like Nowell he had already made maps and in 1561 the queen asked the dean and chapter of Durham to allow him two years' paid leave to survey her realm, a request they could not well refuse.[6] But Rudd was of an age when he was unlikely to be fit enough to undertake the unavoidable travelling and fieldwork that was expected of him and the project came to nothing.

However, John Rudd had an assistant, Christopher Saxton, a local man who came from the hamlet of Dunningley, three miles from Dewsbury. He had been born about 1544, or perhaps a year or two earlier, and had probably been known to Rudd from his boyhood. Rudd certainly trusted him sufficiently to send him to Durham in April 1570 to collect his stipend.[7] It has been suggested that he was by then already working on Rudd's survey of England and he was certainly well enough trained and qualified for his name to be put forward in 1573 to carry out the survey of the counties so long planned by Lord Burghley.

Saxton was apparently commissioned by the Crown initially and in 1574 was rewarded with a grant of arms and a royal grant of property to compensate him for his 'grand charges and expenses'. Shortly afterwards, Burghley contrived an arrangement for more secure official support by other means, like other Elizabethan ventures in which the Crown readily encouraged private entrepreneurs without directly investing too much of its own limited resources. In Saxton's case Thomas Seckford, Burghley's contemporary at Gray's Inn in 1540 and in 1573 surveyor of the Court of Wards and Liveries, of which Burghley was master, became his paymaster. Both Burghley and Seckford are portrayed together in a group painting of the Court in session in 1585.[8] For his compliance, Seckford was allowed to display his arms, uniquely and more ostentatiously, along with the royal arms on Saxton's maps.

In that year his first county maps were printed. As the proofs came off the press a copy was sent to Burghley for his approval and retained for his own use. They are annotated in both his hand and that of one of his secretaries (Plate 1). On the map of Herefordshire a secretary has written the names of the county gentry, including the family name of Cecil beside the place-name Alltyrynys, together with the names of their relatives and neighbours, Skidmore (or Scudamore) of Kentchurch, Parry of Newcourt and Poston and Baskerville of Eardisley and Pontrilas.[9] His master was concerned to know privately where throughout the country influential loyalist protestants and potential catholic dissidents had their homes. Saxton had worked his way westward and northward from south-eastern England to complete the English counties in 1577, the date of his map of Herefordshire (Plate 2). Burghley must have been well impressed with Saxton's rapid progress, for which he was rewarded immediately by the queen with the grant of a monopoly for the production and sale of his maps for ten years. Two years later, although the maps had been on sale individually, he had complete sets bound together for sale as a volume.[10]

The maps included all the principal geographical features of the county, the rivers and larger streams, the lines of hills, drawn like tumpy molehills, trees crowded together to mark the forests and woods of Mocktree, Bringewood, Deerfold and Haywood, the wooded hills above Woolhope and along the Golden Valley, here called for the first time 'The gilden vale'. Following continental precedents the site of the villages are marked by the conventional sign which was to become universal, a little open dot, which Saxton embellished with a church spire or tower. Hamlets are indicated by a square. Important isolated buildings were also marked, a few chapels, the chief castles from Brampton Bryan and Wigmore in the north to Goodrich and Pembridge (in Welsh Newton) in the south. In addition, the seats of the principal gentry were indicated by a circular fence to mark their deer parks, as at Eardisley, Holme Lacy and Newcourt in Bacton. Bridges are marked but not roads, a surprising omission to modern eyes, but Burghley was more concerned to learn the lie of the land, the situation of places and their relationship one to another than the routes between them.

The general accuracy of the map in achieving these objectives is impressive. It is particularly so if one recalls that this was the first such survey and that even if Saxton had studied Humphrey Lhuyd's map of Wales he still had personally to cover the county, and indeed the rest of the country, on horseback and on foot. We know from Elizabeth I's instruction to the Welsh magistrates in 1576 to provide him with guides and assistance that in England also he would have been accompanied by local men, who would have taken him up the prominent hills, towers and other viewpoints. But Saxton, racing round the country in his busiest year, cannot have had time to wait for clear days and most of his survey must have relied upon hard riding and legwork. Proof of personal knowledge, whether from his own observation or first-hand information, may be found not only in his note of the landslip on Marcle Hill but also in his drawing of a small lake as the source of the river Dore. The Dore rises from a scatter of streams and brooks, which join into a strongly flowing stream at the head of its valley. Today there is no trace of a lake, partly because the line of the former Golden Valley railway runs through its site. But the Dorstone tithe map of 1840 names a field there called Pool Meadow and a short distance downstream the coppice of the Bell Alders, in which the Golden Well is situated, remains a marshy enclosure.[11]

The spelling of place-names is, of course, erratic. In Saxton's time there was neither standardised spelling nor approved pronunciation and it is not surprising that a Yorkshireman should occasionally mishear or misunderstand what he was told. Local pronunciation has also changed. The neighbouring mid-Herefordshire parishes of Beggarsweston, Yarcle, Stokedye and Taddington are all recognisable if not so often heard now, but the older pronunciation of Snodhill as 'Snowdell' seems to be entirely forgotten as does 'Lidbury' for Ledbury. Gone also are the Welsh renderings of 'Llanihangell Dewlas' and 'Llanyhangleescley', recorded by Saxton but long anglicised as Dulas and Michaelchurch Escley. To have collected and placed correctly so many place-names was itself a notable achievement. Only occasionally does he stumble, for instance in mishearing 'the Worldes end' for Alders End in Tarrington and in duplicating the name 'Fowemynd Chapel' for both the little medieval chapel at Mynyddbrydd near Dorstone and, mistakenly, the nearby chapel at Urishay.

Saxton's maps are important for providing the earliest detailed picture of England and Wales. In addition to their value as historical evidence they have the further attraction of being decorative. The title of the Herefordshire map is written in a cartouche reminiscent of a church wall monument, flanked by figures of Adam and Eve with the royal arms above. In another corner are the arms of the promoter, Thomas Seckford with his motto, newly changed in 1576 to *Industria Naturam Ornat*, ('Industry enhances nature'). In the bottom left-hand corner dividers bestride the scale bar, which bears the name of the Flemish engraver, Remigius Hogenberg, behind which is displayed a scroll with Christopher Saxton's name. The map is set in a border like a picture frame in which are written the four cardinal points, but it lacks a grid of latitude and longitude, which had appeared on some earlier smaller maps of the British Isles. It is, therefore, rightly attractive to collectors.[12] It has been remarked that the decorative features on Saxton's maps are inconsistent with his style as demonstrated in his later estate maps.[13] Though workmanlike, these are distinctly plain and unadorned. This prompts speculation that another artist drew the decorative features. In Tudor England, continuing the practice of medieval monastic and chancery scribes well known to both Burghley and Seckford, Crown grants and other outstandingly important documents were written by officials but subsequently decorated with illuminated initial letters and other details by specialist limners and painters at the expense of the recipient. Other early engraved maps, printed in black and white, were hand-coloured by specialist painters.[14] It would have been entirely within this tradition for Saxton to draw the working part of the map and for another artist to have added the decorative matter.

Saxton's maps were not merely the first of their kind but the basis of most county maps published over the next century and a half. Once his ten-year monopoly had expired and in the absence of copyright laws, not introduced until 1709, there was nothing to prevent other map-makers and publishers plagiarising his pioneering work, thus saving themselves time and expense in surveying. Some introduced new detail or presented the same information in a new style. Others, or perhaps their Flemish engravers, committed copying errors, which in turn were compounded by their imitators. Perhaps most confusingly, the original copper plates, not only Saxton's but other map-makers' as well, were handed down or sold on by the publishers and re-used time and again. It is therefore essential for historians to determine which issue of a map they are relying upon for evidence, as it may be fifty, a hundred or more years out-of-date. Collectors need to be equally wary. County maps were originally published both in atlases, which have since been broken up by printsellers, and for sale individually. In either case the bibliographical information about their issue and dating, which would have been printed on the atlas's title page, has been irretrievably lost.[15] Nowadays most maps are bought, or found in libraries and record offices, as single items. Even if the map has an imprint, giving the name of the map-maker and engraver it may be difficult, even impossible, to determine which issue of an atlas it has come from.

At about the same time that Saxton was riding the country with his surveying tools, William Camden (1551-1623), schoolmaster, antiquary and historian was similarly engaged, collecting material for his *Britannia*, first published in Latin in 1586. There is no evidence that their paths crossed, but Camden was sufficiently impressed to have the sixth and last Latin edition of *Britannia* illustrated by maps, 43 of which (out of 50) were copied from Saxton and engraved in 1607 by William Hole. They are of interest for two reasons. First, they are less scarce than Saxton's original maps, because of the size of the print-runs of the successive English editions of *Britannia* containing these maps in 1607 and

1637. Second, Camden added antiquarian features. In Herefordshire, for example, the regional name Archenfield is written across the south-west of the county and the Roman town of *Ariconium* is marked, though mistakenly identified as Kenchester. It was not until 1732 that the antiquary, John Horsley, correctly identified *Magnis* as Kenchester; *Ariconium* was the name of the iron-working settlement near Weston-under-Penyard. William Hole also introduced a few copying errors, which later copyists were to compound.

An earlier and more curious version of Saxton's pioneering work may be seen in four large tapestry maps of Gloucestershire, Oxfordshire, Warwickshire and Worcestershire designed as wall-hangings. They were manufactured in 1588 and the 1590s by Richard Hyckes or Hicks of Barcheston, near Shipston-on-Stour, for Ralph Sheldon's new house of nearby Weston Park. Unlike Saxton's maps, and indeed unlike all the county maps before the Ordnance Survey in the early nineteenth century, these maps were not confined to the county boundaries but spilled over to fill the whole rectangular shape of the tapestries. The map of Worcestershire consequently also covers much of east Herefordshire. Although Hyckes's conventional signs for villages and woodland are more pictorial than Saxton's, the spellings of place-names and the reference to the landslip on Marcle Hill betray Hyckes's source. Hyckes, however, introduced some new features including some rather fanciful roads and more bridges.[16]

THE EARLIEST PLAN OF THE CITY

The publication of Saxton's county maps in his mid thirties represented the peak of his achievement and fame. Thereafter he pursued his career as a competent but not especially outstanding estate surveyor, mostly in his native Yorkshire and elsewhere in the north of England, until his death about 1611. Meanwhile, John Speed was completing his *Theatre of the Empire of Great Britain*, which was part of his larger *History of Great Britain*. Unlike Saxton, Speed was not himself a surveyor. His whole project was based heavily on previously published work, the history on Camden and the maps on Saxton, the term 'Theatre' being borrowed from the *Theatrum* of Abraham Ortelius 1570, published in English in 1606. Speed's was the first atlas of the whole of the British Isles and was published by John Sudbury and George Humble of London but engraved in Amsterdam by Jodocus Hondius. The map of Herefordshire is dated 1610, but the volume's title page is dated 1611 and it was apparently not published until 1612.

Despite the lack of originality, Speed's maps are important historically and justly popular as visually attractive. The most important new cartographical features are the inset plans of the county towns (Plates 10, 11). Drafted more accurately in 1606 (see p. 95-6), this is the earliest plan of Hereford, clearly recognisable as the present city centre within the walls with suburbs straggling out along the main roads leading from the town gates. The medieval and Tudor streets, churches and principal buildings are named, including the pre-Reformation friaries. Here, too, are seen the buildings soon to be demolished during or after the Civil War like the castle, the cathedral's chapter house and the two suburban churches of St Martin and St Owen. Within the county Speed marked in approximate outline the boundaries of the hundreds, those ancient Anglo-Saxon administrative and judicial subdivisions of the counties, which may be traced through successive changes in local government from hundred to petty sessions divisions, poor law unions and 1894 rural district councils until in Herefordshire the creation of the present unitary authority in 1998. Within the frame of the city map is a simple compass rose and a scale bar measuring 'A Scale of Pases', an indication that at Hereford, as in some other towns, he personally paced its streets to measure them.

Speed embellished his map with a strapwork inner frame and title cartouche, the coats of arms of the Crown, the city and the medieval barons who had been earls of Hereford, William fitz Osbern (created 1067), Robert 'Bossu' (*c*.1140), Miles of Gloucester (1141), Henry de Bohun (1200), Henry of Bolingbroke (1384, later Henry IV), and 'Stafford' (1380).[17] As in some other counties there is a drawing of a battle scene, the Yorkist victory of Mortimers Cross in 1461. The drawing, which was probably not by Speed himself, focuses on the meteorological phenomenon of the appearance in the early morning mist of three suns at sunrise before the battle. In the bottom corners are two drawings of a map-maker, seated at a cloth-covered table with book, globe, dividers and protractor. The drawings, which are found only on the map of Herefordshire, must be intended to represent John Speed at work.[18] Altogether it is a very useful and appealing map of the county.

The *Theatre* was, however, a large and heavy volume, like Saxton's, the double-page spread of both maps being about 20 by 15 inches (500 x 380 mm.), excellent for consultation on a table or for displaying to impress friends but otherwise inconvenient. Already, in 1599, a young engraver, also from Amsterdam, Pieter van den Keere, had begun a set of 44 county maps, including Herefordshire, used in 1617 to illustrate a pocket-size edition of Camden's *Britannia*. The copperplates of his maps, measuring about 5 by 3 inches (120 x 80 mm.) were acquired by George Humble, who with his uncle John Sudbury had published Speed's *Theatre*, and reissued, rather misleadingly as *England Wales Scotland and Ireland Described and Abridged ... from a farr Larger Voulume Done by John Speed*.[19] It does not, in fact, owe a great deal to Speed, for the selection of place-names is idiosyncratic and their spelling is unreliable. As a complete little atlas it is fun but a single map is only a curiosity.

It was, however, by no means the most curious of contemporary maps as publishers and engravers sought new ways of presenting the same information to a hungry market. John Bill's map of Herefordshire in 1626, though a poor copy of Saxton, was the first to mark a grid of latitude and longitude in its margins, the meridian being measured not from Greenwich but from the Azores. A Greek astronomer, Eratosthenes of Alexandria, had introduced the concept of longitudinal meridians in the 3rd century BC, which Ptolemy developed about AD 150. Ptolemy's zero meridian ran through the Canaries, then the limit of the known world. As the late medieval voyagers explored westward the zero meridian was also moved west to the Azores or Cape Verde islands.[20] Another novelty was a set of playing cards based upon Saxton's map of England and Wales, produced as early as 1590 by William Bowes and copied in 1635 by Jacob van Langeren for the first pocket road book. In this, the top left diagonal half of the page contained a triangular table of road distances between towns of the kind first published in 1607 in an almanac by John Pond and still found in road atlases today.

The Civil War unloosed a demand for accurate maps. Saxton's atlas was republished in 1642 by William Web, an Oxford bookseller thought to have been a Royalist sympathiser. On the Parliamentarian side Van Langeren's road book came out in a new edition with larger maps in 1643, probably published by Thomas Jenner, a Parliamentarian, who the following year and again in 1657 published books of place-names with maps 'Usefull for Quarter-masters' (Plate 12). In the same period in 1645-6 the rival houses of Joannes Blaeu and Jan Jansson, both Amsterdam publishers, unleashed the first of their series of new atlases of the world including maps of the English counties (Plates 4, 13). Their maps of Herefordshire are similar, beautifully drawn and delicately engraved with baroque title cartouches incorporating swags of harvest crops (Blaeu) or sheep and farm workers (Jansson). Both have scale bars to which a surveyor and cherubs are pointing. Both also are derived from Speed, not only the working map but also the coats of arms of the earls of Hereford. Jansson's preliminary map, pre-1646, even had a reduced and much simplified inset plan of the city of Hereford taken from Speed. There is little to choose between them. Both make desirable decorative pictures. Both were long out-of-date when published.

The copying of Saxton and Speed continued. At a time when the estate surveyors in Herefordshire and elsewhere were in their prime, from the late seventeenth to the late eighteenth century, producing exact and beautifully drawn large-scale maps, the printed county maps were unimaginative and often increasingly inaccurate representations of the county as it had appeared in Elizabeth I's reign. The printing presses spewed out Blaeu's maps until 1672, when a fire destroyed the copperplates, and Jansson's until 1724. They were also published, with re-issues of Speed's maps, by the booksellers John and Henry Overton, father and son, in a succession of atlases from the 1670s to 1750s. Late in the seventeenth century the maps of Richard Blome, John Seller and Robert Morden, each claimed as new, were still basically copies of Speed (Plates 6, 15, 16). Morden's county maps for Edmund Gibson's enlarged and popular edition of Camden's *Britannia* in 1695 nevertheless became a new base for the early eighteenth-century copyists like Herman Moll, John Cowley and Thomas Hutchinson.

The single most innovative and influential new atlas of maps in this period was *Britannia* published by John Ogilby in 1675 (Plate 5). It has been suggested that Ogilby may have been inspired by Matthew Paris's strip map of the pilgrim route from London to Rome, *c.*1250. Whether this was the case or not, the format was startlingly new to his contemporaries, for this was a book not of county maps, but a road atlas of the principal roads in England and Wales. Its publication coincided with the rapid expansion of the Post Office and the general post, running on the six post roads that came into existence after the Restoration in 1660. This was also a period of growing standardisation of weights and measures. Despite the slow introduction of statutory measures local variations prevailed in different parts of the country. Herefordshire acres were smaller than statutory acres, Herefordshire miles longer than the statutory miles defined for London and its environs as 1,760 yards by act of Parliament in 1593. By deciding to measure the main roads uniformly John Ogilby did more to spread the use of the statutory mile than any act of Parliament. He did no surveying himself but employed surveyors, to travel the roads with a 'perambulator'. This was a large wheel designed to measure statutory miles which, depending on circumstances, was either pushed or attached to a pony trap. It was a device taken up by other surveyors later. The maps of the roads were drawn in a series of scrolls or strips, a simple linear map of each main road in a form that was to be copied by his imitators in the horse-and-carriage age and survives in the route maps of the motoring associations and road atlases of the present day.[21]

Four main roads crossed Herefordshire in Ogilby's time — London to Aberystwyth (basically the A44 but running via Presteigne rather than Kington), Bristol to Chester (the A466 and A49), Gloucester to Montgomery (A40, A49 and then by B roads to Presteigne), and Hereford to Leicester (the A4103 to Worcester). The maps showed all the landmarks that might be useful to the road traveller, the hills up or down, bridges and fords, road junctions and distances, principal roadside inns and ponds for refreshing travellers and their horses, the country houses near the roadside; the market towns and villages through which the traveller passed were shown in miniature plans. Each scroll had a small compass rose so that the traveller would recognise the changes of direction that he was taking. And, for the first time, Ogilby introduced to the general public the scale of one mile to one inch, an awkward fraction of 1:63,360 perhaps, but instantly memorable and readily measurable on the page by a finger joint. It was a scale that was to be adopted by the eighteenth-century county map-makers and by the Ordnance Survey for three hundred years. The only drawback was the size of *Britannia*, a massive volume suited for the library, not the carriage. As a practical route-finder for the traveller it was useless. Strangely, it was not until 1719-20 that the first of a long series of smaller road books, based on Ogilby, began to pour onto the market created by the growing number of travellers and private coaches on the extending network of turnpike roads.

Ogilby's road maps led other map-makers to mark roads on their county maps. First was Robert Morden, whose set of playing cards in 1676 and his 'smaller maps' of 1701 mark some main roads without much attempt at accuracy. He also marked roads on some of his larger county maps, published in Edmund Gibson's enlarged edition of Camden's *Britannia* in 1695, but there are none on the map of Herefordshire. From the date of Herbert Moll's edition of the smaller Morden map in 1708, however, it was normal for the main roads to be shown, reflecting the growing amount of travel and traffic, even though some were woefully inaccurate. It would be a rash traveller who relied upon them to find a crossing of the river Wye.

Although no new nationwide survey had been undertaken since Saxton's in the 1570s small cartographical improvements crept in. Also, such was the competition among the map-publishers that each sought to obtain sales by demonstrating some novelty. Although most of these little changes are unlikely to interest the local historian they may attract the collector wishing to chart the progress of technology and fashion. To boost sales local patrons were sought and their arms displayed on the map, as on Richard Blome's maps of Herefordshire in 1673 and 1681. John Seller's map of 1694 was the first to be based on a zero meridian running through London at St Paul's cathedral. Morden's maps of 1695 and 1701 were unique in displaying three scales of great, middle and small miles to indicate the local computed miles and the statutory mile calculated from a scale of 60 miles (or more precisely about 69½ miles) to one degree of longitude (see p. 75). Morden also gives both the degrees of latitude and longitude, based on the St Paul's meridian, and the minutes of time west of London in the borders of his maps. Moll in 1708 was the first to indicate towns which returned a Member of Parliament, while Thomas Jefferys and Thomas Kitchin went further in 1748 by also noting fair and market days (Plate 17). Their map was the first to use the Greenwich meridian where the Royal Observatory had been established in 1675.

THE MID-EIGHTEENTH-CENTURY WATERSHED

The shortcomings of all these maps based on Saxton were well known and early in the eighteenth century there was a movement towards entirely new surveys of individual counties at a larger scale founded on the techniques of triangulation and measurement in common use among estate surveyors. Ten counties were mapped at a scale of one inch to one mile or more between 1699 and 1759, the year in which the Society for the Encouragement of Arts, Manufactures and Commerce, after some years' consideration, launched its influential scheme offering premiums for the best county maps at a scale of one inch to one mile.[22] Amongst these ten earlier ones was the map of Herefordshire in 1754 by Isaac Taylor of Ross (Plates 19-25).

Taylor's map set a new standard for maps of Herefordshire. It was a detailed source of information from an entirely new survey by triangulation and measurement of distances, with latitude and longitude marked in the margins. It remained supreme until Henry Price published the results of the Ordnance Survey's work in 1817. In two tables at the foot of the map Taylor describes the features marked and their conventional signs; how, for example, he differentiated between roads crossing enclosed lands and those crossing open country, how the main roads are distinguished from the minor ones, how mileages are written in both computed and statutory miles, how churches with spires, churches with towers, chapels and ruined chapels are each drawn differently. Archaeological sites are marked, and so are industrial features like mills and forges. And, at last, Saxton's molehills are replaced by a newer way of showing relief, by shaded hachures, like woolly caterpillars, with the finely engraved lines drawn to indicate the steepness of the hill slope.

The whole map is overlaid with a grid of letter and number references to pinpoint the seats of the gentry who had subscribed to his publication. Outside the map were two long panels of 288 shields in which the coats of arms of subscribers were displayed, with cross-references to their country seats marked on the map. Sadly, Taylor was over optimistic in the number of supporters he could attract. In the copy of the map in Hereford City Library only 40 of the empty shields have been filled. The side panels were updated at intervals. The British Library copy of the map has 58 shields fully completed and 20 with only crests or names; in the 1786 re-issue of the map the figures had risen to 71 complete and 9 incomplete and another copy, in private hands, has 79 completed arms and 13 with names only. Clearly the sales were slow, despite the map's outstanding qualities and its modest price of 12 shillings for the four sheets. The practice of displaying the arms of patrons or potential subscribers dates from the earliest English printed maps, beginning with Saxton, who showed the arms of the queen and Seckford, and Speed, who included all the arms of the Oxford and Cambridge colleges in the margins of the maps of those counties. John Harris's map of Kent 1719 displayed 152 coats of arms of the county gentry and one of the most influential textbooks for surveyors, William Leybourne's *The compleat surveyor*, which went through five editions between 1653 to 1722, advised them to feature their client's arms on estate maps.[23]

Taylor's is a remarkable and wonderful achievement, measuring 1175 by 962 mm. (about 46 by 38 inches) without the side panels, which were often cut off. Isaac Taylor, however, was no creative originator. The scale, conventional signs, draughtsmanship and the side panels are all taken directly from the outstanding survey of Warwickshire 1722-25 by Henry Beighton (1687-1743), published in 1728. Beighton's widow Elizabeth re-issued his map in December 1750.[24] It would seem likely that it was this re-issue which sparked Isaac Taylor's interest. He had already published a couple of town plans of Oxford and Wolverhampton, both surveyed in 1750 and among the earliest large-scale town maps following John Rocque's maps of Bristol 1742, Exeter 1744, Shrewsbury 1746 and London 1746.[25] The plan of Oxford was rather heavily drawn and engraved by G. Anderton for publication by W. Jackson of Oxford on 29 October 1751. By contrast, Thomas Jefferys of London, a first-class engraver and map publisher, engraved the plan of Wolverhampton, which is of a high quality. From these Taylor, who by 1754 was based in Ross, turned his attention to a county map of Herefordshire. Afterwards he went on to map other counties in the south and west of England in a similar manner.

Questions as to the accuracy of Taylor's maps are often raised. On his map of Hampshire (1759) critics noted that it is difficult to distinguish between his main and minor roads and that some roads just seem to peter out. The map of Dorset (1765) was criticised by his contemporaries for both its standard of accuracy and weak engraving.[26] Historians, however, valuing the diversity and detail of the information he recorded, have been kinder, typified in a recent appraisal 'that Taylor's portrayal of the rural landscape has an evocativeness, due in a large measure to the liveliness of his map language'.[27] Only local historians with a thorough knowledge of their locality in the mid eighteenth century and access to other local sources can judge Taylor's accuracy. In Herefordshire his classification of country houses is reckoned to be sound but there are doubts about his delineation of minor roads.[28] In Gloucestershire, which he mapped in 1777, a close study suggested that places and roads were recorded with reasonable accuracy, but that streams were represented only approximately by a wriggly line. An apparent omission of the main road running north from Newent to Dymock and the Herefordshire boundary, marked on an earlier large-scale estate map of 1775, was explained by highway archives; the road had only been cut through about 1774-5, presumably after Taylor had surveyed that part of the county.[29]

On his Herefordshire map some similar explanation may account for the road in Stretton Sugwas which he marks in 1754 but which is not shown on an estate map made by Meredith Jones of Brecon three years later. Perhaps Meredith Jones had done his fieldwork earlier, or perhaps the route was already planned when Taylor was surveying. Less explicable is his 'Road to Brecon' running more or less due west from the Golden Valley at Wilmaston over the hills to Craswall. It would seem likely that here he had relied not on his personal observation but upon hearsay, perhaps of a traditional route taken by drovers in that direction. Map-makers did occasionally mark proposed routes to demonstrate how up-to-date they were. For example, the British Library's copy of the re-issue of Taylor's map in 1786 has been overprinted with the routes of the Herefordshire to Gloucester and Kington to Stourport canals although neither project was even authorised by Parliament until 1791.

The map, perhaps influenced by Speed, also contained a small inset plan of the city with its principal landmarks intended to serve as a guide to strangers to find their way about the streets. For greater detail 'he Refers to his large Plan on a Scale of twelve perches to one Inch', which he had evidently prepared to follow his plans of Oxford and Wolverhampton before he was diverted to the larger task of mapping the whole county. In the event it was three years before the plan of Hereford was printed in 1757. It too is an excellent map and among the earliest large-scale town plans (Plate 28). It is drawn on a scale of 1:2376, or 26.6 inches to one mile and every building within the city centre and suburbs is marked. The principal ones are shown with impressive accuracy. The Wye bridge is shown with its five arches though a close examination of the cathedral's chapter house reveals that Isaac Taylor drew the ruined but still standing building as octagonal, not decagonal. Even small landmarks are noticed, like the turnpike gates, the Bowling Green and the weighing machine in St Peter's Square, but outbuildings and gardens are probably drawn less exactly. The map exhibits Taylor's trademarks, seen on his earlier plans of Oxford and Wolverhampton, of a flowery dedication to the city authorities and inset drawings of views and prominent buildings, which are also a feature of his manuscript map of Anthony Sawyer's estate at Canon Pyon about 1760.

Taylor's county map had an immediate influence on the best county atlases published in London. Emanuel Bowen and Thomas Kitchin published *The Large English Atlas* in 1755, which set a new standard. The map of Herefordshire, though based on the eighteenth-century copyists of Saxton and Speed, included some features and place-name spellings from Taylor, whose drawings must have been passing through Kitchin's workshop at the same time. Unfortunately, they made a series of copying errors, for example in the Golden Valley area alone, the spellings of 'Backo' for Bacho, 'Arthur's Stoke' for Arthur's Stone' and 'Claunston' for Chanstone. These and similar mistakes were in turn perpetuated in maps of Herefordshire published by Bowen in *The Natural History of England* in 1760 and by Kitchin in *England Illustrated* in 1763. In the second half of the eighteenth century there was a proliferation of monthly magazines containing series of maps of the counties. They are of varying merit and it is clear that the London engravers were copying from each other, with some diminishing accuracy, rather than referring to the more reliable sources now available. Saxton's molehills and source of the river Dore survived into the 1780s, roads wandered erratically off course, place-names spellings remained unchecked.

It was, of course, a period of much change, as industrial areas expanded, turnpike roads were improved and canals built, though Herefordshire was little affected. It was also a period of greater travel so that road-books, like the many re-issues of John Senex's *An Actual Survey of all the Principal Roads of England and Wales*, 1719, derived from Ogilby, and pocket atlases, like Joseph Ellis's *The New English Atlas*, 1765, were popular. There was, therefore, massive publication of county maps and to secure sales the publishers resorted to eye-catching gimmicks. Kitchin's map of Herefordshire in 1754 has a title cartouche featuring a riverside scene, perhaps imaginary, perhaps of Ross, with the fruits and produce of the county arrayed in the foreground, and Emanuel Bowen's of 1755 has a different scene in the same *genre*. In the 1760s both Kitchin and Bowen designed title cartouches within rococo frames. Later, simpler frames became fashionable. Kitchin's map of Herefordshire in *England Illustrated*, 1763 distinguishes not only the parliamentary boroughs and market towns, but rectories and vicarages, whilst the maps in Bowen and Kitchin's *The Royal British Atlas*, c.1764, are surrounded by notes on the county, a practice taken to extremes in Bowen's map of 1767.

Of all the so-called 'new' maps and re-issues only two are of importance for their historical evidence. First is Emanuel Bowen's 'Accurate Map of the County of Hereford' published in *The Royal English Atlas* about 1764 and re-issued about fifteen years later (Plate 7). It was closely based upon Taylor and at 408 by 510 mm. (about 16 by 23 inches) it may be regarded as Taylor at a more manageable size. Second is 'Herefordshire' in *Cary's New and Correct English Atlas* of 1787, sometimes called 'the small Cary' to distinguish it from the maps drawn at a larger scale by E. Noble and engraved by John Cary for Richard Gough's edition of Camden's *Britannia* in 1789. The 'small Cary' was based on Taylor but was the first genuinely new survey for a generation. Relief is shown by hachures, and roads, bridges and ferries are clearly shown. By omitting the hundreds (although they still had a relevance for tax and judicial purposes) and woodland Cary achieved a clean and uncluttered map. It was a commercial success and remained in print for forty years. Its clarity, detail and modest size of 262 by 210 mm. (a little over 10 by 8 inches) maintain its popularity among collectors and make it a convenient choice for reproduction in books (Plate 8).

The small Cary 1787, the large Cary of 1789 (Plate 9), and C. Smith's map of Herefordshire 1802 are best examined side by side and cross-checked against each other. The chances are that deficiencies in one of them will be noticed, which may be corrected by one or both of the others. But as a general rule it is prudent to treat all the other late eighteenth-century maps with caution and check each for authenticity against the one-inch-to-one-mile maps of Taylor in 1754 and Henry Price in 1817.

New standards of accuracy, coupled with a restrained draughtsmanship eschewing decorative features were introduced by the Ordnance Survey. The origins of the Ordnance Survey lie, like Saxton's survey for William Cecil two hundred years before, in precautions for the defence of the realm following the Jacobite rebellion in 1745-6 and in a joint Anglo-French scheme in 1784 to establish a triangulation system linking London and Paris. Its foundation is generally taken to date from the establishment of the Trigonometrical Survey of the Board of Ordnance in 1791. Under the pressure of the war with France, the Board's surveyors, a mix of military engineers and civilian land surveyors, began work on the English survey along the south coast. Kent was published in 1801 and the surveyors steadily worked their way westward and northward. Their survey and drawings for Herefordshire, done at a scale of two miles to one inch, were prepared between 1812 and 1817. In accordance with the practice at the time, they were made available for private publication at the scale of one inch to the mile. Herefordshire was first published as a county map by Henry Price of Hereford in 1817 and after further revision in the field was published in the rectangular sheets of the Old Series by the Ordnance Survey itself in 1831-33. This lies well outside the limit of this study, but the point to make here is the extent to which the government's Ordnance Survey 'First Edition' built upon the practices of the private map-makers, particularly in Herefordshire the eighteenth-century surveyors, Isaac Taylor and, more especially, John Cary and Henry Price.

1. *Christopher Saxton 1577. Proof copy annotated for William Cecil, Lord Burghley.*

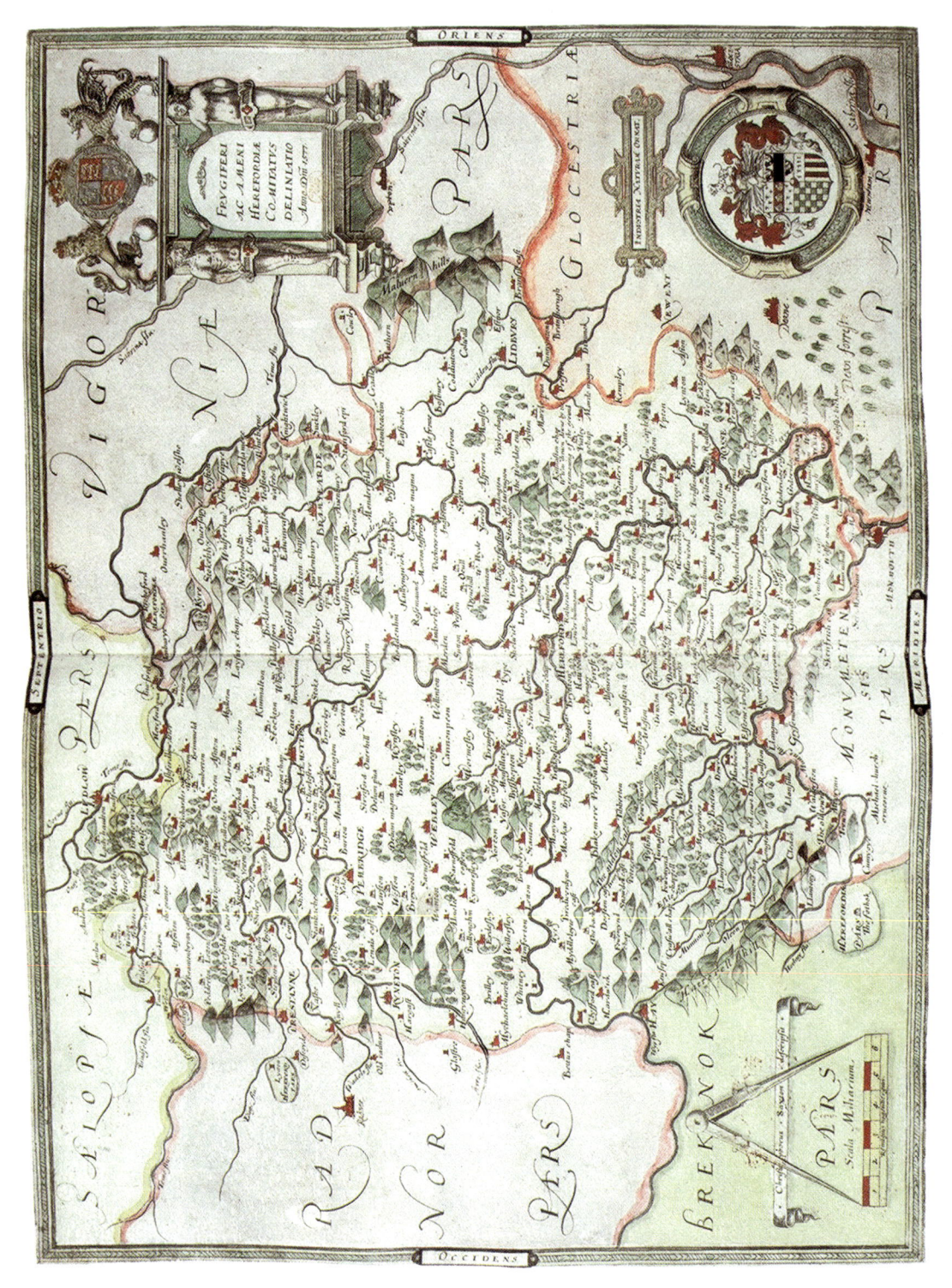

2. Christopher Saxton 1577. The earliest county map of Herefordshire.

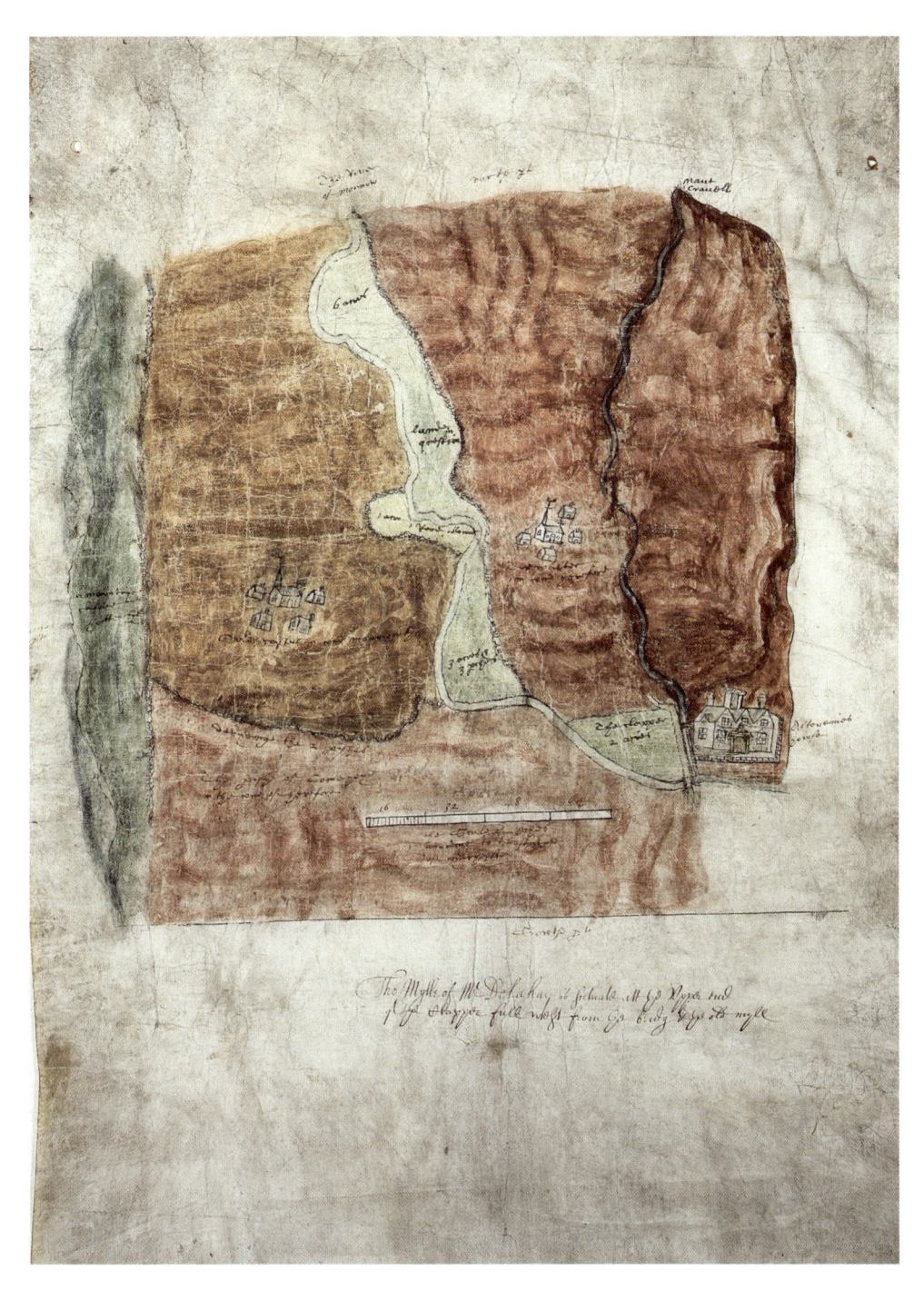

3. *Alltyrynys, the Cecil family's estate at Walterstone c.1600.*

4. Joannes Jansson 1646 (see also Plate 13).

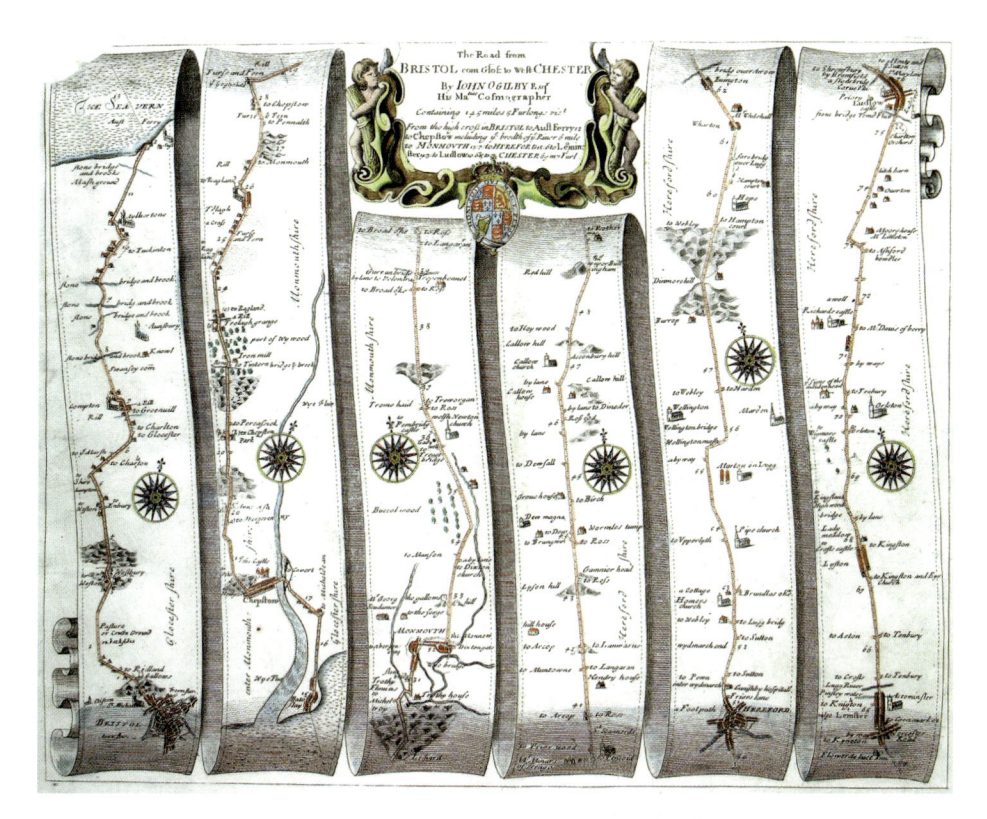

5. *John Ogilby 1675.*

The road from Bristol to Chester. Ogilby's use of the convenient scale of one inch to one mile led to its wider adoption for county maps and by the Ordnance Survey.

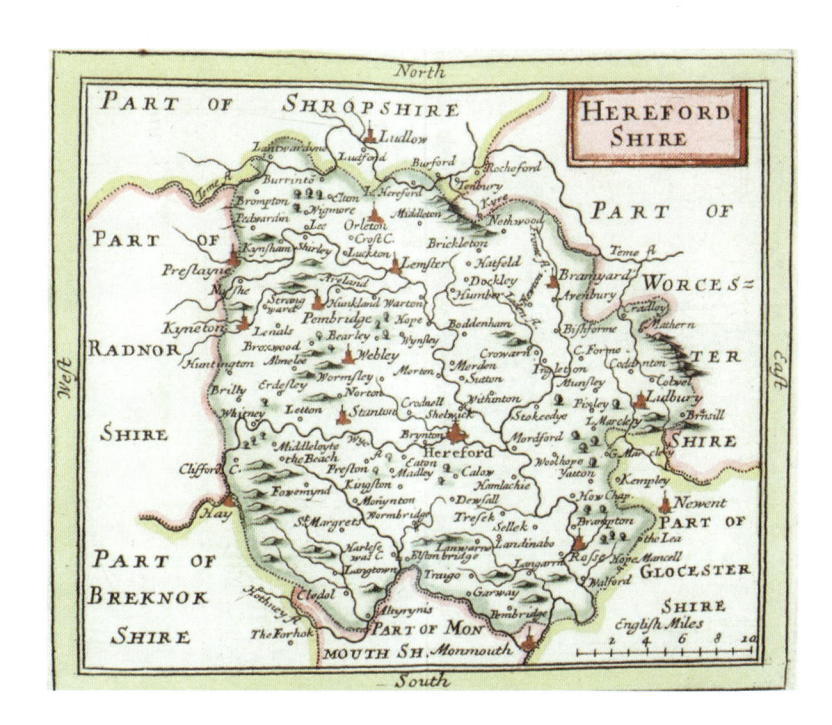

6. *John Seller 1694.*

Reduced from Speed.

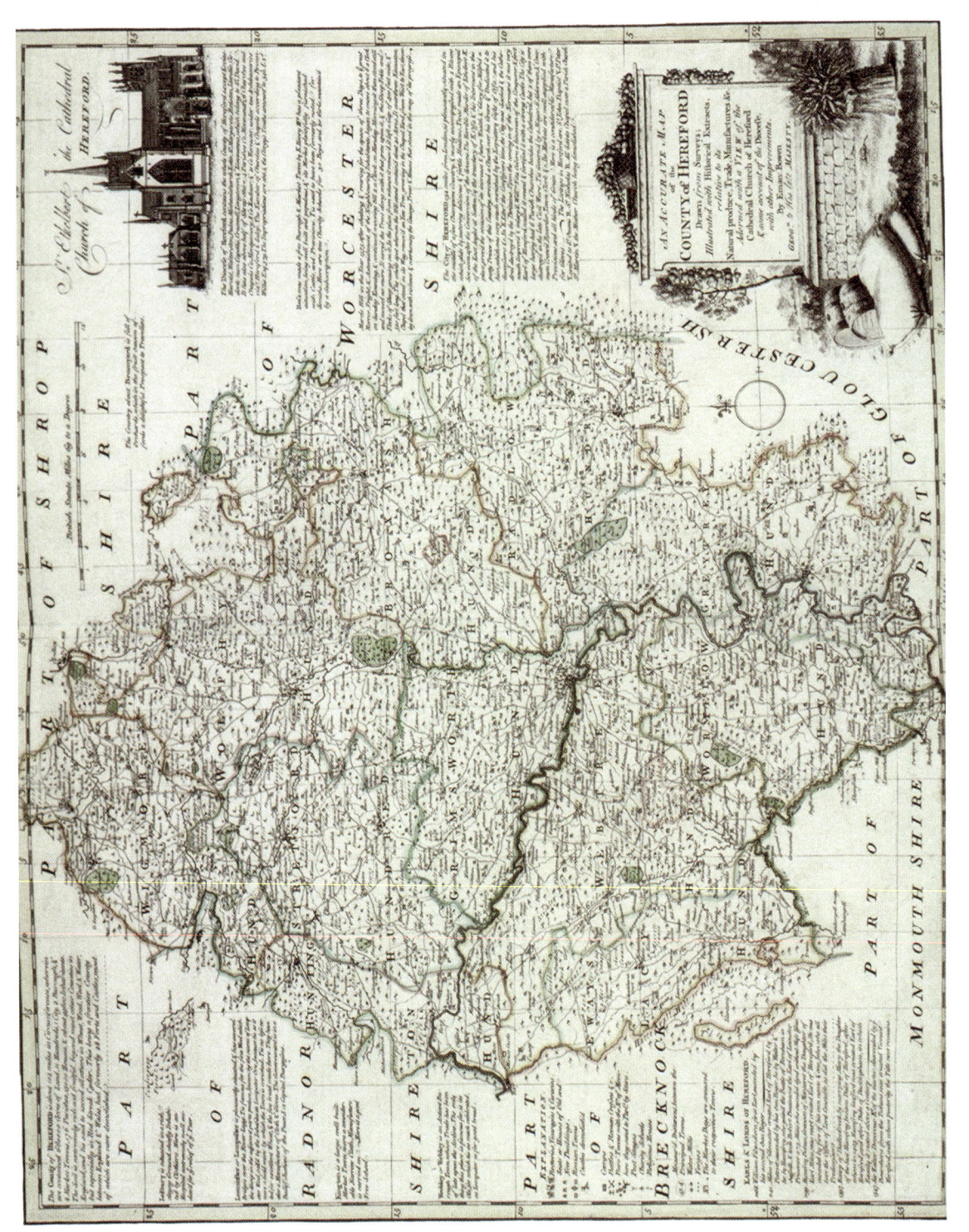

7. Emmanuel Bowen c.1762-64, based on Isaac Taylor 1754.

8. *John Cary 1787 (the 2nd edition of 1793). 'The Small Cary'.*

9. E. Noble, engraved by John Cary 1789 (the 2nd edition of 1805). 'The Large Cary'.

4 Estate Maps

The printed maps provide an overall view of Herefordshire. Manuscript maps, the overwhelming majority of which are estate maps, bring small selected areas under the microscope. Whereas the printed county maps are commonly of very small scales, with the one exception of Isaac Taylor's 'one-inch-to-one-mile' map of 1754, the estate and other manuscript maps are mostly drawn at scales of one inch to three or four chains, that is 26.6 inches to 1 mile (1:2376) or 20 inches to 1 mile (1:3168).[1]

The origins and functions of these large-scale manuscript maps were quite different from those of printed maps. The most distinctive distinction, of course, is the uniqueness of estate maps. Normally only one copy was made, hand-drawn and usually hand-painted. Despite the differences in scale, multiplicity of copies and purpose there are practices common to both in the methods of surveying and equipment used and in the styles of draughtsmanship and decoration. Some surveyors, as distinct from the map-publishers, produced both printed and manuscript maps.

SURVEYS

Written descriptions of property are recorded in Herefordshire from the sixth century. Charters of the Herefordshire lands of the diocese of Llandaff, copied into the early twelfth-century Book of Llandaff, set out the boundaries of the places described. Clearly, a party walked the bounds to identify the streams, hills, prominent stones and other landmarks from point to point. The bounds of Malvern Chase, adjoining the bishop of Hereford's manors of Colwall and Eastnor, were similarly ridden by the verderers in 1584 and the traditional perambulation of bounds of the city of Hereford was still led by the mayor every third year in the nineteenth century, though it has since become a rare token event.[2] More commonly the survey was obtained by inquiry of reliable witnesses, whose evidence was then written down. The best known survey of this kind was William the Conqueror's Domesday Book, ordered by the king at the Christmas meeting of his council at Gloucester in 1085. The county courts were called to provide the information about his new kingdom. Throughout the middle ages local landowners carried out similar surveys through their manorial courts. By Tudor times this verbal evidence could be supplemented by the manorial court rolls and other records. After searching the episcopal archives Swithin Butterfield, steward to the bishop of Hereford, compiled a huge and magnificent volume of surveys of the bishop's estates in 1578-81. A century later the dean and chapter of the cathedral, seeking the recovery of their properties after confiscation during the Commonwealth, used similar methods well into the eighteenth century when renewing leases of their farms.

By that time, of course, landowners were familiar with maps. It is, therefore, a little surprising and a sign of both the cost and Herefordshire's habitual backwardness, that the dean and chapter should not have employed a surveyor to map their estates systematically after the Restoration. The practice of estate map-making had taken root in Kent and Essex in

Elizabeth I's reign and spread out from south-east England during the seventeenth century. It was slow to reach Herefordshire. The earliest map of land in Herefordshire dates from 1577 and only five maps before 1600 and twenty-one before 1700 have been traced. In comparison, Devon has 60 dated before 1700 and Kent 187.[3]

The reasons for this contrast are two-fold. The first is the remoteness from the capital and south-east England. All Wales and Westmorland are similarly poorly represented in distribution maps of surveyors at work.[4] The second is the pattern of landownership in the county. The Crown's property interests in Herefordshire were limited, which makes it the more significant that of the five earliest maps one should be of that part of south Herefordshire adjoining the royal Forest of Dean, a second should be of Bringewood Chase, which came to the Crown with the accession of Edward Mortimer as Edward IV in 1461, and a third should be of William Cecil's ancestral home of Alltyrynys in Walterstone (Plate 3). There were also no large corporate estates after Elizabeth I seized the bishop's estates in 1559 until Guy's Hospital acquired full control over the 16,000-acre estates of the Dukes of Chandos in 1754. The Scudamores of Holme Lacy and the Harleys of Brampton Bryan, were among the more important local landowners interested in the management of their estates in the seventeenth and eighteenth century, but they exhibit a contrasting attitude towards maps. Twice, at the beginning and towards the end of the eighteenth century, the Scudamore family estates were surveyed in series of maps but the Harleys relied almost entirely on written surveys and valuations. Neither family wielded the wealth and power of the noble families elsewhere in England. Herefordshire's squires residing on their relatively small estates had less need of maps than absentee landlords or stewards managing larger estates.

The Purpose of Estate Maps

The earliest maps were often compiled to support other written evidence. The maps of Bringewood Chase in 1577, Marden in 1582 and Alltyrynys a few years later were each drawn up in connection with legal disputes. All are diagrammatic rather than accurate surveys. The map of Alltyrynys was drawn at a variable scale, with the lands in dispute adjoining Alltyrynys house and Trewyn mill at the centre of the map shown both larger and more accurately than the more distant landmarks of the surrounding countryside; Walterstone village, Oldcastle and Hatterrall Hill were included at a smaller scale merely to put the site of Alltyrynys into its context. All three maps also demonstrate that, unlike the printed county maps, estate maps are primarily concerned with boundaries. It is much simpler and more intelligible to draw the boundaries of a plot of land than to describe them in writing. The point is underlined by the oldest map in the Herefordshire Record Office, the late sixteenth-century map of Brilley Common, which marks and measures only the bare outline of the irregular bounds of the common.

It follows that the other features marked are all subordinated to this prime function of showing the extent of the landowner's property. As owners are not concerned with their neighbours' properties only their own property is marked, though sometimes in the complex intermingling of strips in open fields the surveyors indicated the general outline of the whole field. More often, as in the maps of Hellens in Much Marcle 1741, Richards Castle 1743 (Plate 35) and Hillhampton in Ocle Pychard 1791 only the owner's scattered strips are shown, like islands in the sea. Landowners, unlike travellers, did not need to know whether their lands went uphill or down, only where they began and ended and, secondarily, how they were identified, what was their extent and quality, and what was their value or rental. To achieve this purpose early estate maps served as a novel illustration to the familiar written survey, so much so that surveyors sometimes thought it necessary to explain the function and conventions of a map to their clients. In the seventeenth century the use of maps became commonplace but right through to 1800 estate maps usually included a reference table or were accompanied by a separate book or sheet, setting out the written details of the lands surveyed. The written survey was never entirely superseded by a map alone. The survey, with other related records where they survive, remains essential to understanding the map.

The cost of mapping an estate was such that a landowner needed a powerful motive to commission a surveyor. The decision was prompted by special circumstances. Legal disputes continued to be a strong reason. In 1784-5 the treasurer of Hereford cathedral and their tenant and neighbouring landowner, Richard Aubrey, agreed to appoint independent surveyors from London to settle amicably their differences over the ownership and boundaries of unfenced scattered strips in the open fields of Breinton.[5] Most frequently, though, landowners wanted to forestall such trouble. The acquisition of property was commonly the occasion to arrange for it to be mapped. Just as William the Conqueror had ordered Domesday Book to be made to find about his newly won kingdom, so the first step of a new estate owner, whether by purchase, marriage or inheritance, was to find out exactly what he had acquired.

Thomas Foley, the ironmaster, bought the Stoke Edith estate for his second son Paul in 1670, triggering a series of estate maps dating from 1680. Robert Chaplin, a London merchant who owned the Shobdon estate, had William Whittell of Bodenham survey part of the estate in 1705, the year that he sold it to his fellow Londoner, Sir James Bateman. Within three years Bateman, who remained in London for much of his time, had the other parts of the estate surveyed by Whittell (Plate 31). Another rich Londoner, Jacob Tonson, bought the Hazle estate near Ledbury. He was the publisher of Pope, Addison and Dryden, who called him 'the cheerfullest, best, honest fellow living'. Others were less flattering, among whom Thomas Hearne, the antiquary, described him as 'a great, snivelling, poor spirited whigg and good for nothing that I know of'. Immediately after buying the estate in 1720 Tonson had it surveyed by Charles Price, an accomplished London surveyor, instrument-maker and map-publisher (Plate 33). Corporate owners were equally keen to have their estates mapped. Three years after gaining control of the Brydges family's former estates in 1754 Guy's Hospital had them surveyed by Meredith Jones.

Similarly, inheritance or marriage might be followed by an appraisal of newly acquired assets. George Carpenter, 2nd Baron Carpenter, succeeded his father in 1731; he had the Homme estate in Weobley surveyed two years later (Plate 34). Richard Payne Knight inherited the Downton estate from his grandfather Richard Knight, the ironmaster, in 1772. He built Downton Castle in 1773-78 and then turned his attention to landscaping the park in 1780, the year that James Sherriff surveyed the lands of his whole estate (Plates 52-54). The Hon. Charles Howard (11th Duke of Norfolk 1786) married Frances, the heiress of the Scudamores of Holme Lacy, in 1771 and immediately commissioned the Irish surveyor Richard Frizell to survey the estate. In all these cases the maps were intended to form the basis for the owners and their stewards to assess the value of their property, monitor their rentals and plan improvements.

The first and simplest improvement was the clearance of woodland or conversion of waste land into enclosed arable. The map of Brilley Common about 1590 was probably drawn in connection with a proposal for its enclosure. The partial felling and enclosure of Bringewood Chase and Mocktree Forest went on for centuries. The 1577 map of

Bringewood shows past enclosures and felling in progress. A map of 1662 records further enclosures in Bringewood and the felling of Mocktree, although the former forest remained common until its enclosure in 1803 (Plate 30).[6]

The second and more complex improvement was the enclosure of open arable fields and meadows, in which landowners possessed scattered unfenced strips of lands, intermingled with those of their neighbours across the whole of the huge fields. The system of open-field agriculture persisted over much of lowland Herefordshire but the hilly and wooded nature of the county had resulted in piecemeal clearance and ownership from an early date. Whereas in most of Midland England the enclosure of the open fields is associated with late eighteenth and early nineteenth-century enclosures authorised by private acts of parliament, in Herefordshire enclosure proceeded by private arrangements. These are mostly unrecorded and difficult to date. When Guy's Hospital commissioned Meredith Jones to survey Stretton Sugwas in 1757 he portrayed a landscape of open arable fields (Plate 36). The Hospital's estate papers do not reveal whether his map was intended as a preparation to enclose the fields, but enclosure had occurred by 1794 when the area was re-surveyed by James Cranston (Plate 63). In the eighteenth century only three parliamentary inclosure awards were enrolled with the clerk of the peace, the earliest in 1780. Two were primarily for the enclosure of wood or waste and the only typical act of parliament for the enclosure of open arable fields was that for Much Marcle in 1797. In addition, the act for paving and lighting the city of Hereford in 1774 also authorised the enclosure of Widemarsh and Monkmoor commons.

Estate maps might also be commissioned in connection with other forms of estate improvement to raise the value and rental of the estate, but this intention is rarely revealed on the map itself, only in associated estate papers. The survey of the Moccas estate in 1772, with its varied arrangement, cross-references and pencil annotation of the tenants' names and holdings was clearly intended and used as a working tool (Plates 41, 42). Maps recording exchanges of lands to consolidate holdings occur most commonly when one of the parties was a corporate body like the dean and chapter of Gloucester in 1774, the borough of Leominster in 1793 or the parish of Weobley in 1799. On the whole, maps give a snapshot of an estate at a specific moment and the changing picture of adjusting the boundaries of tenants' farms, altering field fencing or changing land-use over a period of time has to be extracted or deduced from letters, leases, rentals, accounts and other estate papers. The correspondence of the estate stewards, John Pye in 1727 and Benjamin Fallowes in 1753 refer in passing to maps made of the Mynde and Shobdon estates, but not until later in the eighteenth century was a new breed of land surveyors commissioned specifically to advise on improvements, like Nathaniel Kent at Kyre Park in 1772-4 and James Sherriff at Downton in 1780. Finally, the bird's-eye view of an earlier proposed improvement, published in 1610 by Rowland Vaughan of Vowchurch, deserves passing mention. His utopian scheme for a 'commonwealth' in the Golden Valley to provide work for his impoverished neighbourhood apparently did not go beyond building a water mill, but his drawing of the workshops, dwellings and gardens is not far removed in style from the map of Alltyrynys of about the same date.[7]

THE FEATURES OF ESTATE MAPS

It did not take the surveyors long to learn how to meet the requirements of their clients. From the beginnings before 1600 surveyors in south-east England, like Ralph Agas, Israel Amyce and Ralph Treswell found the ways to provide this information neatly on their maps. Their example was followed by their successors with little change until about 1700. There were instructional books from Elizabethan times, like Leonard Digges, *A boke named Tectonicon* (1556) and *Pantometrica* (1571), Ralph Agas, *A preparative for the platting of landes and tenements for surveigh* (1596), John Norden, *The surveyors dialogue* (1607) and William Leybourne, *The compleat surveyor* (1653, 4th edn 1679). Towards the end of the seventeenth century the surge of interest in the sciences and mathematics, marked by the foundation of the Royal Society, led to an expanding number of teachers of mathematics like John Dougharty the elder (1677-1755) of Bewdley and Worcester. In surveying and map-making there was a growing desire for accuracy. Among a fresh series of textbooks in the early eighteenth century was Edward Laurence's *The Young surveyor's guide* (1716) and *The duty of a steward to his lord* (1727) and a fifth edition of Leybourne's *The compleat surveyor* (1722).

The textbooks and teachers set out the application of geometry and trigonometry for field-surveying, the methods of using plane-table, circumferentor, thedolite, chain and perambulator to measure angles and distances. They gave instruction on protracting measurements from field notebook to a sheet of paper and the drawing of conventional signs, like trees to indicate woodland or shading to indicate hills. They explained how to mix and apply colours and how to design a map, with compass rose, scale, and the client's coat of arms. Taken together with the normal exchanges of professional experience they ensured that estate maps throughout the country possessed common features and characteristics within which surveyors might develop their individual style. The results may be seen in the maps of William Whittell 1705-09, an unsigned map of Marden *c.*1725, John Green's map of Parton in Willersley 1727 and J. Meredith's of The Homme in Weobley in 1733. As Leybourne concluded, 'These things being well performed, your Plot will be a neat Ornament for the Lord of the Manor to hang in his Study, or other private place; so that at pleasure he may see his Land before him, and the quality of all or every parcel thereof, without any farther trouble'.[8] Today, framed estate maps can still be seen hanging in the study of a country landowner or on the walls of the estate office, whilst the fine leather-bound estate atlases are less likely to be in the muniment room of a country house than on the shelves of its library.

By the time that map-makers were working in Herefordshire it was hardly necessary for surveyors to explain how to read a map. Nevertheless, William Whittell in presenting his map of lands in Shobdon and Aymestrey to his client Sir James Bateman in 1708 took the precaution of noting:

> SHOBDON HILL, the little vallet, and ELBATCH VALLET lyeth in the Parish of SHOBDON.
> The rest of the WOODS and LANDS are Situate in the Parish of ALMESTREE
> The Characters distinguishing the Lands are:
> Red M   Signifies Meadow
> Red P   Signifies Pasture
> Arable is distinguish'd by small Red lines in the Form of Ridges
> Woods are Describ'd by small Trees
> Lands mark'd (A) are those late Hardway's
> Lands mark'd (B) are those in Possession of Rich: Cooke.

As late as 1743 John Corbet, a Warwickshire surveyor, still evidently thought it advisable to inform Henry Jordan when he surveyed his estate at Richards Castle[9]. An old-fashioned draughtsman, his wording perhaps suggests his own lack of confidence rather than the presumed ignorance of his client:

Note. The round Red spot by the House is the Ground where the Dovehouse stands. The Black squares the Ground where the barns stand and the Green line that crosses the Road by Mill green Denotes a Gutter that parts Herefordshire and Shropshire. Note. all lands in Common Fields that belong to the Estate and are not Inclosed are expres'd by Letters. Number'd and wash'd over with several Colours and all pieces Inclosed have their names wrote in the middle Number'd and washed only round the Boundary line. Note. wherever any Person has land adjoyning. There their names are wrote which shows the Bounding of every piece or parcel of land whether it be Inclosed or not. Note. The House near Linehalls Eye. Commonly call'd by the name of Cams House. and adjoyning to that House in the Middle of a Field. is about three parts of an acre belonging to the Estate and Ioyn'd on the South side thereof with Cams Land. and on the North with Land of Sq.^{re} Salwey. Note. The Black line draw'd from the Corner of the Fishpool by Mill Green. Denotes a small stream of Water that runs there. And here Note. That the Black line at the bottom of Orlly round. Denotes a Brook that passes Orltons Parish and Richards Castle &c.

Note. all Bridleways are double spac'd
Foot path single
Gates expres'd thus…..I=I
And Stiles thus……….H.

The features shown on Herefordshire estate maps are also similar to those marked on maps throughout the country. The estate boundaries are fundamental, sometimes confirmed by marking notable landmarks just outside them and routinely naming adjoining landowners or properties. Within the estate, the tenants' holdings are commonly distinguished by different colours, usually by a thin coloured line painted round the edge of the fields or strips in open fields. Where only one holding is mapped, appropriate colours may be used instead to distinguish between arable, meadow and pasture. Woodland and orchards are marked with appropriate drawings of trees. Freestanding trees, either in orchards, parks or isolated elsewhere, are frequently drawn with a shadow as if seen in afternoon sunlight. The identification of coppices, sally-beds and hopyards is revealed only by their field-names. Similarly, quarries, clay pits or brickyards, lime kilns, smithies, forges, inns, mills will be indicated by written field-names rather than conventional signs. The field-names, acreage and land-use are either written within the field or set out in a reference table on the map or a separate sheet or volume, depending on the size of the area mapped. In such cases the fields drawn on the map are identified by reference numbers or combined letters and numbers cross-referenced to the table. Few surveyors distinguished the type of fencing around fields or indicated their ownership, but the position of gates is more often shown. Hedgerow trees are usually spaced conventionally, even when isolated large trees within fields are marked.

Only a little less important than the extent and layout of the estate are its homesteads and buildings. From Tudor times until the early eighteenth century these are represented in little drawings of the houses and outbuildings, most often in bird's-eye view (as if seen from above (Plate 43)), perspective (as if seen from ground level at an angle to show two sides of the building) or in profile (as if seen from ground level 'lying on their backs'). Usually the drawings are merely conventional signs but larger buildings, like the principal house or a parish church, may sometimes be shown with hint of realism. Just occasionally, as in Joseph Dougharty's map of Grendon Warren in 1732 (Plate 47) and Samuel Addams's of Brockbury in Colwall in 1758, the drawings may be more precisely accurate.

Other man-made features commonly include roads and bridges, with access tracks less often and footpaths rarely. Among natural features shown are ponds, streams and rivers, sometimes with an arrow to mark the direction of flow.

Relief, though, is very rarely indicated. The mole-hills marked on the printed maps by Saxton and his imitators would have masked the essential features of estate maps. Edward Laurence drew a hump for Shepards Hill at Goodrich in 1717, but not very successfully (Plate 45). Both he and Daniel Williams in 1758 (Plate 50) attempted only a little more successfully to mark the sheer limestone cliffs above the river Wye near Symonds Yat. Hachures or shading, used by the engravers of printed maps to indicate hill slopes, was rarely resorted to by Herefordshire surveyors of estate maps. Two examples are John Lambe Davis's map of the Moccas estate in 1775 and Edward Thomas's map of Rowlestone in 1776. Meredith Jones's map of Aconbury in 1757 has a varied tone of green watercolour for the woods on Aconbury Hill, but this may be the result of applying the wash unevenly rather than a deliberate attempt to suggest the contours of the hill (Plate 49).

The conventional signs changed little and slowly over the whole period from about 1580 to 1800. The illustrative drawings, often rather crude, of animals or figures in the landscape, which appear on early maps in other parts of England, are almost completely absent from Herefordshire maps. The practice of drawing buildings in block plan was introduced from about 1720 and by 1750 had replaced the earlier conventional bird's-eye and other views. From about 1780 a further distinction was made between dwellings, painted red, and outbuildings, painted grey or black, though earlier examples occur on maps of Leominster in 1747 and Bridstow in 1756. These changes, together with the more familiar developments in handwriting styles, are useful in dating undated maps, with the cautionary reminder that elderly or unskilled draughtsmen may continue or imitate old-fashioned conventions a generation after the introduction of innovations. The maps of Mathon in 1772 by John Aird and of the Longtown area by Edward Penry in the 1790s are examples of this time-lag.

Unlike the printed map-makers the land surveyors often failed to include a scale bar on their maps, especially before 1700. The earliest Herefordshire maps with a scale date from the mid seventeenth century, of Wigmore (undated and unsigned) and Mocktree Forest and Bringewood in Downton by William Fowler in 1662 (Plate 30). Until about 1700 the scale was usually measured in perches (one perch was five and a half yards) but from 1720 a scale of chains was almost universal, using Edmund Gunter's chain of 100 links measuring 22 yards, introduced in 1610. At first estate maps were drawn to any scale which the surveyor found convenient, influenced by the size of the area to be mapped, the detail that had to be included and the proposed size of the finished map. In 1721 Edward Moore drew a map of Fawley at the scale of one inch to sixteen perches and it was this scale, which converts to one inch to four chains or twenty inches to the mile (1:3168) which surveyors most commonly used. It was suitably large to mark all the features in sufficient detail on a reasonably sized rolled map and it was not replaced as the most popular large scale until the Ordnance Survey produced the 1:2500, or so-called twenty-five inches to one mile (1:2534) maps in the mid nineteenth century.

Working at these large scales the estate surveyors did not concern themselves with the distinction between statutory miles and the longer local customary miles that had troubled Celia Fiennes and Robert Morden at the end of the seventeenth century. They could not, however, entirely ignore the differences between statutory acres and the 'computed' acres calculated to reflect land values and recognised by local landholders. Writing on 25 October 1726 John Pye of Kilpeck attempted to explain the difference to the London merchant, Richard Symons:

> Sr.
>
> I receaved a letter from you dated the 27th of September which was not answered sooner because I thought to have seene George Smyth [another Herefordshire surveyor] first, he being at Cirencester. Yet I know he gave you an account that the Cockyard Vallett [a coppice] conteined something above 22 statutory acres without measuring the Limestone ground or any other. And the survey which since came to your hand was made (I believe) by Mr Cleare and it wanted 7 perches of 22a statutory the difference being less than an acre might be by George's measureing every nock [nook] more nice, either of which is betweene 13 and 14 acres Customary measure, (viz) 21 foot to the perch which is the Custome in Hereford shire, but in some other places it is 18 feet, in other 24 feet to the perch. But for computed acres I never heard of any for wood before I heard you mencion it. But there is a measure which we commonly goe by in arable land & pasture laid downe from arable and call'd by some Computed acres, but it is most properly called covers, as they doe in Monmouthshier, 3 of which makes 2 statutory. Soe there is noe mistake of it if you consider it ...
>
>   Farewell soe ever wishes he
>   who's more your friend than he can seeme to be
>   Jo: Pye.[10]

Whether the explanation satisfied Richard Symons is not known but the general rule in Herefordshire was, as Pye stated, that for arable land two statutory acres were the equivalent of three computed acres, and for woodland five statutory acres equalled three computed acres.[11] In practice, surveyors rarely went to the trouble of recording the computed acreage. William Whittell of Bodenham gave both measurements in surveying woodland in 1705-08 and Meredith Jones of Brecon on three occasions between 1754 and 1766. In the reference book to his map of Coddington 1760 Jones noted additionally that 'Meadows most commonly are computed by the days Math which comes near the Statute'. The anonymous surveyor of woods at Little Birch gave both measurements as late as 1794.

Following the example of the printed map-makers it was the common practice for the land-surveyors to write the cardinal points of the compass on the edge of early estate maps or later to include a compass rose on the face of the map. However, unlike the printed map-makers they did not always align their maps with north at the top. Estates were frequently of such an irregular shape that maps of them fitted awkwardly on to a rectangular parchment skin or paper page. It was, therefore common for draughtsmen to skew the map to fit the page, turning the compass to the correct direction of north. Or, sometimes, turning it incorrectly. In a few late eighteenth-century examples the maps have a grid superimposed for cross-reference with the tables of field-names. Only very rarely does a pretentious land-surveyor mark the longitude on an estate map.

DECORATION

The finished maps were drawn in ink, using goose quills for normal lettering and drawing and raven quills for the finer work. Watercolour, mixed according to recipes given in the textbooks, was generously applied in bold colours to the decorative features as well as to the working map. These bright colours were toned down in the second half of the eighteenth century until by its last decades little more than a modest grey wash might be applied, perhaps relieved by a restrained use of brown, blue or pink for roads, rivers and buildings. In the same period penmanship became finer and more delicate and the quality of the best maps matches topographical engraving. Until the last quarter of the century most finished maps presented to the landowner were drawn on parchment, and paper was used only for drafts and rough small maps. Parchment is, of course, a more robust material than paper, though inclined to buckle and, disconcertingly for a scaled map, to stretch or shrink according to the dryness or dampness of its storage place. Fortunately, any such warping of large-scale estate maps is not critical. In the later eighteenth century paper became more generally used, large maps being drawn on several sheets mounted on linen or other cloth. Though paper is a less strong material than parchment for the historian it has one valuable advantage — the manufacturer's watermark may help to date an otherwise undated map.

The cartouches for the title of the map and its reference tables, perhaps incorporating the landowner's coat of arms, the scale bar, compass rose, outside border or frame, allowed surveyors the opportunity to embellish the essentially practical estate maps. It was the one area where they could express themselves freely following the dictates of contemporary fashion. In this they were undoubtedly influenced by artistic trends and the practice of the map-makers and engravers.

The heavy title frames, like those laboriously executed by William Fowler (Plate 37) and William Whittell at the turn of the seventeenth and opening years of the eighteenth century, hark back to the strapwork designs of a century earlier. They gave way in the early and mid century to ornate baroque cartouches reminiscent of silverware and picture frames. In the second half of the eighteenth century lighter rococo designs with intertwining trails of foliage and flowers prevailed. These were followed in turn by 'picturesque' scenes worked around the title, which might be written on a stone slab or pediment among ruins, set against a background of trees or river scenes, with a castle or timber-framed cottage to lend a romantic air. Occasionally a figure steps into the scene, a peasant girl or a fisherman perhaps. One of the earliest such scenes, a classical pavilion set by a pair of trees flanking the title, occurs on the map drawn by Thomas Leggett of Brockhampton (Bromyard) in 1769.

A pair of dividers, at first ornate and heavy, commonly bestrode the scale bar, sometimes, as on John Harris's late eighteenth-century maps, with protractor, square, pen and other drawing instruments tucked behind it. Compass roses were a prominent feature of seventeenth- and early eighteenth-century estate maps, brightly coloured and elaborate multiple-pointed stars. William Fowler's 1662 map of Mocktree Forest is dominated by a large compass rose, and one of the two on John Corbet's 1743 map of Richards Castle has thirty-two points. But in line with penmanship generally they became simpler and less flamboyant from the mid eighteenth century. The north point was always given prominence, and sometimes the east point was embellished with a small cross. The outside borders similarly became less decorative until they were often reduced to a mere double-ruled line. Throughout, depending upon the purpose of the map, the client's wishes, and the surveyor's inclination, some maps were drawn without decoration, but with plainly framed titles and minimalist scale bars and compass points.

Decoration was an effective form of advertisement for surveyors. They could demonstrate their own skills of draughtsmanship and at the same time give pleasure to their clients. It is, therefore, surprising that so many maps were not signed. Of those catalogued here only 30% before 1700 were signed. Throughout the whole of the eighteenth century 70% were signed. Some, of course, were intended as no more than drafts. In other cases the maps would have been accompanied by separate books or sheets of reference tables, which perhaps were signed by the surveyor. Because of their different shape and size map and reference tables have often become parted from each other and the map alone, being larger and more attractive, has survived. Even when unsigned some maps can be identified by the distinctive styles of decoration developed by individual surveyors. Benjamin Fallowes spread pale coloured flowers around the borders of his maps (Plate 38), John Pye pasted a strip of paper with an engraved pattern along the edge of the parchment (Plate 44), the Dougharty family frequently incorporated a man's head or a shell at the top of the title cartouche and painted the double-ruled border in yellow watercolour.

These decorative features fulfilled a secondary purpose of estate maps. The maps were intended to be displayed, to show off not merely the ability of the surveyor but the wealth of the client, the extent of his property and his pride in having it mapped attractively. The great leather-bound atlases of Lord Bergavenny's estates in 1718, John Whitmore's Monnington-on-Wye estate in 1771 or Richard Payne Knight's Downton Castle estates in 1780, are superlative examples of this pride. Large parchment maps were intended to be hung for display, like the engraved maps seen hanging in the rooms of Dutch houses in Jan Vermeer's paintings of seventeenth-century Dutch house interiors. In a few cases, like John Pye's map of Sir William Compton's farm of Cwm Maddoc in Garway about 1695, the map remains attached to its wooden roller. Smaller maps were sometimes framed for display. But just as one can imagine Lord Bergavenny in 1718 opening his great estate atlas to show to his aristocratic friends so also may John Lewis have taken equal pleasure in 1794 in laying out on his table Edward Penry's map, no larger than a tea-tray, of his modest 32-acre farm of Neuadd-Lwd in Longtown.

5   The Surveyors

THE PIONEERING COUNTY MAP-MAKERS

The men who mapped Herefordshire mostly came from outside the county.[1] This reflects the fact that Herefordshire is far removed from the richer, innovative and fashionable south-east of England and London where patrons and clients, printers and teachers, instrument makers and surveyors were concentrated. In the period of map-making in Herefordshire, from about 1600, the county town was declining in regional importance. Despite continuity of ownership even the larger estates in Herefordshire, like those of the Scudamores of Holme Lacy, Foleys of Stoke Edith or Coningsby of Hampton Court were of a relatively modest extent and influence in a national context. The only large institutional owner was Guy's Hospital. The general pattern of landownership in Herefordshire revolved round a country house and its surrounding farmland, coupled with the lordship of few manors and a seat on the magistrates' bench. Unsurprisingly, therefore, much of the map-making work in Herefordshire was carried out by surveyors based elsewhere in the region or by well recommended men brought in by the owners of the bigger estates or wealthy mercantile incomers. Throughout the whole period relatively few surveyors were based in Herefordshire. Of the 108 named surveyors identified only 35 certainly or probably had a Herefordshire address. Little evidence can be traced about the lives of most of them.[2]

The making and publication of the printed maps of the counties was dominated by the London book trade. In the early days experienced continental engravers and printers in the Low Countries powerfully influenced this trade. Many Flemish Protestants took refuge in London and four out of the seven engravers of Saxton's maps in 1579 were Flemish, Remigius Hogenberg being responsible for the largest number, including that of Herefordshire. Few of the engravers, printers and publishers left their workshops and offices in the streets around St Paul's cathedral, being content to get their information from earlier maps and other works. Only the true surveyors had to work in the field locally.

Christopher Saxton, a Yorkshireman born and bred, was an unlikely candidate to survey the whole of England and Wales for the Elizabethan government.[3] As he rode round the Herefordshire countryside in the mid 1570s, with his strange accent and official credentials, plotting and drawing his map of the county, he must have been received with curiosity and hostility. The introduction of parish registers in 1538 had been popularly viewed as a preliminary to taxation, and to have a man from the government making enquiries and notes about the neighbourhood cannot have been welcome. Swithin Butterfield, bishop John Scory's steward, aptly described the typical attitude of Herefordshire farmers in warning his successors in 1581:

> You that be officers looke well upon the Recordes & Court Rolles, For Customes which are for the lordes benefitt: For you shall finde that the tenaunts will encroache (as they have donne) Too much upon the Lorde For to Frustrate his Customes and enlarge their owne.[4]

Two centuries later Nathaniel Kent felt threatened by a tenant of Uvedale Price's estate at Foxley, 'who intimated that he was a Man which it would be imprudent for M:r Price to have any dispute with'. The local inhabitants encountered by Saxton would have been suspicious of any stranger with a notebook wandering by their homes and property.

John Speed and the agents employed by John Ogilby would have attracted less attention, spending little time in only small parts of the county. Speed (1552-1629) was the son of a Cheshire tailor, who qualified as a freeman of the Merchant Taylors' Company. The younger Speed, having made his way to London, also became a freeman of the Company and obtained patronage to pursue his interests in writing history. His atlas was to him no more than a supplement to his *History of Great Britain*.[5] His map of Herefordshire is closely based on Saxton's and his fieldwork was mainly directed at exploring the streets of Hereford before drawing his map of the city. It did not take him very long. Hereford within the walls is small and can be crossed on foot in ten or fifteen minutes. In 1606 in the course of a long journey between the county towns of Winchester and Carmarthen he was in Monmouth on 26 August, Hereford on Monday 1 September 1606 and Brecon on 3 September. We may suppose that he made himself known to the mayor or town clerk and the dean, and that he may have mounted the tower of the cathedral, All Saints church or the castle to obtain a bird's-eye view of the city. His day or two in Hereford would have been ample time to measure the streets by counting his paces and to plot the principal landmarks, drawing these at four times the scale of the printed map with some semblance of accuracy but filling in the rows of town houses merely as conventional signs.[6]

At the time when Speed was pursuing his historical interests in the capital, John Ogilby's father, like other Scotsmen, came down to London from Scotland at the beginning of James I's reign. Ogilby himself (1600-1676) had a lively and varied career as poet, translator, Master of the King's Revels in Ireland, printer and publisher. *Britannia* was 'his last great dream'.[7] The surveyors that he employed must have been a strange sight as they traversed the country's main roads pushing the great 'perambulator' wheel, clocking the measured miles and noting the countryside they were passing through and the bridges they was crossing. Even allowing for Herefordshire's notoriously bad roads, their journeys within the county — between London and Aberystwyth (within Herefordshire, 24 miles), Bristol and Chester (43 miles), Gloucester and Montgomery (35 miles), Hereford and Leicester (14 miles) — would probably not have totalled more than seven or eight uncomfortable days.

THE FIRST HEREFORDSHIRE SURVEYORS

The local estate surveyors did not have to cope with such difficulties of travel or hostile curiosity. They enjoyed the protection of their employers and many came from that class of lesser gentry and estate stewards familiar with land-management and dealing with tenants and suitors to the manorial courts. Their reputations became known locally and were passed on in correspondence or by word of mouth between neighbours and relatives, at the sporting and social events attended by the gentry, at assizes and quarter sessions. Some of these links can be deduced from the names of the clients of individual surveyors and the rare references to them in estate papers.

The first native Herefordshire surveyor about whom something is known was John Pye of Kilpeck, a member of the gentry. He was the nephew of William Pye of Thruxton, who in his will of 1681 had left him the manor of Thruxton in trust to fund charitable bequests. This branch of the family was related, if distantly, to the Pye family who owned The Mynde estate in Much Dewchurch from the fourteenth to the early eighteenth century. On his death in 1731 John Pye himself owned a range of property in the countryside between Much Dewchurch, Allensmore and Treville. He was a lawyer and in this capacity was active as a land surveyor between 1695 and 1704.[8] The second Viscount Scudamore, in organising his affairs between making his will in 1694 and his death three years later, engaged him to survey his Holme Lacy estates and about the same time Sir William Compton of Hartpury in west Gloucestershire had him survey the farm of Cwm Madoc at Garway, which lies adjacent to the estate of the Scudamores of Kentchurch Court. Whether this undated commission came through the Scudamore family can only be guessed at, but a web of patronage through the Scudamores, Sir William Compton and Pye's other work in Gloucestershire and Wiltshire can be discerned. In 1700 the 11th Duke of Kent, who had estates in both Gloucestershire and Herefordshire, had employed him and he subsequently mapped part of the duke's estate at Goodrich in 1704.

Pye was an accomplished surveyor and an advanced draughtsman for his time. He drew buildings in the new style that did not become prevalent until the 1730s, in block plan rather than either in the bird's-eye view or perspective favoured by Tudor and Stuart surveyors and map-makers (Plate 44). His scale bars, with dividers astride them, were also attractively drawn, though it may be noted that his scales were measured in perches, an old-fashioned practice, rather than chains. He bequeathed all his law books to his grandson, Arnold Russell, later lord of the manor of Thruxton, together with his prized copy of the Revd Francis Gouldman's Latin-English-Latin *Dictionary* (1664).

He also advised Richard Symons, the London merchant who bought The Mynde estate in 1727.[9] In a letter to Symons he expressed confidence in the accuracy of a younger local surveyor, George Smyth, compared with that of an outsider, 'Mr Cleare'.[10] At the time that Pye wrote this letter George Smyth was at the start of his career. His earliest known commissions were both for the measurement of woodland, in 1723 to survey Lady Scudamore's woods on the Holme Lacy estate, which establishes his link with John Pye, and then in 1726 for Richard Symons. In 1728 he was working near Symonds Yat and in 1733 at Llangarron, before doing larger and better executed surveys at Bromyard, Madley and Much Marcle in 1741 and 1742. He appears to have been the surveyor of an undated estate map of Uffculme in Devon and possibly of two estate maps in Norfolk in 1732 and 1734. The presence of a George Smyth as agent to the Scudamores of Kentchurch in 1752-54 raises the possibility that he was a local man connected with the Scudamore estates.

The 'Mr Cleare' referred to in Pye's letter to Richard Symons was probably Thomas Cleer, who was active all over south-east England between 1685 and 1706. He published a map of Norwich in 1696 and turned up in Herefordshire in 1700 to map the Ford estate, presumably for Thomas 1st Lord Coningsby of Hampton Court, whose estate included most of Ford. Cleer was in Essex in 1701 and 1703, so what caused him to return to Herefordshire so long afterwards in 1726 is curious, but his London connections could have brought him to Symons's notice.[11]

Another of Pye's Herefordshire contemporaries was William Whittell. He was the younger son of the vicar of Bodenham and both his elder brother and (probably) his grandfather were also parsons in north-west Herefordshire. He was baptised at Orleton in 1684 and signed himself as 'of Bodenham' during the short period that he was map-making between 1705 and 1709. He has not been traced after his father's death at Bodenham in 1710.[12] He appears to have worked only near his home for the owners of the Shobdon estate. In 1705, Robert Chaplin, a London merchant, employed him to survey his estate at Livers Ocle shortly before he sold it and his whole Shobdon estate to his fellow Londoner and future Lord Mayor, the financier, Sir James Bateman. Almost immediately, in 1708-9, Bateman commissioned him to survey the rest of the estate. His draughtsmanship, like William Fowler's, was heavy and laboured, harking back to a seventeenth-century style influenced by the printed map-makers. His habit was to draw a human face in the frame of the title cartouche (Plate 31). It would be nice to think that this was a self-portrait, but it is more likely to have been merely conventional ornamentation.

Thomas Croft at Garway in 1698 and Job Gilbert at Weston-under-Penyard in 1709 are otherwise unknown. They may have been local men. The Croft family of Croft Castle is well known in Herefordshire; the surname Gilbert is less common in the county but a family of that name occurs in the mid eighteenth century in Kentchurch and Llancillo. All were competent enough to have made other maps elsewhere.

The Early Surveyors Brought into Herefordshire

Despite the presence of some local, adequately qualified surveyors, major landowners whose main estates lay elsewhere in the country, preferred to employ men with whom they were already familiar. Prominent among these were three leading surveyors brought into Herefordshire by important absentee landowners.

There was William Fowler, whose work is recorded all over the west midlands. He was born about 1610, lived near Wolverhampton and in 1662, two years before his death, was employed by Lord Craven of Hampstead Marshall in Berkshire to survey Mocktree Forest and Bringewood Chase (Plate 37). He liked to decorate his maps and the title cartouche of this map has a hefty and rather crude frame, lightened by the little drawing of his working instruments of plane table with compass, circumferentor for sighting angles, a tripod, set squares and two shelves loaded with books including three entitled 'Euclid', 'Records' and 'Ramus' (Giovanni Ramusio, a sixteenth-century Venetian travel-publisher).

Edward Laurence or Lawrence (1674-1739), who surveyed Credenhill (a map that has not survived) and the large manor of Goodrich for Henry Grey, Duke of Kent, was a surveyor of truly national repute. He was born and died at Stamford but spent much of his life in London and worked all over the country. He was an accomplished draughtsman, well known for his books of maps, an agriculturist and author, friend of William Stukeley, the antiquary, and of the London instrument-maker, Jonathon Sisson. The calf-bound volume of his Goodrich survey in 1717 contains ten maps, which, it must be said, are not elegant. The title cartouches and the representation of both buildings and the cliffs above the river Wye near Symonds Yat are all of naïve artistry. He re-used one of the maps, that of the land surrounding Goodrich Castle, to demonstrate a model estate map in the second edition of his book *The duty and office of land steward* (1731), where it is fictionally disguised as 'A survey of Dun Boggs Farm in the manor of Haversham, in the county of Hereford' (Plates 45, 46).

His accounts are a rare survival for a Herefordshire survey, which followed similar work on the duke's estates in west Gloucestershire:

	£	s	d
Expenses in Credenhill & Goodrich Manners			
To a man & horse Removing from Huntley to Weston		2	6
To a man for 24 Days Assistance at Credenhill	1	4	0
To a man and two Horses Removing from Credenhill to Goodrich		6	0
To a man for his Assistance for 31 Days at Goodrich	1	11	0
To Horse hire and Passage over the Wye whilst at Goodrich		13	6
To Carriage of Boxes to Gloucester & from thence to London		12	6
To Coach hire & Expences to London from Gloucester	1	9	6
To 39 weeks Board wages from Jan 1st to 7ber [September]: 29th 1717	13	13	0
To Coach Hire and Expences from London to Rest		11	0
To 21 Weeks Board wages whilst making up the Books of Gloucestershire And Hereford shire	7	7	0
Paid for Binding the Book of Gloucestershire	1	15	0.

    The total bill came to £29 5s. 0d. Surveying was a job for two, with an assistant needed to hold one end of the chain or the measuring poles, as well as leading the horse laden with the precious instruments and draft papers, clothes and other paraphernalia needed in the field. This assistance was required for four weeks at Credenhill and five weeks at Goodrich, not counting Sundays and days of bad weather, but Laurence himself was away from home for thirty-nine weeks, including his time travelling and, presumably, for his work in Gloucestershire. The whole area which he mapped comprised 5,778 acres in Herefordshire and 2,772 acres in Gloucestershire. According to the terrier attached to the Herefordshire maps this included 900 acres of tenanted lands and at the end of his accounts he noted that he had been obliged to survey 1,000 acres of 'free land', not belonging to the duke, because 'it lay so intermixt with other of your Graces Lands'. He also claimed that he had spent 'a Considerable part of my time' making abstracts of leases, manorial documents and other writings in order to compile new and detailed rent rolls. This would explain the discrepancy between the weeks he spent away from home compared with those in the field with his assistants.[13] The favoured time in the year for out-of-door work was the spring and early autumn. The weather was not likely to be too cold or wet and the cultivated land could be walked over without damaging either the grass or cereal crops. During the winter the surveyor could comfortably be at his drawing board and desk.

    An even more elaborate survey and estate atlas was compiled the following year for George Neville, 11th Baron Bergavenny of his huge estates in Monmouthshire, Herefordshire and Worcestershire. It was the work of Benjamin Fallowes of Maldon in Essex. Only the first nineteen folios containing fifteen maps remain in the fine leather-bound atlas, for at least eight maps have been cut out of the volume, perhaps when the estates were sold about 1800. Six of the remaining maps relate to the Longtown and Ewyas Harold area. Whereas Laurence spent nine weeks in the field and ten months overall in surveying the Duke of Kent's estates, Fallowes completed his survey in the three summer months of June, July and August 1718; the Herefordshire estates alone comprised 525 acres.

    A good deal is known about Fallowes. He was a Quaker, living in Essex from at least 1714. He had a short but exceptionally prolific career as a surveyor between 1714 and 1720, in the year 1716 alone carrying out no less than twenty-seven surveys covering over 6,000 acres, mostly for the Western family of Rivenhall, before becoming a full-time farmer. He worked almost exclusively in Essex for two of the largest county landowners and a variety of London gentlemen but it has been noted that 'he was the only local Essex man throughout the century to make a long journey for a commission'. This was to survey the estates of Lord Bergavenny in 1718. He then returned to Essex, where he surveyed two large estates in 1719-20 and took up farming near Maldon until his death about 1730.[14]

His surveys for Lord Bergavenny were carried out competently and the parchment maps are well drawn, if not exceptionally so. The map of Longtown, for example, has an interesting inset plan of the castle, which shows the buildings of the keep, house and shop (perhaps a blacksmith's) drawn in perspective in a conventional rather than accurate fashion (Plate 40).[15] More unusually, he dated the maps precisely by day, month and year, apparently the date of carrying out the field survey. In 1718 his map of the lands between Ewyas Harold and Abbey Dore was dated 30 June, followed by Ewyas Harold castle on 8 July, three further maps of lands east of Ewyas Harold on 5 July, 8 July and 17 July, Longtown on 19 July, and finally another map of lands east of Ewyas Harold on 25 July. Later, the atlas was prepared, the parchment cut to size, the field drawings and measurements pricked or marked out on the parchment, pencilled and inked, the buildings, trees, gates and other features drawn, and the scales, compass roses and title cartouches set out. Then followed the decorative borders that make Fallowes's maps instantly recognisable. Watercolour was applied and in a rather backward-looking style he added brightly coloured leaves, berries and flowers, sometimes as a border, sometimes individual flowers well spaced or placed only at the corners, and in one case supplemented by vividly painted bird (Plate 32). He was probably assisted in the draughtsmanship or painting, for in the title of the map of Longtown the place is spelt 'Long Sown', a simple copying error which surely the surveyor himself was unlikely to have committed. Finally the parchment skins had to be sent off to a bookbinder to make up the volume.

By coincidence, about the time of Fallowes's death a second Benjamin Fallowes appears in Herefordshire as estate steward to William Bateman (1st Viscount Bateman 1725) and his successors at Shobdon between about 1730 and 1777. An extensive correspondence relating to his management of the estate in the mid eighteenth century survives among the Shobdon Court estate papers.[16] He also acted as Lord Bateman's political agent and, being on friendly terms with his employer, he used his position to secure for his son, Benjamin Fallowes III, the post of clerk of the peace of the county in 1752. Benjamin Fallowes III held this influential post until 1797, in which he was succeeded by his son, Benjamin Fallowes IV from 1804 to 1817.[17]

It is not a common combination of names that was handed down in this family and the unknown link between Benjamin Fallowes, the distinctive surveyor of Maldon, Essex, and the Herefordshire family of estate stewards and lawyers came to light in 'An Exact Map of the estate called the Broome Farm lying in the Parish of Eardisland … belonging to the Honourable William Bateman of Shobdon Court'. It has two floral cartouches, an array of thirty-four flowers arranged around its edges and a brightly coloured compass rose and coat of arms. It is signed and dated by Benjamin Fallowes on 28 July 1720 (Plate 38). So, Benjamin Fallowes of Maldon not only returned to Herefordshire at the end of his career as a surveyor but was commissioned to do so by the Lord Bateman who later employed the younger Benjamin Fallowes as his estate steward. It is not too fanciful to imagine that Benjamin Fallowes travelled back from Essex to Herefordshire in 1720 with a son, who impressed William Bateman sufficiently for him to be employed as steward at Shobdon until shortly before his death in old age.

The Professional Surveyors

The late seventeenth-century interest in mathematics, science and estate improvement deepened and spread early in the eighteenth century and there grew up a class of professional men for whom surveying was one of several related activities that provided them with a livelihood around the country.

Among Edward Laurence's contemporaries who worked in Herefordshire was Charles Price. He was a London instrument maker, engraver and publisher, who in 1694 had been apprenticed to John Seller, one of London's map publishers, before setting up in business himself. He was an accomplished surveyor and draughtsman who worked all over England and in Ireland. In 1720 Jacob Tonson, a fellow London publisher, commissioned him to survey his newly acquired estate of The Hazle, south of Ledbury (Plate 33). He returned ten years later to survey the adjacent Vineyard estate, shortly before his bankruptcy (1731-2) and death in 1733. He had lost none of his skills in the interval; the map of the Vineyard is even more beautifully drawn than those of The Hazle.[18] Robert Whittlesey worked all over southern England as well as his single survey at Holme Lacy in 1729, whilst J. Meredith of Shrewsbury and Joseph Dougharty of Worcester were better known locally. Meredith is recorded only in Shropshire and Herefordshire, where in 1733 he produced a splendid map of The Homme estate in Weobley for Lord Carpenter (Plates 34, 39). Joseph Dougharty was a member of a prolific family of map-makers in Worcester. The practice was started by his father, John Dougharty, an Irishman from Dublin who migrated as a teacher of mathematics first to Bewdley and then to Worcester. His two sons, Joseph (1699-1737) and John (1709-72) continued teaching and surveying until the death of John Dougharty the elder in 1755. Although they had many connections with the Herefordshire and Worcestershire gentry, not least with the Foleys, the only map within Herefordshire that has been traced is that of Grendon Warren in 1732 by Joseph Dougharty for Thomas Coningsby, a cousin of Lord Coningsby of Hampton Court. In his usual way it is neatly and precisely drawn, with clear details down to the three-storeyed and jettied farm house with four gables and smoking chimneys (Plate 47). John Dougharty the elder charged the dean and chapter of Worcester 3*d.* an acre for surveying in 1727 and some four years later the three Doughartys received 4*d.* an acre for surveying the 6,554-acre Hanbury Hall estate in Worcestershire. By 1751 the younger son, John Doharty, was charging 6*d.* an acre for enclosed lands and 8*d.* for lands divided into small pieces in open fields and, he added in a letter that year to the bursar of King's College, Cambridge, 'when the division run small, no money can be harder earn'd'.[19]

Later in the century the maps of other Worcester surveyors who worked occasionally in Herefordshire show that they were strongly influenced by the Doughartys. William Hull of Kempsey surveyed an estate at Bosbury in 1775 and John Aird, a teacher of mathematics in Worcester surveyed estates in Wacton (1775) and Mathon (1780) in what was by then a distinctly old-fashioned style reminiscent of John Doharty and his father between 1730 and 1755. Other writing masters who were part-time surveyors in the middle of the century were John Ward, who occurs between about 1725 and 1760 and Thomas Ward, who occurs in the 1760s, both of Leominster.

In Hereford the schoolmaster, John Bach, was also busy surveying in the 1760s, venturing into landscape design at Stoke Edith in 1766. He was not afraid of promoting himself, but his talents do not appear to have matched his claims. He advertised in the *Hereford Journal* in 1770 that 'In his Maps are given perspective Views of Gentleman's Seats and Farm Houses, together with every other Embellishment necessary to render them superior to most other Maps hitherto executed in this Country'.[20] The claim is exaggerated. His only map that has a view of a house is of a farm at Much Birch for John Williams of Worcester in 1768, which has an inset drawing of a proposed farm house that appears not to have been built

(Plate 51). His most extensive survey was of the Monningon-on-Wye estate of John Whitmore in 1771, for which he compiled an estate atlas in a heavily ornate style of garish colours that owed more to classroom maps than the elegant rococo penmanship which was then fashionable.[21]

Also in the older tradition was Meredith Jones of Brecon, who was active throughout South Wales and the Marches between 1743 and 1757.[22] In Herefordshire he was employed by both Guy's Hospital and the bishopric of Hereford. His surveys and draughtsmanship are sound without being outstanding. His map of the manor of Aconbury for Guy's Hospital in 1757 contains a large inset drawing of Aconbury Court, which the Hospital, having no need of a country mansion, demolished later in the eighteenth century (Plate 49).[23] John Green of Bridstow likewise included a little inset drawing of Wilton castle, near Ross, in a map of 1755 (Plate 48). He was the third of four surveyors of that name, three of whom had connections with Bridstow and Ross. They appear to be members of the same family, gentleman-surveyors capable of mapping a small area reasonably accurately but drawn in an old-fashioned style. The other Herefordshire surveyor, whose maps also included inset drawings of major buildings, was Isaac Taylor of Ross.

Isaac Taylor of Ross

Isaac Taylor was the only surveyor from Herefordshire with a reputation outside the county. His achievements as a map-maker of printed town plans and one-inch-to-one-mile county maps, already noticed (pp. 25-27) were supplemented by his work as an estate surveyor, for which he advertised in the titles of several of his maps. His stylistic trademarks included well-drawn inset views of scenes and buildings, lavishly phrased dedicatory titles for the printed maps and good, though not outstanding, penmanship as an estate surveyor.

The date and place of his birth have not been traced.[24] He appears first in 1750 as the author of printed town plans of Oxford and Wolverhampton, and on his first county map, that of Herefordshire in 1754, he signed himself as Isaac Taylor of Ross. He did not consistently use this title — for example, about 1780 he signed an estate map of Madley merely as J. Taylor of Ross — but Ross was his home for the rest of his life, even though his work took him away for long periods. It is important not to confuse him with his contemporary namesake, perhaps a few years younger, Isaac Taylor of Worcester, born in 1730, the son of a brass founder and engraver.[25] Isaac Taylor of Worcester went to London in 1752, where he obtained employment with Josiah Jefferys of Brentwood, Essex, brother of Thomas Jefferys, the map engraver and publisher. He married his employer's daughter in 1754, worked as a land-surveyor 1754-58, but was best known as an engraver and undistinguished watercolourist, eventually returning to London and dying at Edmonton, Middlesex in 1807. To add further to confusion, he named his younger son Isaac, whose son, also Isaac, was a map-engraver.

Isaac Taylor of Ross published the first one-inch-to-one-mile map of Herefordshire in 1754 and, apparently had ready for publication a large scale plan of the city of Hereford. For some reason, perhaps as a consequence of the few local gentry coming forward to subscribe to the county map, the engraver Thomas Kitchin did not also engrave the map of the city, for which Taylor went to Richard Benning, a competent but less well known engraver. That map was not published until 1757 when Taylor was already working as an estate surveyor in Hampshire, where two surveys have survived, and was preparing his county map of Hampshire, published in 1759, for which he also chose Benning as the engraver. He came to regret the choice, for he was put to unexpected trouble and expense in travelling to London to oversee the work and the standard of the engraving was disappointing.[26]

There is no known portrait of him but his appearance at this time may be revealed in the dedicatory title of the plan of Hereford city in 1757. This depicts the prominent figure of an untidily dressed young man, standing to draw or write on a large sheet of paper spread out before him (Plate 29). It is tempting to believe that the drawing is a self-portrait. With the Hampshire map ready for the press, if not published, he returned to Ross, where just before Christmas 1759 he married Eleanor Newman. His dissatisfaction with the London engravers was such that he now set about training himself as an engraver and successfully produced a map of the Isle of Wight taken from his map of Hampshire.[27] He also found a little work locally as an estate surveyor, mapping the estates of Anthony Sawyer at Canon Pyon, probably about 1760, of the dean and chapter of Hereford at Preston-on-Wye in 1764 and surveying the course of the river Wye in 1763. However, he was already looking farther afield again with the intention of preparing another one-inch-to-one-mile county map.

His first two county maps, Herefordshire and Hampshire, had been published before the Society for the Encouragement of Arts, Manufactures and Commerce, formed in 1754 and better known by its later shortened name of the Royal Society of Arts, announced in 1759 its scheme to award premiums of not more than £100 for the best county maps published at a scale of one inch to one mile. He had the experience and qualifications to win a premium and he prepared a map of Dorset for this purpose. It was a good map, published in 1765, but contemporary critics found fault with his accuracy and with the quality of engraving, done by Taylor himself, and the Society did not award him a premium. Their decision was controversial and scholars, though remaining critical of his draughtsmanship, now look on his work more favourably.[28] Taylor was not alone in failing, for of the twenty-three surveyors who submitted maps between 1759 and 1801, only thirteen were given awards.[29]

At about the same time he and Eleanor suffered the joy of the birth and then the grief of the infant death of a daughter, Mary Newman Taylor, early in 1765. A year later, another daughter, Elizabeth, was born, but she too died in infancy in 1768 at the age of only two and a half. Eleanor had to bear a lot of this personal tragedy alone for her husband spent much of the five years between 1765 and 1770 employed in surveying the estates principally in Dorset and Devon of the Sturt family of Crichel before returning to prepare the county map of Worcestershire, published in 1772. Between 1773 and 1777 he was back in Dorset to survey the estates of the Erle-Drax family of Charborough Park. As before, he again overlapped this commission with preparing the county map of Gloucestershire, published in 1777.

It was unusual for a country land-surveyor like Taylor, not close to the London map-trade, to undertake so much work of this nature. By mingling his preparation of county maps for publication with estate surveying for individual clients it is difficult to estimate how long it took him to carry out either task. Each county map appears to have taken about five years from beginning the survey to publication.[30] This estimate is borne out by apparent inaccuracies in the maps of Herefordshire and Gloucestershire that may have been due to topographical changes, particularly noticeable in the routes of roads, occurring during the years between survey and engraving. As already noted, in 1754 he had imitated the methods and style of Henry Beighton, but by the date of his last map of Gloucestershire in 1777 he had reduced the range of his conventional signs and restricted the outpourings of his dedication cartouches. In other respects he adhered to his style of twenty years earlier. Although between 1764 and 1777 he produced some one hundred and twenty estate maps the long time that it took him to complete the Sturt and Erle-Drax estate surveys can only be attributed to the fact that he was also working on the county maps of Dorset, Worcestershire or Gloucestershire.[31]

The two types of maps offered him both variety and a balanced income. The printed maps promised some long-term continuation of income but demanded an outlay of capital for engraving and publishing. Estate surveying, though an unpredictable form of employment, involved less expense and offered quicker returns. The large commissions in Dorset gave him the security to continue preparing the county maps. Together, the two contrasting sources of income enabled the Taylors to live in some comfort in a large house at 54-55 High Street in Ross, situated at the west end of the town just below The Prospect (Plate 26). The property included the tall, four-storied house with stable, brewhouse and an adjoining house, untenanted and empty in both 1776-77 and 1780.[32] The buildings were insured for the sizeable sum of £400 in 1780 and the contents of household and other goods for £100, which included one itemised luxury, a harpsichord insured for £25.[33] By that year Taylor was back at home, acting as a local estate surveyor and agent and advertising his services in 1780 for the sale of property at Wilton and Peterstow.[34] His map of Herefordshire was re-issued in 1786, perhaps to provide income in his old age or infirmity, for he died in 1788 and was buried in Ross churchyard on 17 June. As he apparently did not leave a will and the death of Eleanor Taylor has not been traced it has not been possible to discover who may have arranged for and benefited from the re-issue of his maps of Worcestershire and Gloucestershire in 1800.

The Late Eighteenth-century Surveyors

The number of surveyors working in Herefordshire increased markedly in the second half of the eighteenth century, but it was uncommon for most of them to receive more than two or three commissions and it is remarkable that about a third of their maps were still unsigned. John Harris of Wickton and Joseph Powell of Bridgnorth were exceptional in their activity and, as earlier, there was the same mixture of national experts, competent local surveyors and rough land-measurers, joined now by a new class of engineers and professional surveyors specialising in public works.

In the first category are three important surveyors, Richard Frizell, John Lambe Davis and Nathaniel Kent. The Frizells were a dynasty of Irish surveyors. After working extensively in Ireland with his father and brother, both named Charles, Richard Frizell moved to England and between 1771 and 1780 was employed by Charles Howard, subsequently 11th Duke of Norfolk, to survey the Holme Lacy estate, in which Howard had acquired an interest by his marriage with the Scudamore heiress, Frances. At first Richard was assisted by his brother Charles, who soon returned to Ireland never to work in England again. Later Richard Frizell became agent to the 1st Marquess of Ely. John Lambe Davis similarly came from a family of agents to a grand estate, that of the Dukes of Bedford at Woburn Abbey. At the time that Richard Frizell was at Holme Lacy, Davis was surveying the Moccas Court estate for Sir George Cornewall. Nathaniel Kent, the London-based agriculturist and estate agent, completed a thorough three-year survey of the Kyre Park estate, which straddled the Herefordshire-Worcestershire border, by producing a well-organised two-volume estate atlas for his client, Edmund Pytts in 1774. He then moved on to survey the Foxley estate in Yazor for Uvedale Price. He returned to Herefordshire twelve years later, at the peak of his reputation, to value the Hampton Court estate in 1786-87 and the Hellens estate at Much Marcle in 1787-90.

All three of these distinguished surveyors produced accurate and neatly drawn working surveys, remarkable for their practicality and lack of ostentation at a time when fine penmanship and rococo decoration was fashionable among map-makers of both printed and estate maps. Frizell's thirty-five drawings of the Holme Lacy farms were starkly simple, the time-consuming work of penning title cartouches and compass roses being eliminated by using engraved stamps for these repetitive features. They would seem to have been fair-copy drafts from his field-notes, neatly and unerringly executed on paper, which bears the prick holes for copying the lines of the field boundaries. He later drew four larger maps of some of the estate on parchment but still used the same stamps and added little by way of other ornamentation. His large estate atlas compiled five years afterwards, in 1780, was based upon these earlier drawings, with occasional lapses in spelling and a notable absence of scale bars. The atlas contains additional remarks about the condition of farms and their potential improvement, suggesting that he must have revisited Herefordshire to update his observations, and he added surveys of property in Buckinghamshire and Gloucestershire. Davis's and Kent's volumes of the Moccas (Plates 41, 42) and Kyre Park estates were clearly designed as working tools. They are conveniently smaller, respectively of quarto and octavo sizes (that is somewhat over and under A5), thick, strongly bound volumes showing the signs of being well used. In both surveys the tables of lands and tenants are laid out for easy reference, classified with cross-references. Kent added advisory remarks for the improvement of the farms on the estate. They are in striking contrast to Frizell's great rolled parchments to be hung in the estate office or his massive 'coffee-table' estate atlas designed for display in the library.

An example of one such estate atlas, well surveyed and beautifully drawn with rococo cartouches and 'picturesque' views is the large atlas of forty-two maps of the Downton Castle estate drawn by James Sherriff for Richard Payne Knight in 1780 (Plates 52-54). Knight was Uvedale Price's contemporary and partner in promoting the 'picturesque' landscape. He published *The landscape: a didactic poem* in 1794, the same year that Price published *Essays on the picturesque*. James Sherriff, an accomplished surveyor and land agent from Birmingham, who worked all over the midlands and southern England, set out to provide Knight with a volume that his client would himself appreciate and enjoy showing to his friends. The working maps and reference tables are of the highest standard, but it is Sherriff's title cartouches which give the volume its special quality. He illustrated it with a frontispiece of 'A south west view of Downton Castle' and some of the maps have delicate title cartouches of rococo frames with foliage, flowers and trees, ruined castles, follies, and stone slabs.

Two Herefordshire men produced maps of a similar quality. John Harris of Wickton Court in Stoke Prior was baptised in Stoke Prior parish church on 26 July 1753, the son of John Harris, the steward of the Hampton Court estate, which owned Wickton. His first map, containing all the essential features of an estate map but rather poorly drawn, was made for Lady Frances Coningsby of Hampton Court when he was only nineteen. As it bears the page number 241 it must have been torn out of a large volume of estate surveys compiled under his father's tutelage. Between 1772 and 1781 he was living at Newton, south of Leominster and a short distance from Stoke Prior, but on his father's death in 1784 he moved into Wickton Court.[35] He married Jane Eaton at Leominster on 14 July 1786, perhaps rather hurriedly as four months later she bore him their eldest son Samuel, baptised at Stoke Prior on 12 November.[36]

Although he then appears to have concentrated his attention on his own farm or estate management at Hampton Court he continued to carry out surveys for other clients. Two of his last commissions in 1784 had been for the dean and chapter of Hereford but his relationship with them appears to have been erratic. They did not appoint him the next year when they were seeking a surveyor but in 1786 he accepted a commission to survey an estate at Ledbury belonging to St Katherine's Hospital, a charitable foundation within the cathedral's jurisdiction. Then in 1793, when he was acting for Guy's Hospital, the dean and chapter made a grumpy response to the proposal that he should act for both parties in arranging a lease of the Hospital's tithes of Llangarron. Over the next few years they ignored him in preference for a Mr Wainwright in 1795 and David Pain in 1797 only to relent in 1798 in resolving to employ him 'or any eminent surveyor' to value the demesne lands at Upton Bishop.[37] He was still at work as a surveyor in 1809, when he was chosen to apportion the four parts of the auctioned Belmont estate in Clehonger among the consortium of successful bidders. After acting with James Cranston as an inclosure commissioner at Bodenham in 1811-13 they were joint agents for the Duke of Norfolk in arranging a subsequent mortgage in 1815.[38] He died in 1829, leaving other members of the family to continue the business.

Over a dozen of his maps have survived. His style can be seen to develop from his first rather crude penmanship of a nineteen-year old to the confidence of an accomplished draughtsman and the minimalist requirements of a parliamentary inclosure award. Early in that career he produced an extraordinarily tiny map of Red Hill, Bredenbury in 1777, measuring a mere 120 x 160 mm. (about $4^{1}/_{2}$ x $6^{1}/_{4}$ inches), yet containing all the relevant features of a full-blown estate map — an accurate survey marking field-names and land-use, gates and roads, a title, precise date and name of the surveyor. It admittedly lacks the name of the client and a north point, but the map is correctly aligned with north at the top. Later, he cultivated a delicate penmanship as finely drawn as an engraving, enhanced with a soft grey wash of watercolour. Reference tables are portrayed as scrolls with curling corners laid on the larger surface of the working map. Rococo title cartouches abound, with flowers and foliage, 'picturesque' trees and tablets, occasionally incorporating figures like the horned cow and bagpiper on the map of the Yoke estate in Hope-under-Dinmore in 1778.[39] His scale bars are surrounded by pictures of drawing instruments and, with a lighter touch, on his two maps for the dean and chapter in 1784 he drew the north points as a flag standard and a half-coiled snake (Plates 55-58).

His younger contemporary, David Pain of Lugwardine, was an equally proficient draughtsman. The map of Pipe and Lyde for the dean and chapter in 1797 no longer survives among the cathedral's archives, but two others of estates in Weobley and King's Pyon for John Peploe Birch in 1795 and 1797 demonstrate Pain's skills and his taste for decorating the title cartouches with imaginary 'picturesque' scenes. He worked later, about 1819, with Joseph Powell of Sutton, who is thought to have been the son of Joseph Powell 'senior' of Bridgnorth, a more familiar figure in eighteenth-century Herefordshire. Joseph Powell senior worked in the four neighbouring counties of Herefordshire, Gloucestershire, Shropshire and Worcestershire between 1765 and 1801. Powell's maps, like those of John Harris, show the development of his style but over a much longer period. His first map of Thornbury in 1765 was crude and heavily drawn but within ten years he was producing neat working maps with decorative borders and elaborate cartouches. After a further ten years his maps of Preston-on-Wye, Blakemere and Bosbury in or about 1791 were embellished with 'picturesque' pastoral scenes (Plate 61). But despite these artistic flourishes, which on the Bosbury map also included a scale bar with dividers and protractor, on the carefully drawn working part of the map he was already colouring the little block plans of dwelling houses in pink watercolour in what was to be the universal style of the nineteenth-century surveyors.

Some of the other surveyors beginning work late in the century also belong to the history of nineteenth-century map-making. These include James Cranston of King's Acre, Hereford, whose map of his own neighbourhood, the Stretton

Court estate belonging to Guy's Hospital (Plate 63), was workmanlike but less of artistic merit than of historical interest in comparison with Meredith Jones's map of Stretton thirty-four years earlier. He worked with John Harris on the inclosure of Bodenham in 1811-13 but is better known as the founder of a dynasty of nurserymen who were eventually succeeded by the existing Wyevale Nurseries in 1930. Another surveyor whose career began in the closing years of the eighteenth century was William Galliers of Presteigne and Leominster. His earliest known map was dated 1799, the first in a family business that continued for the next half-century.

THE ENGINEERS

Navigation and canal schemes in the final years of the eighteenth century brought some of the promoters and engineers to Herefordshire to investigate potential routes in the county. The well-known canal engineer Robert Whitworth surveyed the whole 58-mile course of the river Wye from Hereford to Tintern in 1779 for a scheme to improve the navigation of the river, returning in 1796 with a more specific improvement at New Weir, near Symonds Yat.

Meanwhile schemes for canals to link Herefordshire and the river Severn had obtained parliamentary approval in 1791. Josiah Clowes, engineer of the Thames and Severn canal through Gloucestershire was responsible for the route of the Gloucester and Herefordshire canal on which work had started when he died in 1795. He was succeeded as the engineer of both canals by Robert Whitworth, who redesigned and shortened the first section of the Gloucester to Herefordshire canal as far as Ledbury, opened in 1798. Thomas Dadford junior designed the route of the unsuccessful Stourport to Leominster and Kington canal in 1789.

THE LAND-MEASURERS

John Harris of Wickton styled himself on one of his maps as 'land-measurer', but though a respectable enough term at the start of his career it later became a derogatory description of the untrained men whose maps are marked by inaccurate surveying and crude draughtsmanship.

There were many such surveyors who knew the rudiments of measuring distances and angles but lacked the skills of penmanship to produce a well presented map. Samuel Addams made a satisfactory survey of the manor of Brockbury in Colwall in 1758 but his attempt to draw an inset bird's-eye view of Brockbury Court and its farm buildings exposed his inability to command the rules of perspective. In the same year Daniel Williams made two maps of Admiral Thomas Griffin's estates in the lower Wye valley. The one is of the Doward Wood with rather childish perspective drawings of the cottages and lime kilns near the river bank. He found it even more difficult to represent the cliffs of the Seven Sisters rocks towering above the river (Plate 50). The other map shows the forge and associated buildings at New Weir, near Symonds Yat, where the houses are again drawn crudely. To compound his faults the compass is pointing in the wrong direction. He was aware that a map needed a scale and a north point but was still gripped in a belief that houses should be drawn as seen and not in plan, and that the river should be highlighted by adding the embellishments of a swimming salmon and a trow being hauled upstream by a team of seven men. One wonders what moved Robert Bright of Brockbury, one of the more important residents of Colwall, and Admiral Griffin to choose these men as their surveyors, and whether they were disappointed with the results. Admiral Griffin, in particular, would have been familiar with maritime charts drawn to a high standard of accuracy and legibility.

Daniel Williams's out-of-date style was exceeded by Edward Penry, who appears to have been a member of the family of that name from Longtown. His maps of farms in Longtown in 1794 (Plate 64) and in Michaelchurch Escley, undated but for the same owner, John Lewis, have a colouring and patterns of scattered flowers very similar to the better executed maps of Longtown and Ewyas Harold drawn two generations earlier by Benjamin Fallowes of Maldon. The inescapable conclusion must be that either Edward Penry or John Lewis had seen or knew of Benjamin Fallowes's earlier maps, liked them and copied their style in the belief that this was correct. The limitations of Penry's skill were exposed by another client, Tomkyns Dew, who had inherited the Whitney Court estate in 1780. He commented bitterly in 1783 that 'This Map or Plan of the Estate at Whitney & Clifford Herefordshire is very Bad & Not at all Correct'.

Even more extraordinary, and from the same part of the county, are four undated and unsigned maps and a reference table of farms in Longtown. The anonymous surveyor knew the rudiments of his craft, writing a terse title on each map, giving tenants' names, some landmarks, plot numbers and the acreage of the lands mapped. They are roughly drawn but, given the clue of the names of each farm, the fields are identifiable. They have no scale or north point and they have been written crudely in a late eighteenth-century hand on poor quality paper pasted for strength onto scraps of discarded parchment deeds (Plate 62). One of the parchment pieces is dated 22 August 1691. They appear to be the work of a countryman, whether the landowner or his agent, sketching out the lands of his tenant farmers for his own use, perhaps either just before or after their sale.

In some respects they are more significant than the professionally crafted maps of James Sherriff or John Harris, proof that in the second half of the eighteenth century maps had become such familiar and everyday objects that their value for identifying and describing property was widely recognised. It would not be long before the Ordnance Survey's teams were in the field throughout Herefordshire, imposing their well regulated standards of surveying and draughtsmanship upon the whole country.

The days of both the independent map-maker and the unqualified land-measurer were coming to an end.

References — Part One

1. Introduction (pp. 3-4)

1 E. Gibson, *Camden's Britannia, newly translated ... with additions and improvements*, London 1695, p. 574.
2 T. Chubb, *A descriptive catalogue of the printed maps of Gloucestershire, 1577-1911*, Bristol & Gloucestershire Archaeological Society, 1912; D.P.M. Michael, *The mapping of Monmouthshire*, Bristol 1985; P.D.A. Harvey and H. Thorpe, *The printed maps of Warwickshire 1576-1900*, Warwick 1959; F.G. Emmison, ed., *Catalogue of maps in the Essex Record Office 1566-1855*, Chelmsford 1947, with *Supplements*, 1952-68, 4 vols.; H.M. Thomas, ed., *A catalogue of Glamorgan estate maps*, Cardiff 1992; M.R. and M. Rowe, eds., *Devon maps and map-makers;manuscript maps before 1840*, Devon and Cornwall Record Society, new ser., vols. 44 and 45, (2002).
3 T. Chubb, *The printed maps in the atlases of Great Britain and Ireland, a bibliography, 1579-1870*, London 1927, reprinted 1966; R.A.Skelton, *The county atlases of the British Isles. A bibliography 1579-1703*, Folkestone 1978; D. Hodson, *County atlases of the British Isles, vol. 1, 1704-1742*, Tewin 1984; *vol. 2, 1743-63*, Tewin 1989; *vol. 3, 1764-1789*, London 1997.
4 The author may be contacted privately through either his publisher or The Society of Antiquaries, Burlington House, Piccadilly, London W1J 0BE.

2. Landscape of Herefordshire (pp. 5-15)

1 [W.] Marshall, *The rural economy of Glocestershire; including its dairy: together with the dairy management of North Wiltshire and the management of orchards and fruit liquor in Herefordshire*, London, 1789, vol. 2, p. 221.
2 W. Ravenhill, 'Christopher Saxton's surveying: an enigma' in S. Tyacke, ed., *English map-making'* London 1983, p.115.
3 B. S. Smith, *A history of Malvern*, Leicester 1964, p. 15.
4 The original charter is at the Town Hall, Hereford, the full text being published in the Historical Manuscripts Commission, *Thirteenth report*, London 1892, Appendix pt. 4. It is illustrated in E.M. Jancey, *The royal charters of the city of Hereford*, Hereford 1973, 1989.
5 F. A. Youngs, *Guide to the local administrative units of England*, Royal Historical Society Guides and Handbooks, vol. 2, pp. 122, 128.
6 M.A. Faraday's forthcoming edition of *Herefordshire taxes in the reign of Henry VIII*, WNFC 2005.
7 J. Buchanan-Brown, 'The natural history of Herefordshire: John Aubrey's projected tract', *Trans. WNFC*, vol. 49, pt. 3 (1999), p. 383; for field-names, see the parish fasicules of the Woolhope Naturalists' Field Club, *Herefordshire field-names survey*, 1987-93.
8 The Herefordshire part of the parish of Cwmyoy was transferred to Momouthshire in 1866.
9 For example in 1683, 1793 and 1896. T. W. M. Johnson, 'The diary of George Skyppe of Ledbury', *Trans. WNFC*, vol. 34, pt. 2 (1953), p. 58; D. Whitehead and R. Shoesmith, eds., *James Wathen's Herefordshire 1770-1820*, Almeley 1994, unpaginated.
10 I am grateful to Dr Paul Olver for revising these paragraphs on the geology of the county.
11 M.A. Faraday, ed., *Herefordshire militia assessments of 1663*, Royal Hist. Soc., Camden 4th ser., vol. 10 (1972), pp. 21, 24.
12 E.B. Wood ed., *Rowland Vaughan his booke, published 1610*, London 1897, pp. 31-33. Vaughan's claims about the population and poverty of the Golden Valley is not corroborated by the figures in M.A. Faraday, *Herefordshire militia assessments of 1663*, Royal Historical Soc., Camden 4th ser., vol. 10 (1972).
13 L.T. Smith ed. *The itinerary of John Leland in or about the years 1536-1539*, pt. 6, London 1964, p. 47.
14 R. Shoesmith, *Hereford history and guide*, Stroud 1992.

15 M.D. Lobel, ed., 'Hereford', p. 9 in *Historic towns,* Vol. 1, London 1969; Jancey, *op. cit.*, in note 4, p. 23; Shoesmith, *op. cit.* in note 14, p. 76.
16 D.A. Davies, 'Plague, death and disease in Herefordshire 1575-1640', *Trans. WNFC*, vol. 43, pt. 3 (1981), pp. 307-314.
17 J. Hillaby, 'The parliamentary borough of Weobley, 1628-1708', *Trans. WNFC*, vol. 39, pt. 1 (1967), pp. 104-151.
18 E.L. Jones, 'Hereford cattle and Ryelands sheep: economic aspects of breed changes, 1780-1870, *Trans. WNFC*, vol. 38, pt. 1 (1964), pp. 36-48.
19 E.L. Jones, 'Agricultural conditions and changes in Herefordshire, 1660-1815', *Trans. WNFC*, vol. 37 (1961), p. 36.
20 PRO, C2/Eliz/P2/56.
21 P. Cross, 'A 1577 plan "A platte of part of the Chase of Bringewood and of certayne groundes adjoining leased to Mr Walter of Ludlowe". A re-evaluation of the landscape', *Trans. WNFC*, vol. 48, pt. 3 (1996), pp. 573-81.
22 E. Taylor, 'The seventeenth century iron forge at Carey Mill', *Trans. WNFC*, vol. 45, pt. 2 (1986), pp. 451-67.
23 J. van Laun, '17th century ironmaking in south west Herefordshire', *Journal of the History of Metallurgy Society*, vol. 13, pt.2 (1979), pp. 55-68.
24 D.G. Bayliss, 'The effect of Bringewood forge and furnace on the landscape of part of northern Herefordshire to the end of the seventeenth century', *Trans. WNFC*, vol. 45, pt. 3 (1987), pp. 721-29.
25 See, for example, *Victoria County History of Gloucestershire*, vol. 10 (1972); D. Whitehead, 'Some connected thoughts on the parks and gardens of Herefordshire before the age of landscape gardening', *Trans. WNFC*, vol 48, pt. 2, p. 216; R. Newman, 'The effect of orcharding and the cider industry on the landscape of west Gloucestershire 1600-1800', *Trans. WNFC*, vol. 44, pt. 2 (1983), pp. 202-03; Brampton Bryan, deeds and papers, tenancy agreements among bundles 2, 31, 73, 81, 82, 85; Gloucestershire RO, D326/21.
26 Marshall, *op. cit.* in note 1; J. Clark, *General view of the agriculture of the county of Hereford*, London 1794.
27 Marshall, *op. cit.* in note 1, p. 287.
28 [Bromyard and District Local History Society], *A pocketful of hops. Hop growing in the Bromyard area*, Leominster 1988, p.5. In Kent the close in which hops are grown is called 'a garden', in Herefordshire 'a yard', the West-Country word still commonly used for a garden in the southern states of the U.S.A.
29 Clark, *op. cit.* in note 26, pp. 44-51.
30 *A pocketful of hops*, *op. cit.* in note 28, p. 5.
31 L.T. Smith, *op. cit.* in note 13, p. 47; J. N. Jackson, 'Some observations upon the Herefordshire environment of the seventeenth and eighteenth centuries', *Trans. WNFC*, vol. 36, pt. 1 (1958), p. 29; Jones, *op. cit.* in note 19, p. 37.
32 Marshall, *op. cit.* in note 1, p. 224. The spelling 'inclosure' was standard in the eighteenth century and is still commonly, if pedantically (as here), used when referring to parliamentary inclosure acts and awards.
33 Jones, *op. cit,* in note 19, p. 37, citing H. L. Gray, *English field systems*, 1915. Jones dismissed as pointless further research on open fields in the county, but since then more local archives have become available for research.
34 This evidence is also found in tithe apportionments and maps of *c*.1840, which cover nearly every parish in the county.
35 Clark, *op. cit.* in note 26, p. 57-58.
36 C. Watkins, 'The parliamentary enclosure of Much Marcle', *Trans. WNFC*, vol. 443, pt. 3 (1981), pp. 315-31; W.K. Parker, 'Opposition to parliamentary enclosure in Herefordshire, 1793-1815', *Trans WNFC*, vol. 44, pt. 1 (1982), pp. 79-90.
37 Clark, *op. cit.* in note 26, p. 27-28.
38 E. Heath-Agnew, *A history of Hereford cattle*, London 1983.
39 *Whitehead 2001.*
40 C. Morris, ed., *The illustrated journeys of Celia Fiennes c.1682-c.1712*, London 1984, pp. 189, 230.
41 Marshall, *op. cit.* in note 1, p. 223.
42 Marshall, *op. cit.* in note 1, pp. 224; Clark, *op. cit.* in note 26, p. 51.
43 A. Crow, *Bridges on the river Wye*, Hereford 1995.
44 V.R. Stockinger, ed., *The rivers Lugg and Wye navigation. A documentary history 1535-1951*, Almeley 1996, pp. 62-77.
45 I. Cohen, 'The non-tidal Wye and its navigation', *Trans. WNFC,* vol. 35, pt 2 (1956), pp. 83-101.
46 Stockinger, *op. cit.* in note 4, pp. 83-107, 131-150.
47 Jackson, *op. cit.* in note 31, p. 38.
48 Whitehead and Shoesmith, *op. cit.* in note 9.
49 I. Cohen, 'The Leominster-Stourport Canal', *Trans. WNFC*, vol. 35, pt. 3 (1957), pp. 267-86; D.E. Bick, *The Hereford & Gloucester canal*, Newent 1979, pp. 8-17.
50 A copyist possessed of patience and an ultra-violet lamp might succeed in recovering some of the evidence of these earliest large maps of the whole city within and w i t h o u t the walls.
51 Shoesmith, *op. cit.* in note 14, pp. 64-74.

52 Marshall, *op.cit.* in note 1, pp. 231-32.
53 S. Daniels and C. Watkins, eds., *The Picturesque landscape. Visions of Georgian Herefordshire*, Nottingham 1994, pp. 89-90; S. Ireland, *Picturesque views of the river Wye*, London 1797, pp. v-vi.
54 D. Whitehead and R. Shoesmith, *op, cit.* in note 9.

3. Printed County Maps (pp. 17-29)

1 P.D.A. Harvey, *Mappa mundi. The Hereford world map*, London 2002; pp. 7-11; V.I.J. Flint, 'The Hereford map: its author(s), two scenes and a border', *Trans. Royal Hist. Soc.*, 6th ser., vol 8 (1998), pp. 19-44.
2 Matthew Paris's map and the Gough map are reproduced in Harvey, *op. cit.* in note 1, pp. 36-37.
3 Mercator compiled his map from about 1550, greatly aided by information from an anonymous friend in England. P. Barber, 'The British Isles' in M. Watelet ed., *The Mercator Atlas of Europe circa 1570-1572*, English edn., Pleasant Hill, Oregon 1998, pp. 43-47 and facsimiles. Lhuyd's map and the others are reproduced in *Moreland & Bannister.*
4 *DNB; Robinson*, p. 322;
5 *Delano-Smith & Kain*, pp. 61-66; P. Barber, 'England II, Monarchs, ministers and maps 1550-1625' in D. Buisseret, ed., *Monarchs, ministers and maps: the emergence of cartography as a tool of government in early modern Europe*, Chicago and London 1992, pp. 57-98.
6 *Bendall*, vol. 1, p. 21.
7 *Delano-Smith & Kain*, pp. 66-71; P.D.A. Harvey, *Maps in Tudor England*, London 1993, pp. 54-65; I.M. Evans and H. Lawrence, *Christopher Saxton, Elizabethan map-maker*, Wakefield and London, 1979, pp. 1-11.
8 At Goodwood House, Sussex. The painting is published and described in F.W. Steer, 'The Court of Wards and Liveries in session', *Journal of the Society of Archivists*, vol. ii, no. 9, (April 1964), pp. 400-402.
9 BL., Royal MS. 18.d.iii, fol. 95.
10 The term 'atlas' was not used until 1595 by Gerard Mercator of Duisburg in the Rhineland.
11 A small (blue) open circle became the conventional sign for a spring. Saxton does not use this sign elsewhere on his map of Herefordshire where the source of the Dore is shown as a larger irregularly shaped small lake at The Bage in Dorstone, upstream of the Golden Well.
12 Reproductions are sold in the British Library shop.
13 *Ex inf.* Peter Barber, Head of Map Collections, British Library.
14 *Hodgkiss*, p. 20.
15 In some cases, beginning with Saxton, the maps were also sold loose by the publisher.
16 H. Turner, 'The Sheldon tapestry maps belonging to the Bodleian Library', *Bodleian Library Record*, vol. 17, no. 5 (April 2002), pp. 293-311. The original tapestry map of Worcestershire is on long-term loan to the Victoria & Albert Museum, London. A photograph is in Hereford City Library.
17 Robert 'Bossu' of Beaumont, Earl of Leicester was granted the earldom of Hereford by King Stephen about 1140 at the time of civil war; Miles of Gloucester was created earl by Queen Matilda in 1141. Thomas of Woodstock (1355-97) received a Crown rent from the county and his Stafford family descendants, Dukes of Buckingham, styled themselves earls of Hereford but the title was never officially recognised.
18 N. Nicolson and A. Hawkyard, *The counties of Britain. A Tudor atlas by John Speed,* London 1988, reproduces Speed's maps and contains a portrait.
19 Published as E.G.R. Taylor, *An atlas of England and Wales*, King Penguin Books 61, London 1951.
20 *Moreland & Bannister*, p. 63.
21 J. Ogilby, *Britannia 1675,* Facsimile with an introduction by J.B. Harley, Amsterdam 1970; K.S. Van Eerde, *John Ogilby and the taste of his times*, Folkestone 1976; D. Hodson, 'The early printed road books and itineraries of England and Wales', Ph.D. Thesis, Exeter University 2000, typescript, 2 vols. A copy will shortly be available at BL, Maps 237. a. 44.
22 *Delano-Smith & Kain*, pp. 84-91; P. Hindle, *Maps for historians*, Chichester 1998, pp. 15-16; D. Smith, *Maps and plans for the local historian and collector*, London 1998, p. 82.
23 *Hodgkiss*, p. 19; W. Leybourne, *The compleat surveyor: or the whole art of surveying of land ... and an appendix of practical observations in land surveying by Samuel Cunn*, 5th. edn., London 1722, pp. 113-14.
24 P.D.A. Harvey and H. Thorpe, *The printed maps of Warwickshire 1576-1900*, Warwick 1959, pp.19-35, 93-95. John Senex had earlier imitated Beighton in his map of Surrey (1729).
25 *Hodgkiss*, p. 77.
26 See p. 49.
27 *Delano & Kain*, pp. 96-97.
28 Ex inf. David Whitehouse (country houses and parks), Heather Hurley (roads) and Ron Shoesmith (Hereford city).
29 Author's unpublished notes on the history of Oxenhall presented to the editors of the Victoria County History of Gloucestershire, 2001.

4. Estate maps (pp. 31-40)

1. More precisely, the scale of Isaac Taylor's map was 69 1/2 miles to 1 degree (1:64360), not one inch to one mile (1:63360).
2. R. Johnson, *The ancient customs of the city of Hereford*, Hereford 1868, p. 29.
3. M.R. Ravenhill and M.M. Rowe eds., *Devon maps and mapmakers: manuscript maps before 1840*, Devon and Cornwall Record Society, new ser., vol. 43 (2002), part 1, pp. 3,5.
4. *Bendall*, vol. 1, figs. 7-20.
5. DCA, D649.
6. D.G. Bayliss, 'The effect of Bringewood Forge on the landscape of northern Herefordshire to the end of the seventeenth century, *Trans. WNFC*, vol. 45, pt.3 (1987), pp. 721-29; P. Cross, 'A 1577 plan "A platte of part of the Chase of Bringewood and of certayne groundes adjoining leased to Mr Walter of Ludlowe". A re-evaluation of the landscape', *Trans. WNFC*, vol. 48, pt. 3 (1996), pp. 573-81. The Mocktree inclosure act was passed in 1799 and the award made in 1803 (HRO, T74/438 and Q/RI/26). I am grateful to David Lovelace for pre-publication sight of his study of Bringewood Chase, 2003.
7. E.B. Wood, *Rowland Vaughan his booke published 1610*, London 1897, frontispiece.
8. W. Leybourne, *The compleat surveyor: or the whole art of surveying of land ... and an appendix of practical observations in land surveying by Samuel Cunn*, 5th edn. London 1722, pp. 113-14.
9. HRO, F76/III/22, Survey of Mr. Henry Jordan Estate at Richards Castle, 1743.
10. NLW, Mynde Park papers, 1466. The punctuation has been modernised and the abbreviations extended. Editorial comments are in square brackets..
11. The same customary perch of 18 feet was used in Devon and Dorset. In Herefordshire the customary acre for hops was 1/2 a statutory acre, containing 1,000 plants: J. Thirsk ed., *The agrarian history of England and Wales, vol.5, pt. 2 1640-1750*, Cambridge 1985, p. 826 and G.E. Mingay, *ibid.*, vol. 6, *1750-1850*, Cambridge 1989, p. 1117.

5. Surveyors (pp. 41-54)

1. By the time that this book is published the revised and enlarged *Oxford dictionary of national biography* should be available. Meanwhile, *Delano-Smith & Kain*, and *Bendall* have throughout been the fundamental sources for background information.
2. In the whole period up to 1850 and out of a total of 13,744 entries, *Bendall* recorded only 13 surveyors with addresses in Hereford and a further 26 elsewhere in the county. These figures may be compared with Gloucester (25), Shrewsbury (28) and Worcester (28). Similar numbers to Hereford were recorded in other towns far from London, such as Carlisle (13), Durham (15) and Plymouth (18). *Bendall*, vol. 1, pp. 1-2 and 197-256. In the course of preparing this volume an additional 20 surveyors, not recorded by *Bendall*, have been identified.
3. I.M. Evans and H. Lawrence, *Christopher Saxton, Elizabethan map-maker*, Wakefield and London 1979.
4. HRO, Church Commissioners' archives, Survey of the bishopric 1577-81, fol. 170v.
5. N. Nicolson and A. Hawkyard, *The counties of Britain. A Tudor atlas by John Speed*, London 1988, pp.10, 15-16.
6. S. Bendall, 'Draft town maps for John Speed's *Theatre of the empire of Great Britain*', *Imago Mundi*, 54 (2002), pp. 30-45; Merton College, Oxford, D.3.30.
7. K.S. Van Eerde, *John Ogilby and the taste of his times*, Folkestone 1976.
8. *Robinson*, pp. 93-6 (including pedigree), 304; *Whitehead 2001*, p. 363; National Library of Wales, Mynde Park papers, 964 (will 1729, proved 1731); *Bendall*, vol. 2, 419 (P382).
9. *Robinson*, p. 94 says incorrectly that Symons bought the estate 'about 1750'. According to *Whitehead 2001*, p. 363, it was sold to the tenant James Brydges 1st Duke of Chandos in 1723, who sold it on in 1727 to Richard Symons. The Mynde Park papers at the National Library of Wales indicate his interest in the estate began a little earlier.
10. For a transcript of the letter, see p. **38**.
11. *Bendall*, vol. 2, p. 101 (C264); A.S. Mason, *The 18th century land surveyors of Essex*, Chelmsford 1990, pp. 32-33.
12. HRO, Bodenham, Orleton and Yarpole parish registers.
13. *DNB*; E. Laurence, *The duty and office of a land steward*, London 1731, frontispiece; Bedfordshire RO, l26/1023.
14. Mason, *op. cit.* in note 11, pp. 48, 50.
15. P. Ellis, 'Longtown Castle: a report on excavations by J. Nicholls, 1978', *Trans. WNFC*, vol. 49, pp. 80-81.
16. HRO, G39/III/E1-600.
17. E. Stephens, *The clerks of the counties 1360-1960*, p. 100. He was also bailiff of Leominster in 1762 and 1776: J. Price, *An historical and topographical account of Leominster and its vicinity*, Ludlow 1795, pp. 69-70.

References — Part One

18 Hodson, vol. 1, pp. 179-182.
19 B.S. Smith, 'The Dougharty family, eighteenth-century mapmakers', *Trans. Worcs. Arch. Soc.,* 3rd ser., vol. 15 (1996), pp. 251-2; Cambridge, King's College archives, MON/60(1).
20 *Hereford Journal,* 29 November 1770. I am grateful to David Whitehead for this and other references in the *Hereford Journal.*
21 For his landscape garden design, see Whitehead, *Trans. WNFC,* 1980, pp. 186-7 and plate VIII; *Whitehead 2001,* pp. vii, 192.
22 His maps are well known to Welsh archivists and librarians but he has not attracted a biographer.
23 *Whitehead 2001,* p. 2, which includes an engraving of the house in 1787 looking rather less imposing.
24 The name is not uncommon in the early eighteenth century, but no relevant entries have been found among baptismal records. The sources searched unsuccessfully include the Ross parish registers 1715-35 and the IGI Index for Herefordshire and other counties with which Taylor was associated. See also p.**196**.
25 He signed all his maps, on some occasions using the interchangeable initial J for I.
26 P. Laxton ed., *Two hundred and fifty years of map-making in the county of Hampshire,* Lympne 1976, unpaginated. It includes a facsimile of Taylor's map, with an introduction.
27 Laxton, *ibid.,* quoting a letter from John Hutchins, the Dorset historian, to Richard Gough 1780.
28 Laxton, *ibid.; Delano-Smith & Kain,* pp. 89, 97.
29 *Hodgkiss,* p. 61.
30 Benjamin Donn, who won the first award for his map of Devon, spent five and a half years in its composition.
31 Estimates vary of the number of his Devon and Dorset estate maps. There are about twenty-one maps of the Erle Drax estate and twenty-five of the Sturt estate (*ex inf.* Paul Laxton, 2.December 2003), but the total is about one hundred and twenty (*Bendall* data sheet). See *Delano-Smith & Kain,* p.264, note 159; Laxton, *op. cit.* in note 26; M.R. and M.M. Rowe, *Devon maps and map-makers: manuscript maps before 1840,* Devon and Cornwall Record Society, new ser., vol. 45, pt. 2 (2002), p. 403; *Bendall* (T049) and the data sheet for Isaac Taylor.
32 HRO, L78/3, Ross churchwardens' accounts.
33 London, Guildhall Library, Sun Life Insurance Co. archive, policy no. 432165, 15 Aug. 1780. (I am grateful to Heather Hurley of Hoarwithy for this reference).
34 *Hereford Journal,* 17 November 1780.
35 *Bendall,* H144.
36 HRO, Leominster and Stoke Prior parish registers.
37 DCA, 7031/5, fols. 296r, 308v, 336v, 344r. Wainwright appears not to have carried out his commission and was therefore replaced by Pain. No maps for any of these commissions have been traced.
38 HRO, C38/16/1; HRO, R8/1/1. An inclosure act for Bodenham was passed in 1801-2 and the award followed ten years later (HRO, M61/72).
39 Although in 1609 Worcestershire was famed for its bagpipers and the place-name Bagpipers Tump occurs at Mordiford, it is unlikely that John Harris was harking back to these folk memories.

Part Two

Carto-Bibliography of the Printed Maps of Herefordshire before 1800

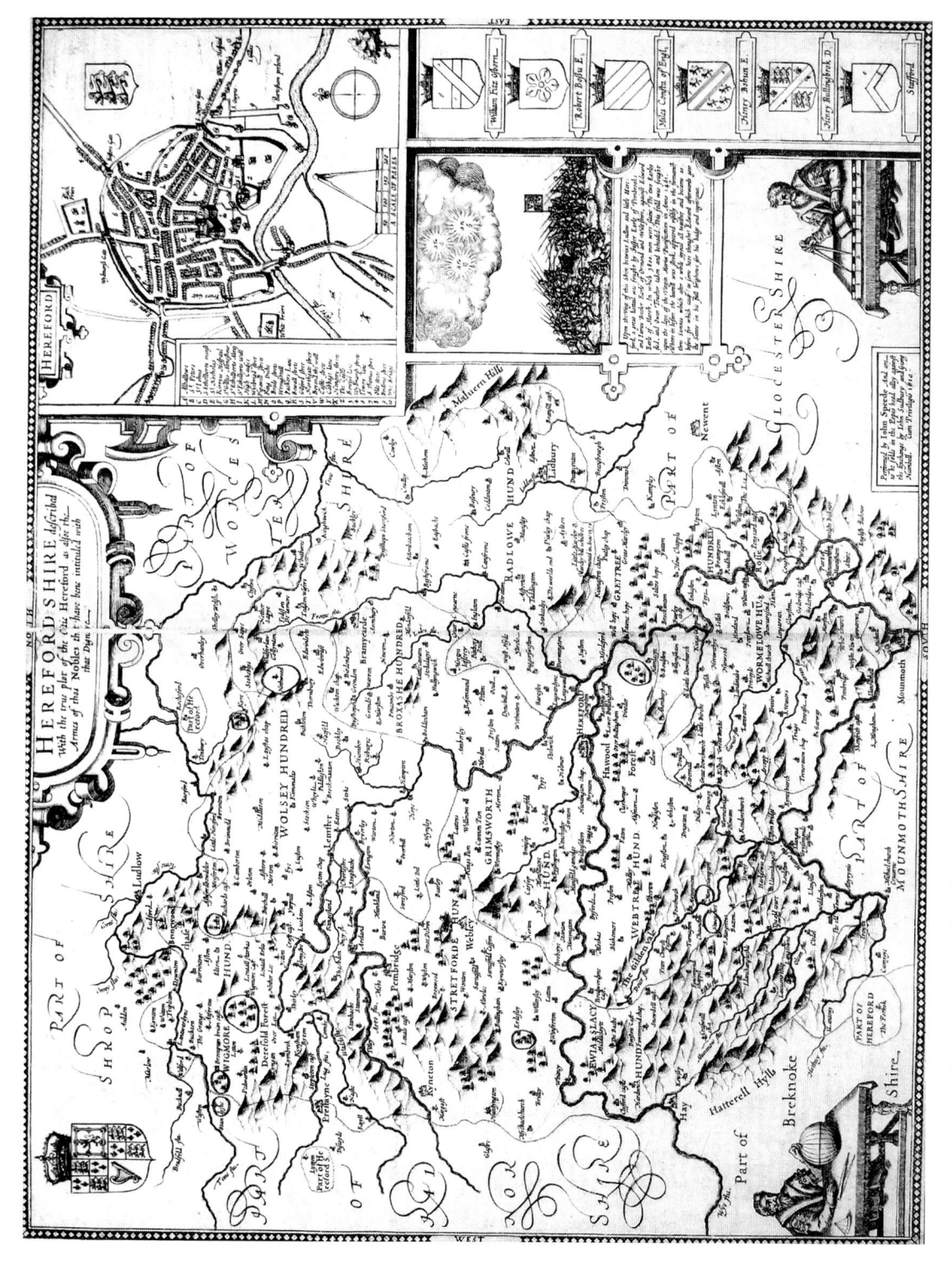

10. John Speed 1610. The county map is based on Saxton; the city was measured by Speed.

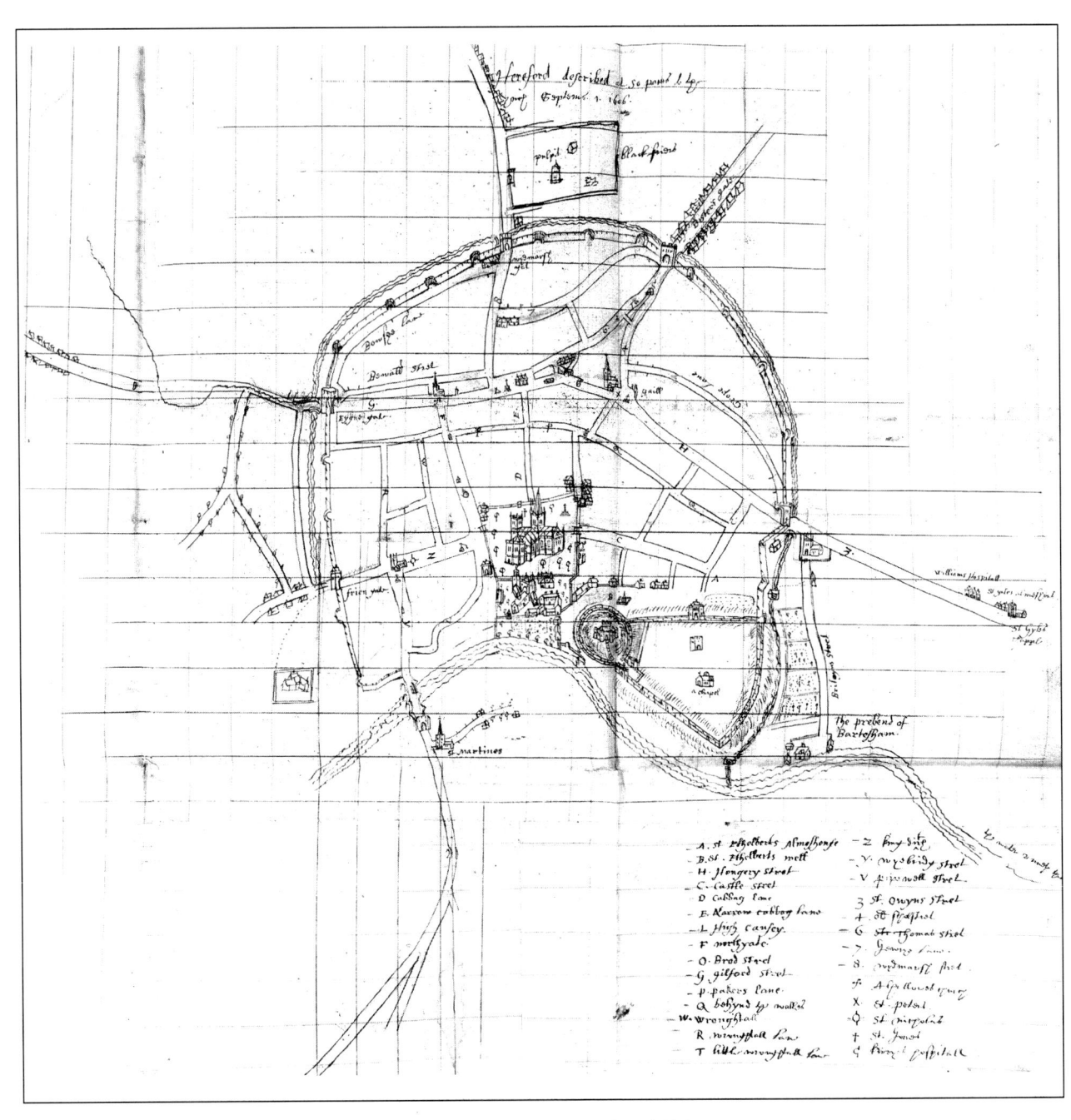

11. *John Speed, 1 September 1606.*
Speed's more detailed draft for the earliest map of Hereford city.

12. *Thomas Jenner's 1643 issue of Jacob van Langeren 1635.*

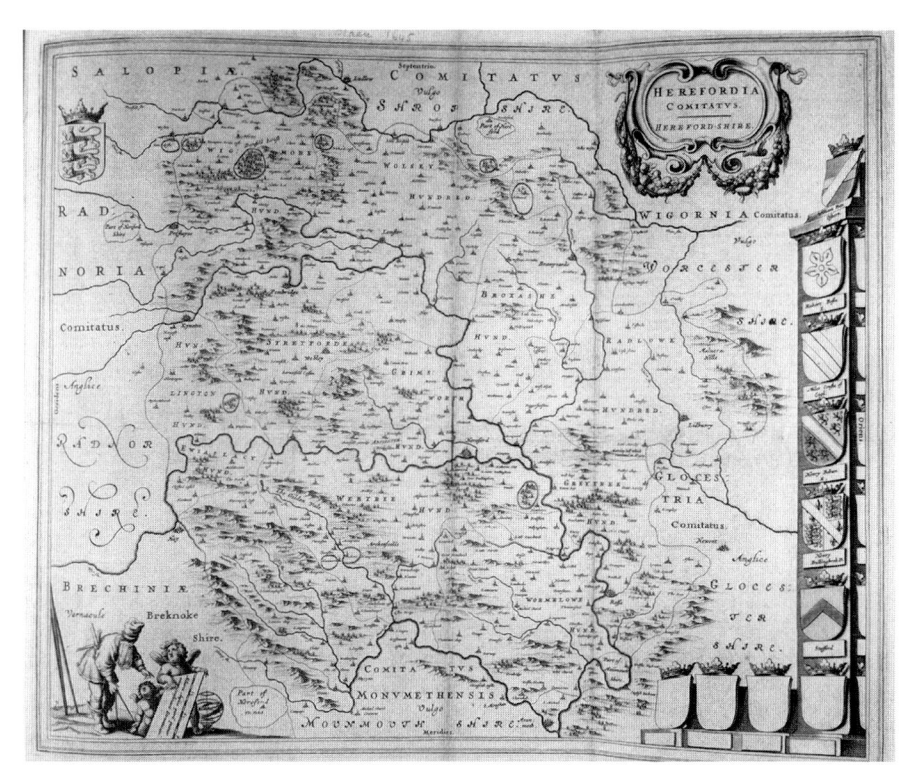

13. *Joannes Blaeu 1645 (the Spanish edition of 1659). The rival Amsterdam map-publishers, Blaeu and Jansson, produced similar maps based on Speed (see also Plate 4).*

14. *Unidentified French map c.1700. This late seventeenth-century map, based upon Speed, is not recorded by British bibliographers.*

15. *Richard Blome 1673.*

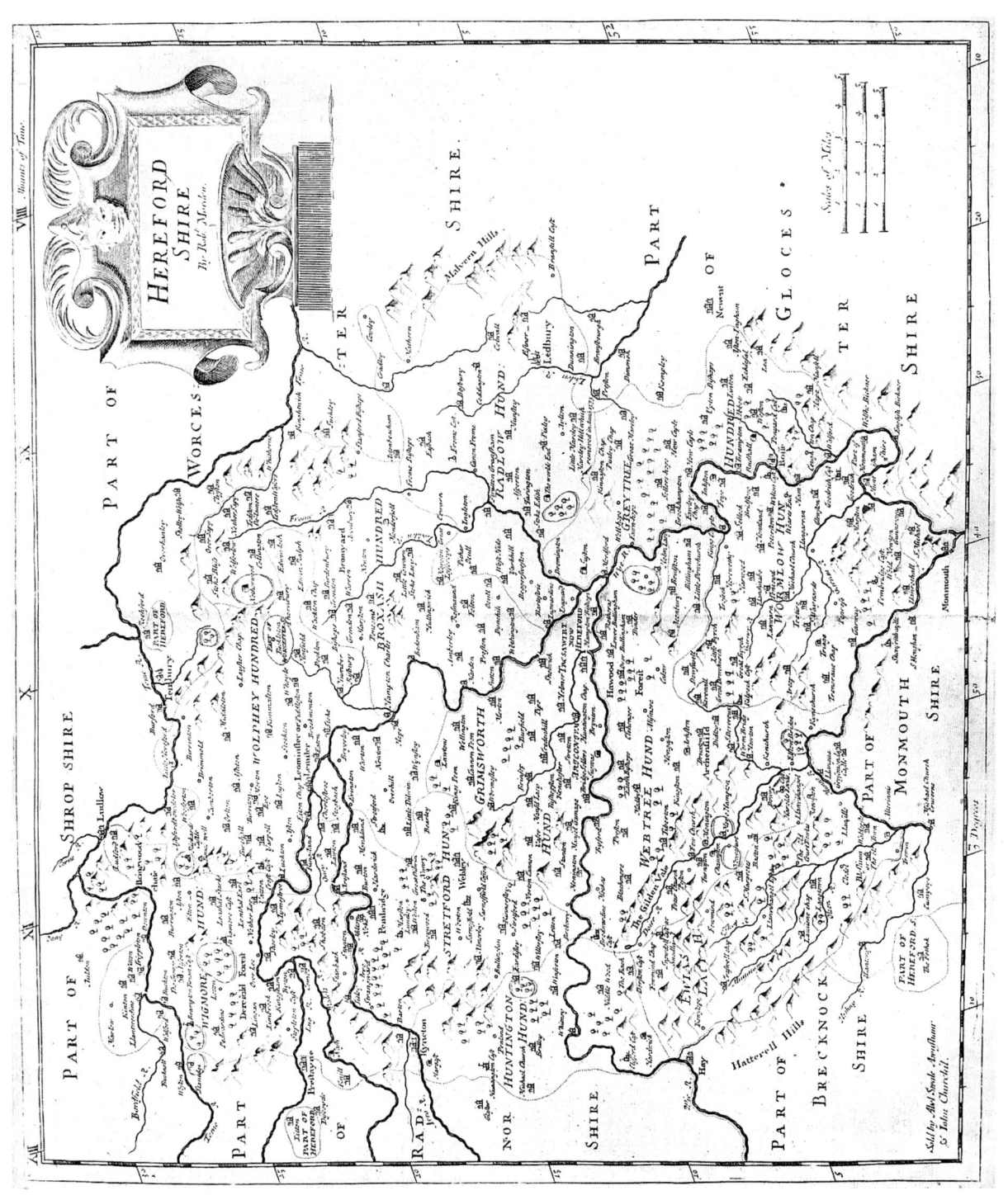

16. Robert Morden 1695.

A Map of HEREFORD-SHIRE.

HEREFORDSHIRE Contains one City, 2 Boroughs, and five other Market Towns. It sends 8 Memb.rs to Parliament; 2 for the shire, two for the City of Hereford, and four for the other two Boroughs.

Hereford City sends 2 Members to Parliament. Market W. F. & S. Fairs W. after Easter, May 19, June 20, and October 9.

Webley sends 2 Members Market Th. Fairs Holy Th. & Trinity Th.

Leominster sends 2 Members. Market F. Fairs Feb. 2. May 2. June 29. Aug. 24. October 28. and Tues. before Midlent

Rofs Market Thurs. Fairs Ascension Day. Corpus Christi Day. July 20. and 25. August 15. Sept.r 14. October 30 and November 29.

Lidbury Market Tues. Fairs May 1. June the 11th and September 21.

Bramyard Market M. Fairs Sunday.

March 25. and May 25.

Kyneton Market Wednesday Fairs July 22. Sept.r 13. Wed: before Easter and Whit–Monday.

Pembridge Market Tuesday Fairs May 1. Nov.r 6. and 11.

a.a.a. Part of Herefordshire
b. ——— Worcestershire
c. ——— Monmouthshire

17. *Thomas Jefferys and Thomas Kitchin 1748.*

18. *Alexander Hogg 1784.*

The Carto-Bibliography

Nearly all the printed county maps of Herefordshire described here were published in atlases of county maps. The two notable exceptions are the maps by Isaac Taylor of the county in 1754 and of the city of Hereford in 1757. This bibliography concentrates upon the features portrayed by the map-makers rather than the fine details of printing and publication. First editions have been listed in some detail, but subsequent impressions, re-issues and revised editions have only been noted briefly, though sufficiently, I hope, to alert historians and collectors to the number and variety of issues of the map before them.

Most of the atlases referred to may be consulted in the Map Room of the British Library or in other libraries of national importance. However, in some cases the publishers also sold county maps singly and, especially since the 1930s, copies of many atlases have been broken up by printsellers to provide loose maps for collectors. It is such loose copies that are now found most commonly in the printsellers' shops. Taken out of the context of their atlas loose maps present particular problems of identification. Collectors will be familiar with these difficulties. The map plate alone may not provide sufficient evidence to identify the edition or issue of the map in question.

Fuller details of every impression and edition of atlases containing a map of Herefordshire can be found in the excellent bibliographies of Chubb, Skelton and Hodson (see p. xiii), upon which I have drawn heavily in compiling this selection of Herefordshire maps. With such reliable bibliographies readily available it would have been unnecessarily redundant to examine every atlas or map listed here. For the most part, therefore, I confined my personal inspection to those loose maps to be found in the public collections in Hereford, the locations and references of which are noted at the end of each entry.

The description of the maps is intended to provide sufficient information to satisfy general readers, guide historical researchers and serve as an initial point of reference for map-collectors. The name of the 'surveyor' or, if that is unknown, the name of the engraver or publisher, is given first, with the title of the Herefordshire map, the title of the atlas in which it was published and the reference in the bibliographies mentioned above. A brief description of the map follows, concluding with its scale, size and the location of local copies. A note of subsequent editions and issues completes the entry.

The scales vary widely, usually being expressed by the map-makers as 'A Scale of x miles' on scale bars of differing dimensions. The English measures in use at the time have been retained, rendering these scales, following Chubb, in miles and inches, for example '6 miles (to $2^{7}/_{8}$ inches)', whilst also, for the sake of uniformity, presenting them as a fraction, in this case as 1:132230. Because of the difficulty in measuring the small scale bars precisely the fractions have been taken to the nearest round figure and must be regarded as approximate.

The size of the maps has been measured in millimetres. Measurements are taken to the borders of the map (not the page or the plate, which may sometimes contain a title or imprint outside the borders of the map). In a few cases, noted in the entries below, there were irreconcilable differences between the measurements of a map given by the bibliographers of the atlases and a loose copy of apparently the same map inspected in a Hereford library.

The bibliographers cited above understandably measured the atlases as books, that is, measuring the maps first by the height of the page and then by its width, sometimes supplying only an average measurement for all the county maps within an atlas. This practice is endorsed by the Anglo-American Cataloguing Rules for Cartographic Materials and is followed in this book throughout.

1577

Christopher Saxton (*c*.1542/4-1610/11).
FRUGIFERI AC AMENI HEREFORDIAE COMITATUS DELINIATIO Anno D(omi)ni 1577. (Plates 1, 2).

Published in *An Atlas of England and Wales*, London 1579.[1] The maps were engraved from 1574, Herefordshire being dated 1577. In the original 'atlas' in 1579 they did not have plate numbers. Of the four variant arrangements of the maps in the atlas, Herefordshire was first in two settings and seventeenth in two. In Lord Burghley's proof copy Herefordshire was nineteenth. They were also sold loose both before and after publication of the 'atlas'. (Chubb I-IV, Skelton 1).

The pioneering series of maps of English counties, upon which most county maps for the following two centuries were based. Conceived by William Cecil, Lord Burghley, Elizabeth I's Lord Treasurer, funded by Thomas Seckford, Master of Requests to the queen and carried out by Christopher Saxton of Dunningley near Wakefield in Yorkshire. Dutch and Flemish engravers and printers were prominent in the production of all atlases and maritime charts in the sixteenth and seventeenth centuries. Although Saxton's maps were probably engraved and printed in London, only three of the seven engravers were English. The others were Flemish, of whom Remigius Hogenberg engraved the most – nine including that of Herefordshire.

The title of the map of Herefordshire (top right) is on a tablet like a church wall monument with the royal arms above and supporters of Adam and Eve holding an apple. Scale bar of miles with ornate dividers and a scroll *Christophorus Saxton descripsit* above the bar and *Remigius Hogenbergius sculpsit* beneath (bottom left). Coat of arms of Thomas Seckford (bottom right) with his post-1576 motto *Industria naturam ornat*. Cardinal points are written in Latin in the decorative border.

The map is remarkably accurate and filled with detail, of which the only notable omissions are roads and the names and boundaries of the hundreds. It shows the whole county, including the small detached areas in Rochford (Worcs.), Litton in Cascob (Radnor) and 'The fathok' (Ffawyddog) in Cwmyoy (Monmouth), but not the more distant area of Farlow (Shropshire). (These areas were lost to Herefordshire in nineteenth-century boundary changes, for which see F.A. Youngs, *Guide to the local administrative units of England. Volume II: Northern England*, Royal Historical Society 1991, pp. 121-137 for Herefordshire, and other pages for relevant places in adjoining counties).

It marks towns, villages, castles and country houses with distinctive conventional signs, which are not explained in the atlas. Following continental precedents towns and parishes are marked by small open circles, with a representation of buildings in profile; for towns a group of buildings with three spires, Hereford's central spire bearing a Maltese cross; and for parishes a church with a west tower and spire. Castles are marked by an open circle flanked by two towers and country houses by either a small square with door and windows or a gabled house in perspective. Parks, indicating the principal seats of the gentry, are shown by a circle of paling. Rivers and streams are named and some bridges shown; the river 'Doier' (Dore) rises from a small lake, one of Saxton's features repeated by copyists for at least 150 years. Forests, indicated by drawings of trees, included 'Moktre', 'Derefolde' and 'Hawood' (Haywood) forests and 'Bringewood chase'. Relief is indicated by 'molehill' or 'sugar loaf' drawings for hills, with afternoon shading on their eastern flank. A whole range of these is drawn to mark 'Hatterell hill' (the Black Mountains) and the Malvern Hills.

It names the valley of the river Dore as 'The Gilden Vale', apparently the first published use of this name for the Golden Valley. It notes the landslip in 1575 on Marcle Hill, 'Kinnaston chap(el): W^{ch} was dreven downe by the remoueing of the ground'. The spelling of place-names indicates sixteenth-century pronunciation, for example 'Lemster' (still the pronunciation of Leominster), 'Snowdell' (Snodhill), 'Taddington al(ia)s Tarrington' (Tarrington) and 'Yarcle' (Yarkhill). Alders End in Tarrington is written as 'the Worldes end', a corruption that was not only copied from Saxton in printed maps but survived in an estate map of *c*. 1700 (HRO, E3/459). He gives the Welsh names 'Llanyhangleeskle' for Michaelchurch Escley and 'Llanihangel Dewlas' for Dulas. The place-name 'Fowemynd chappell' at (Mynyddbrydd in Dorstone) is duplicated by error, apparently for Urishay chapel, and Yarsop appears as 'Carsop', errors that were repeated by Saxton's copyists.

SCALE: 6 miles to 2$^{7}/_{8}$ inches. (1:132230).

SIZE: 372 x 502. (Skelton 410 x 517), (Hodgkiss 381 x 464).

REFERENCE: HCL, 3 copies (2 damaged at edges) and 2 copies of the reproductions published by the British Library.

Other issues:

1579

Chubb notes six impressions of the maps, the popular counties being repeatedly reprinted. Many of the maps contain minor alterations, though apparently not Herefordshire. Sets were made up for binding from the stock of the various impressions, which, being printed on separate batches of paper, had differing watermarks. The earliest of these watermarks feature designs of crossed arrows and bunches of grapes. (Chubb I- VI). Skelton comments that sales were affected by the publication of Speed's *Theatre* in 1611, in which most of the county maps were derived from Saxton.

1588
Four tapestry maps of Warwickshire (dated 1588), Worcestershire, Gloucestershire and Oxfordshire (all undated), designed as wall-hangings for Ralph Sheldon's new house at Weston Park near Shipston-on-Stour, Warwickshire. They were woven under the direction of Richard Hyckes and his son Francis and based on Saxton's county maps of 1579. Though clearly owing much to Saxton there are also marked differences. The maps are not confined to the county boundaries, but in order to fill the rectangular tapestries included parts of neighbouring counties. The map of Worcestershire also covers east Herefordshire. Towns and villages are marked by small drawings, some roads are shown and a few beacons, though not the Worcestershire Beacon on the Malvern Hills. In the borders are extracts peculiar to the 1590 edition of William Camden's *Britannia,* suggesting that the tapestries may not have been completed until well into the 1590s. Later versions of the maps of Worcestershire and Oxfordshire were made in the mid-seventeenth century, more probably in the 1660s than the earlier ascription to 1647. (H.L. Turner, 'The Sheldon tapestry maps belonging to the Bodleian Library', *Bodleian Library Record*, vol xvii, no. 5 (April 2002). The Worcestershire tapestry owned by the Bodleian is now on loan to the Victoria & Albert Museum, London.

REFERENCE: HCL, photograph (reduced).

1590, *c*.1605
(William Bowes).
Playing cards, signed *W.B. invent. 1590*. A set of county maps based on Saxton's general map of England and Wales. Plates engraved as 60 cards on 4 plates, intended to be cut out and pasted on backing card. The first example of playing cards in this form. (Skelton 2).

1607
(William Hole).
See below (p. 66-67). Distinguished by its smaller size).

1642
FRUGIFERI HEREFORDIAE COMITATUS...DELINIATIO, with the arms of Charles I. Published in *The Maps of all the Shires in England, and Wales. Exactly taken and truly described by Christopher Saxton ... Now newly revised, amended, and reprinted ... for William Web,* 1645. Arms of Charles I but not the initials C.R. No plate numbers; Herefordshire was sixteenth in the atlas. Web, an Oxford bookseller,is thought to have been a royalist sympathiser. (Chubb VII, Skelton 27).

1665 (1689), 1693.
THE COUNTY OF HEREFORD, RESURVEYED & ENLARGED. Anº: 1665. The title tablet (top right) flanked by Adam and Eve, with C.R. above the royal arms. Arms of Milo Fitzwalter, Henry Bohun, Hump. Stafford Earl of Hereford, Hen. Bollingbrok Duke of Hereford (left side) and Mortimer Earl of Hereford (beside the scale bar, bottom left). Beneath the scale bar the attribution to Hogenberg as engraver remains unchanged.

Although unmistakably based on Saxton 1577, the map has been extensively revised. Hills appear as steep tumps, especially Hatterrall Hill and the Malvern Hills. Additions include the hundreds, the principal roads (as in Ogilby 1675) and *Ariconium* (at Kenchester); but Archenfield has been omitted. The place-name Fowemynd occurs twice, as in 1577, but 'Uris-hay' has incorrectly been added near the northernmost one. A plan of Hereford, based on Speed, replaces the arms of Seckford (bottom right). It gives the names of the city gates and has a key to the principal landmarks, with a note 'The Scotch Army laying Seige to this Citty of HEREFORD in the yeare 1644, occasioned the demolishing of the suburbs therof'.

Prepared for an edition which was not published (Skelton 80). Eventually published in ALL THE SHIRES OF ENGLAND AND WALES *described by Christopher Saxton ... by Philip Lea'*, 1689, with the title unchanged but some additional symbols (Skelton 110).

Published again, with roads added in *THE SHIRES OF ENGLAND AND WALES ... by Philip Lea*, 1693. (English and French editions) (Skelton 112, 113).

SCALE: 6 miles to 2¾ inches. (1:138240).

SIZE: 373 x 500.

REFERENCE: DCA, B.5/306.

***c*.1732**
(George Willdey).
Philip Lea died 1700 and, following the death of his widow Anna *c*.1730, the plates were acquired by Willdey. He died 1737 and his son Thomas in 1748. (Hodson 183).

***c*.1749**
(Thomas Jefferys).
Jefferys was an engraver who turned to publishing *c*.1750. Bankrupt in 1766 he was saved by Robert Sayer, who he was then obliged to take into partnership. (Hodson 184).

*c.*1772
(Cluer Dicey). (Hodson 185).

1. ' Atlas' is a misnomer. Saxton sold the maps loose as they were engraved and published. It seems that it was only on completion of the last counties that he decided to put sets together for binding in a volume. The title 'atlas' was first used in 1595 by Gerard Mercator, a Fleming then living in Duisburg in the Rhineland, for his *Cosmography of the World*.

(1599), *c.* 1605

Pieter Van den Keere (*c.*1571-1646).
HEREFORDIA COMITATUS.

In a collection of 44 miniature maps of the British Isles, three of which (Warwickshire and Leicestershire; Radnor, Brecon, Cardigan and Carmarthen; northern Scotland) were dated 1599. Based on Saxton and engraved by Van den Keere, probably in Amsterdam. A limited number of places shown, determined not so much by their importance as a desire to fill empty spaces. 'Arcop' is marked for Orcop and duplicated midway between Bredwardine and Hay, probably a misreading and misplacing of Saxton's 'Kewsop' (Cusop). Rivers and some hills are marked, though neither Hatterrall Hill nor the Malvern Hills. (Chubb IX, Skelton 4).

SCALE: 10 miles to 1¼ inches. (1:506880).

SIZE: (Chubb *c.*90 x 120).

Other issues:

1617
Plates re-used by Willem Janszoon Blaeu for his Dutch epitome of Camden's *Britannia, c.*1619. The map of Herefordshire has the page number 412 printed on it. (Chubb X, Skelton 12).

1627
Plates, acquired by George Humble (d.1640), who with his uncle John Sudbury published Speed's *The Theatre of the Empire of Great Britaine* in 1611. Forty of those reprinted by Van den Keere were issued, with others, in 1627 in the miniature atlas *England Wales Scotland and Ireland Described and Abridged...from a farr Larger Voulume Done by John Speed.* Known as 'the Pocket Speed', subsequent editions were brought out up to 1676 at about the same time as the folio editions of Speed. The map of Herefordshire has the plate number 24 (bottom right). (Published in E.G.R. Taylor, *An atlas of Tudor England and Wales,* King Penguin Books 61, London 1951. Chubb XI, Skelton 17).

SCALE: 10 miles to *c.*1⅛ inches. (1:533558).

SIZE: 83 x 123. (Hodgkiss 76 x 121).

REFERENCE: DCA, B.5/304.

?1632
(G. Humble). (Chubb XI, Skelton 19).

?1646
(William Humble). (Chubb XIII, XIV, Skelton 37).

1662
(Roger Rea the elder and younger). (Chubb XV, Skelton 69).

1666-1668
(Roger Rea). (Chubb XVI, Skelton 82, 83, 86).

1676
(Thomas Bassett and Richard Chiswell). (Chubb XVII, Skelton 93).

(1607)

William Hole.
'FRUGIFERI AC AMAENI HEREFORDIAE Comitatus qui olim pars fuit Silurum. delineatio' (bottom right).

Published, possibly in Amsterdam, in William Camden's *Britannia,* sixth and last Latin edition but the first to include maps. *Britannia* was first published in 1598. Twenty-one of the maps were engraved by Hole, thirty-four by William Kip. (Chubb XVIII, Skelton 5).

No plate number but with the Latin text of pp. 467 and 470 on the back. Ornate dividers bestride the scale bar (bottom left) laid across a compass rose, beneath which is the inscription *'Christophorus Saxton descripsit. W(illiam) Hole sculp'*. Though derived from Saxton 1577, the engraving is finer and clearer than Saxton's and the representation of hills, forests and parishes are different, as are some spellings. The duplicate entry for 'Fowemynd chappell' has been removed and antiquarian features added. The ancient regional name 'Archenfeild' spreads across SW Herefordshire, and the name *Ariconium* is added for Kenchester (not correctly identified as *Magnis* until 1732 by John Horsley, a mistake perpetuated on maps until the end of the eighteenth century). A few copying errors have been introduced, for example 'Mooktre' for Saxton's 'Moktre', 'Canfrome' for 'Canfrome' and 'yᵉ for 'the' in the note about Marcle Hill; Yarsop is further distorted from Saxton's Carsop to 'Crasop'. Dulas is here named 'Llanihangell Dowlas' with a clear space between the two words; later copyists widened the spacing further with the result that eventually the word Dulas, variously spelt, becomes attached to Old Castle in Clodock leaving Llanvihangel by itself, also variously spelt, as the name of the village. Royal arms of James I with the motto *'BEATI PACIFICI'* (top right).

SCALE: 5 miles to 1⁷/₈ inches. (1:168960).

SIZE: 284 x 311.

REFERENCE: DCA, B.5/313. With the Latin text of pp. 470-474.

Other issues

1610
Plates reprinted for the first English edition of Camden's *Britannia* 1610, with the plate number 34 added. No English text on the back. (Chubb XIX, Skelton 6).

1637
Plate reprinted as in 1610, but with some signs of wear. In the top right hand corner is the trace of a partly erased numeral or symbol. (Chubb XX, Skelton 23).

REFERENCE: DCA, B.5/302. HCL. HRO, K38/D/1.

1610

John Speed (1552-1629).
'HEREFORDSHIRE described With the true plot of the Citie of Hereford. as alsoe the Armes of thos Nobles that have been intituled with that Dignity' (top centre). Performed By John Speede ... solde ... by John Sudbury and Georg Humbell ... 1610' (bottom right).

Published in *The Theatre of the Empire of Great Britaine...*, London 1611. Engraved by Jodocus Hondius (1563-1612). (Chubb XXIV, Skelton 7). With pp. 49-50 printed on the back.

The maps in the *Theatre*, including Herefordshire, were mostly based on Saxton 1577 with some derived from county maps by John Norden, William Smith and others. The *Theatre* formed the first volume of a larger work, *A History of Great Britaine*, which Speed compiled 1596-1610. It is the earliest published atlas of the British Isles.

It marks towns, villages, castles, parks and some country houses, forests, woods, rivers with some bridges, streams and hills (by 'molehills'). Some of Saxton's place-name spellings have been amended but the test-names of 'Lemster', 'Cledol', 'Carsop' and the Welsh names remain unchanged, with the words 'Dowles' and 'Llanihangell' for Dulas completely separated (see above, Hole 1607). Bringewood chase is named; Mocktree forest is shown but not named. Notes 'Marcley Hill which removed in Anno 1575' but on p. 49 he gives the date as 1571. Roads are not marked.

The significant new features are the boundaries and names of hundreds, the inset plan of Hereford, the six coats of arms of William Fitz Osborn, Robert Bossu E[arl], Miles Consta[ble] of Engl., Henry Bohun E[arl], Henry Bullingbrok D[uke], Stafford, Hereford, and a drawing, probably by another artist, of the battle of Mortimers Cross 1461 recording the legend of the three suns that appeared in the sky before the battle.

The inset map of Hereford shows the city in bird's-eye view, with the streets named and the principal buildings and landmarks indicated by letter or number references A-Z (excluding J and U) and 3-10. It marks the cathedral with a west tower, central spire and chapter house, the former friaries (Greyfriars being incorrectly named Whitefriars), churches, almshouses and hospitals, crosses in High Town and Chapter House Yard, the castle, town walls and gates, St. Ethelbert's Well, the river Wye and Wye bridge, streams, two mills in the City Ditch upstream from the castle and, outside the city walls, suburbs along the radial roads. Small compass rose. Scale bar with dividers for 'A scale of 200 paces'. Speed measured and drafted the map on 1 September 1606. For the variations between the printed map and the draft, which is more accurate and at a larger scale, see pp. 95-96.

Border, double-ruled and ornamented with strapwork, with the cardinal points written in English.

Descriptive text in English on the back, as also, with some variations in typesetting, in all later issues unless noted otherwise below.

SCALES: County map: 5 miles to 2¹/₄ inches (1:140800). City map: 200 paces to 1³/₈ inches (1:9750 when compared with O.S. map).

SIZE: 380 x 507.

REFERENCE: HRO, K38/D2.

Other issues:

1605-10
Like Saxton's maps, Speed's were also issued in a series of impressions, sometimes before the imprint, engraver's name and date had been inserted. No text on the back. Within this series Chubb notes an undated map of Herefordshire with a plain border. (Chubb XXIII).

1614-1654
1614 (as 1610), 1616 (Latin text) (by John Sudbury and George Humble), 1623, 1627, 1632, 1646 (by George Humble), 1650-54 (repeated issues by William Humble). The 1632 and 1646 issues were printed in large quantities as separate maps to meet demand during the Civil War. (Skelton 7, 36, 55, 57). The 1616 issue has

been published as N. Nicolson and A. Hawkyard, eds., *The counties of Britain. A Tudor atlas by John Speed*, London 1988.

1650, *c*. 1665
(Roger Rea the elder and younger, after acquiring the plates from William Humble). Roger Rea the elder died in 1665, the year of the Plague, and the stock of this issue was largely destroyed in the Great Fire of London. (Chubb XXVI, Skelton 81. Chubb noted an undated map of Herefordshire in a copy of *The Theatre of the Empire of Great Britain*, 1662, the publication of which Skelton amends to 1665).

SIZE: 381 x 508.

REFERENCE: DCA, B.5/309.

1676
(Bassett and Chiswell). HEREFORD. Above the scale are added the arms of 'Walter D'Eureux (Devereux), Visc(ount) Hereford'. Published in *The Theatre of the Empire of Great Britain ... With many Additions never before Extant ... Printed for Thomas Bassett and Richard Chiswell, 1676.* With text of pp. 49-50 on the back of the Herefordshire map. (Chubb XXVII, Skelton 92). Bassett and Chiswell's imprint at the foot of the map is undated. The additions include the place-names *Ariconium* as the Roman name for Kenchester and 'Next Inne' (Nextend) in Lyonshall.

REFERENCE: HCL. HRO, K38/D3.

1676
Similar maps, but without text on the back. Issued for sale separately, possibly before publication of the atlas. It is thought that some sharper looking plates, which had been showing signs of wear, were re-touched for the 1676 publications. Later issues are of poorer quality as the copper plates again became worn.

REFERENCE: HCL.

From *c*. 1690.
Maps printed by Bassett and Chiswell with some by Christopher Browne *c*.1695, and sold as sets or singly from *c*. 1713 by Henry Overton (d.1751) and his nephew Henry Overton. John and Henry Overton's atlases were made up with maps from Speed, Blaeu, Jansson and others. They are listed below at *c*.1670-*c*.1755. Their complex history is unravelled by Hodson 135-138, 142-145.

***c*.1713, post-1716, *c*.1720.**
England Fully Described in a Compleat Sett of Mapps ... Printed and Sold by Henry Overton. (Hodson 135-137).

1743
England Fully Described in a Compleat Sett of Maps ... Reprinted Anno 1743. Printed & Sold by Henry Overton ... Reprints of Bassett and Chiswell with their names replaced by that of 'Henry Overton at the White Horse without Newgate, London'. The main roads are engraved on all the plates. Henry Overton's reprints are usually without text on the back. (Hodson 138).

1770
HEREFORDSHIRE, still with the Henry Overton imprint, appeared in *The English Atlas*, no. 20 printed by C. Dicey & Co. On most of the maps Overton's imprint was replaced by that of C. Dicey. There is no text on the back of the maps. For re-issues of the map of Herefordshire in John and Henry Overton's atlases see below *c*.1670-*c*.1755. (Hodson 145).

See also p.74 under Speed and Blome 1681 and Taylor 1715.

1612

Michael Drayton (1563-1631).
Herefordshyre, Parte of Glocestershyre, Parte of Worestershyre
    Probably engraved by William Hole. London, 1612. One of eighteen illustrations to Drayton's great series of topographical poems on England and Wales. (Chubb XXXIII, Skelton 8).
The so-called pictorial 'maps' portray hills and rivers peopled with legendary and symbolic figures. The Malvern Hills have a shepherd, Marcle Hill a man tumbling down from its summit. Rivers and streams contain naked nymphs and a bearded man, 'The goulden Vale' by a seated girl with a basket of flowers and garlands. Of the few named places, 'Lemster' is represented by a woman spinning. There are no county boundaries and no scale. A cartographical curiosity.

SIZE: 245 x 330.

REFERENCE: HCL.

Other issues:

1613
Inside the top border of the Herefordshire map the number of the following text page 101 has been printed. (Chubb XXXIV, Skelton 9).

1622
(Chubb XXXV, Skelton 13).

1626

(John Bill) (1604-1630).
Hereford Shire.

Published in *The abridgement of Camden's Britan(n)ia. With The Maps of the Severall Shires of England and Wales. Printed by John Bill ... 1628.* (Chubb XLI, Skelton 15).

The atlas is reputedly the rarest of all county atlases and single maps are hard to find.. The maps are rather crude and overcrowded, based on Saxton and Van den Keere, but are of interest as the earliest set of county maps to have a grid of latitude and longitude marked on the outside left-hand (west) and bottom (south) borders. Longitude is measured from a meridian running through the Azores. Cardinal points are also written outside the border. Scale bar with dividers. Description of the county on the back of the map.

SCALE: 10 miles to 1¼ inches. (1:563200).

SIZE: *c*.125 x 95.

1635

Jacob van Langeren.
Hereford shire.

Published in *A Direction for the English Traveller ... Sold by Mathew Simmons ... 1635.* (Chubb XLIV, Skelton 20).

Herefordshire is plate number 19. The first of the roadbooks/pocket atlases, with thumb-nail county maps copied from William Bowes's playing cards of 1590 but about twice the scale in a square of about 40 mm. The diagonal left-hand top part of the page is occupied by another innovation, a triangular table of distances between towns and villages, with the mileages based on John Pond's almanac, 1607. Printed on one side only. No signatures. Engraved by Jacob van Langeren.

SCALE: 10 miles to ¼ inch. (1:2534400).

SIZE: *c*.100 x 100.

Other issues:

1636
Two further editions. (Chubb XLV, Skelton 21, 22).

1643
(Thomas Jenner). With larger maps at a scale of 20 miles to 1 inch (1:1267200). Names of towns in full and names of neighbouring counties; some plate numbers erased; Herefordshire is plate 16. Jenner, the puritan who produced the 'Quarter-master's Map' (see below) for Parliamentary armies, is thought to have prepared this edition with the larger-scale maps. Skelton remarks that the order in which the maps were revised seems to reflect the progress of the Civil War. A variation 1644, of which only one copy is reported, lacks Herefordshire. (Chubb XLVI, Skelton 25). (Plate 12).

1644?
Most of the maps were published by Jenner in *A Booke of the Names of all the Hundreds ... Together with the Number of the Towns, Parishes, Villages and Places of every Hundred ... Usefull for Quarter-masters.* Only one copy is known, which lacks Herefordshire. (Skelton 26). The map does, however, appear in Jenner, *A Book of the Names of all Parishes ... in every Shire,* 1657. (Chubb XLIX, Skelton 62).

1662, 1668
(printed for Thomas Jenner). (Chubb L, LI, Skelton 70, 87, 88).

1677, 1677? 1680?
(John Garrett). (Chubb LII, XLVII, XLVIII, Skelton 98, 99, 101).

1645

Joannes Blaeu (1596-1673)
Herefordia Comitatus. Hereford Shire.

Published in *Guil. et Ioannis Blaeu. Theatrum Orbis Terrarum sive Atlas Novus. Pars Quarta.* Amsterdam, 1645. (Chubb LIX, Skelton 28).

One of the finest atlases, beautifully written and well engraved, but the Latin text comes from Camden's *Britannia* and most of the well-drawn maps, including Herefordshire, are derived from Speed post-1623, which they closely imitate, omitting the city plan and battle scene. The coats of arms and decorative features were rearranged. The title cartouche (top right) incorporates fruit and barley in a baroque frame. The map of Herefordshire has (right) the six coats of arms of William Fitz Osborn, Robert Bossu E(arl), Miles Consta(ble) Of Engl., Henry Bohun E., Henry Bullingbrok D(uke), Stafford, and England; four more are blank. The scale bar (bottom left) shows an elderly surveyor with measuring poles and chain pointing to the scale bar held up by two cherubs. The map does not have Blaeu's name or an imprint. Two numbered pages of text are on the back.

SCALE: Millaria Anglica, 3 miles to 1⅝ inches. (1:116970).

SIZE: 403 x 498. (Y. Beresiner, *British county maps,* Woodbridge 1983, p. 58; 416 x 492).

REFERENCE: HCL, (facsimile, 410 x 497).

Other issues:

1646, 1648, 1662
(Latin). (Skelton 30, 42, 71).

1645, 1646, 1648, 1662, 1663, 1667
(French). Skelton 29, 31, 43, 72, 75, 84). Size: 410 x 497.

REFERENCE: HCL

1645, (1646), 1648
(German). (Skelton 32, 33, 44).

1646, (1647), 1648, 1664
(Dutch). (Skelton 38, 45, 77).

REFERENCE: HCL.

1659, 1662
(Spanish). (Skelton 64, 73). With a scale of 3 English miles '*quorum quatuor unum constituunt Germanicum*' and Spanish text on the back. Size: 408 x 498. (Plate 13).

REFERENCE: DCA, B.5/300.

In 1672, apparently whilst a further Spanish edition was being prepared, a fire destroyed the Blaeu family's premises and the copper plates on which the maps were engraved.

See also John and Henry Overton's atlases, *c.*1670—*c.*1755.

c. 1645, 1646-7

Jan (Joannes) Jansson (1588-1664).
HEREFORDIA COMITATUS vernacule HEREFORD SHIRE Amstelodami Apud Ioan. Ianssonium (bottom right). (Plate 4).

First published undated, but probably 1644, as a loose sheet derived from a pre-1623 Speed. Revised and published in *Ioannis Ianssonii Novus Atlas Sive Theatrum Orbis Terrarum ... Tomus Quartus.* Amsterdam, 1646. The other three volumes of *Novus Atlas* are dated 1647. (Chubb LXX, Skelton 34).

Like his competitor Blaeu, Jansson based his atlas on Speed, though perhaps with more justification, for he was the son-in-law of Jodocus Hondius the elder, the engraver of the maps in Speed's *Theatre*. In the race to publication Blaeu completed his atlas first but Jannson brought out some of his maps ahead of Blaeu in 1644. Amongst this hurriedly produced number was Herefordshire. In the pre-1646 preliminary state of the plate he omitted a scale bar and cardinal points but included a small plan of the city of Hereford, without named landmarks, in a monumental cartouche (bottom right) as well as the title cartouche (top right).

In the version published in *Novus Atlas*, 1646, he excised the city plan and rearranged the decorative features of the title cartouche (bottom right), seven coats of arms, one blank, (top right) and the scale bar (bottom left). He took his maps from more than one issue of Speed, and indeed also from Blaeu, together with the Latin text of Camden's *Britannia*. Like Blaeu he omitted Speed's town plans and battle scenes. It has been noted that Jansson's draughtsmanship is a little more flamboyant than Blaeu's and (in Monmouthshire) that his spelling of place-names is less reliable. The hundreds are marked but not roads. Malvern has been miscopied as 'Halvern Hills'. Archenfield has been reduced to a minor place-name between Druxton (Thruxton) and Didley. He put scales of both English and German miles, but later issues of the Herefordshire map have only a scale of English miles. There are further minor distinctions between the otherwise similar maps. The Herefordshire map has (top right) the arms of Will. Fitz Osborn, Robert Bossu E., Miles Co(n)sta of Engl., Henry Bohun E., Henry Bullingbrok D., Earl of Stafford, and one left blank. The scale bar (bottom left) incorporates a standing figure of a man, with two cherubs holding dividers and a compass, pointing to the scale bar. The cardinal points are written in Latin within the plain border. There is no text on the back.

The map of Herefordshire was first published as a loose sheet, derived from the pre-1623 state of Speed. The date of this earlier version is not known. For the *Novus Atlas* it was revised by giving it a new title and scale bar and adding shields in place of the old title and cartouche. The imprint 'Apud Amstelodami Apud Ioannem Ianssonium' was also added. The hundreds are shown with their boundaries, forests, parks, hills (by molehills), some river crossings, and the note about Marcle Hill. Place-name spelling follows Saxton and Speed; duplicate Fowemynd chapels are marked.

SCALE: Millaria Anglica, 4 miles to $1^{7}/_{8}$ inches. (1:135168).

SIZE: 376 x 500.

REFERENCE: DCA, B.5/305 with Latin text on back.

Other issues:

1659
(Latin). (Chubb LXXVII, Skelton 66).

REFERENCE: HCL (date in ink 1659, later amended by HCL to 1646).

1646, 1647, 1652, 1656
(French). (Skelton 35 39, 52, 61).
REFERENCE: HCL (1646).

1647, 1649, 1652, 1653, 1659
(Dutch). (Skelton 41, 49, 54, 56, 67). **HCL** (1652), measuring 370 x 492 with no text on the back.
REFERENCE: HRO, K38/D4.

1647, 1649, !652, 1658? 1659?
(German). (Skelton 40, 46, 53,63, 68).

After 1694, when the stock of Jansson's son-in-law, Joannes van Waesberghe (d.1681) was sold and dispersed, Peter Schenk acquired the copperplates, from which he and Gerard Valck issued loose sheets of the English county maps, without text on the back.

REFERENCE: HCL (1683).

***c.*1705**
(Latin). Published by Karel Allard in *Atlas Major*, maps unrevised. (Chubb LXXX, Hodson 130).

1714, 1715
(French). (D. Mortier). Herefordshire bears plate number 30 (Hodson 131, 132).

1724
(Joseph Smith). (Chubb LXXXI, Hodson 134).

See also John and Henry Overton's atlases, *c.*1670-*c.*1755.

*c.*1670-*c.*1755

John Overton (1640-1713) and **Henry Overton** (d.1751). HEREFORDIA COMITATUS.

John Overton, bookseller, and Henry Overton his son, print and bookseller, issued a series of eight atlases of existing county maps over the imprint of their names. Some of their stock came from Peter Stent (d. of the plague in 1665), who had acquired material from George Humble, Speed's publisher. They made up sets of county maps into 'atlases' from their stock as demand required, buying in fresh copies as necessary or running off new impressions from the increasingly worn copper plates in their possession. A single atlas might, therefore, be made up with maps from various sources and impressions. The serious collector or cataloguer wishing to establish the provenance of a specific county map is faced with a headache. Similarly, the historian must, before accepting the evidence of these maps, check the imprints and bibliographical entries in Skelton and Hodson to determine its date and reliability. The sources of the Herefordshire maps (Speed, Blaeu, Jansson) are listed below.

*c.*1670	Overton atlas I. Blaeu (Skelton 89).
*c.*1675	Overton atlas II. Jansson (Skelton 91).
*c.*1685	Overton atlas III. Jansson (Skelton 107).
*c.*1690	Overton atlas IV. Jansson (Skelton 111).
*c.*1700	Overton atlas V. Jansson (Skelton 121).
*c.*1713	Speed/Overton atlas I (Hodson 135).
post-1716	Speed/Overton atlas II (Hodson 136).
*c.*1720	Speed/Overton atlas III (Hodson 137). Bassett and Chiswell's imprint replaced by Overton.
1743	Speed/Overton atlas IV (Hodson 138).
post-1716	Overton atlas VI. Speed (Hodson 142).
post-1738	Overton atlas VII. Speed (Hodson 143a).
post-1738	Overton atlas VIII. Speed (Hodson 144).
post-1756-*c.***1770.**	C. Dicey atlases and loose maps using Henry Overton's stock, including Speed's maps, still with Overton's imprint. (Chubb XXXII, Hodson 145).

1673

Richard Blome (1641-1705)
A generall Mapp of the County of HEREFORD with its Hundreds by Ric: Blome (top right). (Plate 15).

Published in *Britannia; or, A Geographical Description of the Kingdoms of England, Scotland, and Ireland ... Illustrated with a Map of each County of England ... The like never before Published*. Printed by Thomas Roycroft, London, 1673. (Chubb XCIX, Skelton 90).

Blome was a heraldic painter and an organiser of illustrated works, which he published on a subscription basis. Contemporary antiquaries had a poor opinion of him and condemned his *Britannia* as theft from Camden and Speed.

Despite his claim, 'the like never before published', his maps are copied from Speed, with the same features but in a less bold style. The hundreds are named. Archenfield remains a minor place-name but the source of the river Dore is not shown as a lake. Bringewood Chase is marked without trees. Copying errors include the separation of the words 'Huntington chap' west of Hereford and the representation of Ledbury as a town larger than Hereford. Has (bottom left) the arms of Herbert Croft, bishop of Hereford, and his son Sir Herbert Croft to whom the map was dedicated. Cardinal points written in English in the plain ruled border. No text on the back. Title cartouche of a stage curtain drawn back. The dedication in a baroque frame, incorporating two heads as supporters, is signed Ric. Blome.

SCALE: 5 miles to 1¾ inches. (1:181030).

SIZE: c.13 x 11 ins/304 x 282.

REFERENCE: DCA, B.5/304. HCL. HRO, K38/D5.

Other issues:

1677
(John Wright). (Skelton 100).

1675

John Ogilby (1600-1676)[1]
Britannia ... or, an Illustration of the Kingdom of England and Dominion of Wales: By a Geographical and Historical Description of the Principal Roads thereof. Actually Admeasured and Delineated ... London, 1675. (Chubb C).

The first measured road-book, containing 102 strip-maps of the main roads, with cross-roads indicated and showing bridges, hills, villages, country seats and other landmarks along or within sight of the roads. There are usually six strips, drawn as scrolls, to each page. The titles (centre top) give a brief summary of the route and distances set in a decorative cartouche with elaborate foliage and the arms of Charles II, sometimes incorporating figures.

The most innovative atlas since Saxton a century earlier, though lacking the spatial relationships of a standard map and being limited to illustrating the network of main roads. Distances in statute miles of 1,760 yards, as authorised by Act of Parliament for the London area in 1593, were measured along the roads by a perambulator. The mileage is marked on the strip-maps, for which Ogilby introduced the scale of 1 inch to 1 mile but in the accompanying text he gives both the measured miles and the 'vulgar computation', which in Herefordshire varied between 2,156 and 2,750 yards; it was usually reckoned one-third longer at 2,346 yards.

Landmarks noted on the maps include miniature plans of towns and villages, destination of side roads, bridges over rivers and streams, hills (drawn as large and sometimes rugged 'mole-hills', right way up for ascent and upside down for descent), type of country traversed (arable, heath), country houses and prominent wayside inns. Each strip has a small compass rose. In the absence of signposts he gave further guidance in the text about road junctions and added comments about the condition of the roads. Reproduced as J. Ogilby, *Britannia, 1675, with an introduction by J.B. Harley*, Amsterdam, 1970.

The maps of roads passing through Herefordshire are:

The Continuation of the Road from LONDON to ABERIST-WITH ... Plate y[e] Second Com(m)encing at Islip ... to Bramyard.
Six scrolls, of which the last is the route from a point just N of Worcester to Bromyard, crossing a mile of 'Furrs & Furn' W of Kingwick (Knightwick), 'Bramiard Heath' and a two-arch bridge over the river Frome approaching Bromyard.

The Continuation of the Road from LONDON to ABERIST-WITH ... Plate 3[d]. & last Com(m)encing at Bramyard ...
Seven scrolls, of which two cover the 24 miles from Bromyard to Combe near Presteigne, marking two stone and one wooden bridges immediately S of Leominster and a ford (?) at Kingsland.

SIZE: c.306 x 465.

REFERENCE: HCL, plate trimmed.

The Road from BRISTOL com Glos. to West CHESTER By Iohn Ogilby Esq[r] His Ma[ties] Cosmographer Containing 145 miles 5 Furlongs. (Plate 5).
Six scrolls, of which three and a half cover the 43-mile length through Herefordshire from Monmouth to Ludlow. Landmarks include Lyson (Lysson) Hill and Gamnier (Gamber) Head south of Wormles (Wormelow) Tump (SO 492303), Callow Hill, Coningsby Hospital and a footpath running from W of Eign Gate to (Black)friars Lane in Hereford, Brundles Oak NE of Holmer church (SO 424505), Dinmore hill, Flower de Luce inn S of Leominster 'vulgo Lemster', 'y[e] signe of the maidenhead' at Orleton (SO 489677).

SIZE: 345 x 432.

REFERENCE: DCA, B.5/308. HRO, K38/D6. HCL, with the plate number 56.

The Road from GLOCESTER to Montgomery.
Six scrolls, of which three cover the 35 miles from The Lea to Combe near Presteigne. Marks stone bridges at Wilton of six arches and Stretford of four arches, a roadside pond at Much Birch, and wooden bridges at Pembridge of seven arches and at Combe.

SIZE: 318 x 439.

REFERENCE: HCL 1698 issue, without a plate number.

The Road from HEREFORD to LEICESTER By JOHN OGILBY Esq[r]

Seven scrolls, of which one covers the 14 miles from Hereford to Storridge. Marks three-arched bridge over the river Lugg, two wooden bridges over the river Frome, Fromes Hill. The county boundary with Worcestershire appears to be marked at Stiffords Bridge in Cradley not at Holywell in Leigh Sinton. Plate number 72.

SIZE: 340 x 445.

REFERENCE: DCA, B.5/307. HRO, K38/D7. HCL.

SCALE: 1 inch to 1 mile (1:63360).

Other issues:

1675, 1698
(reprint). (Chubb CI, CII, CIIa).

1719
The maps were reduced for publication by Thomas Gardner in *A Pocket-Guide To The English Traveller*, the map of the road from London to Aberystwyth being signed Tho[s]. Gardner. (Chubb CXXXVII).

1719
The maps were similarly reduced and engraved by John Senex for publication in *An Actual Survey of all the Principal Roads of England and Wales ... First perform'd and publish'd by John Ogilby, Esq: And now improved, very much corrected, and made portable by John Senex ...* London 1719. (Chubb CXXXVIII).

1720
(John Owen and Emanuel Bowen). (See separate entry below).

1742
(Senex). (Chubb CXXXIX).

1757
(Senex). Printed for John Bowles & Son. (Chubb CXL).

1767
(Thomas Kitchin). A re-issue of 1757, with considerable changes, printed for John Bowles, Carington Bowles and Robert Sayer. (Chubb CXLI).

1767
A re-issue of 1757, French edition. (Chubb CXLIII).

1770
(Kitchin and Carington Bowles). Reprint of 1767 with small changes. (Chubb CXLII).

1775
Jeffreys's Itinerary: or Travellers Companion ... London, Printed for R. Sayer and J, Bennett ... 1775. Based on Senex, whose plates were re-engraved. The strip-maps are enclosed within straight single-line borders instead of being drawn on scrolls. (Chubb CXLIV).

REFERENCE: DCA, B.5/295, HRO, K38/D7 both being plate XXX, containing part of the road from Gloucester to Montgomery and the road from Hereford to Leicester, which appears to be from this edition. Size: 297 x 325.

1775
Reprint of 1775 above as *The Roads of England Delineated: or, Ogilby's Survey, Revised, Improved, and Reduced to a Size portable for the Pocket, By John Senex, F.R.S.* (Chubb CXLV).

1. See also for this and other road books D. Hodson, 'Early printed road books and itineraries of England and Wales', Ph.D. thesis, Exeter University, 2000. A copy will soon be available at BL, Maps 237.a.44.

1676-1717

(Playing cards)
1676
Robert Morden. A set of playing cards of county maps (82 x 57) with *Hereford Sh:* as the king of diamonds. A small outline map, derived from Saxton, as modified by later map-makers, with 33 rather randomly chosen place-names to fill empty spaces. Under the influence of Ogilby these little county maps are the first to show the principal roads, with the distance from Hereford to London given in computed and Ogilby's measured miles.. A portrait of Charles II as the king (top right). A simple compass rose and a scale bar. Gives the measurements of the length, breadth and circumference of the county, and the latitude of Hereford. Re-issued ?1680. See also below, Robert Morden 1680. (Skelton 94, 94a, 95, 102, 103).

SCALE: 5 miles to 5/16 inch. (1:1013760).

SIZE: 93 x 58.

HEREFORDSHIRE MAPS

1676
W. Redmayne. A crudely designed set of no cartographic or historical value. Re-issued 1677 and by John Lenthall 1717. (Skelton 96, 97, Hodson 146).

1717
J. Lenthall. A set (91 x 56), based on Morden. Herefordshire was the king of diamonds. (Hodson 147).

*c.***1785**
H. Turpin (108 x 71, untrimmed), in *A Brief Description of England and Wales*. Based on Morden. (Hodson 273).

1680

Robert Morden (?1650-1703).
HEREFORD SH.

Published in *A Pocket Book of all the Counties of England and Wales*. London, 1680. The same outline map as on the playing cards of 1676, above. (Skelton 103).

Other issues:

*c.***1780?**
(H. Turpin). (Skelton 103).

1681

(John Speed and) Richard Blome (1660-1705).
A MAPP of ye County of HEREFORD wth its Hundreds (bottom left) R(ichard) P(almer) sculp.

Engraved *c.*1667-71 but not published until 1681 in *Speed's Maps Epitomiz'd: or the Maps of the Counties of England, Alphabetically placed*. London, 1681. (Chubb CX, Skelton 104).

Crudely engraved reductions from Speed with a numbered list of the hundreds. It was not updated from the time of engraving. The Herefordshire map was dedicated to and has the arms of John Viscount Scudamore, who died in 1671.

SCALE: 5 miles to 1¼ inch. (1:253440).

SIZE: 187 x 236.

Later issues:

1685
(Samuel Lownes). Only one copy of this atlas is reported. It would appear to have been made up later from a mixture of plates as the Herefordshire map is re-dedicated to Thomas Lord Coningsby, Baron of Clonbrazil, Ireland, a title that was not created until 1693. (Skelton 105).

1693
(Richard Blome) in *Cosmography and Geography ... to which is added the County-Maps of England*. (Skelton 114).

1715, 1717
(Thomas Taylor). Republished with the plate number 18 (top right) in *England Exactly Described Or a Guide to Travellers...Printed Coloured and Sold by Thomas Taylor*. Despite its title the atlas was issued uncoloured. On the map of Herefordshire the arms of Viscount Scudamore were replaced by those of Thomas Lord Coningsby.

Taylor added roads in the 1717 issue. Distances on the maps are given in Ogilby's measured miles, but a new table of distances at the beginning of the atlas is in computed miles. (Chubb CXXXVI, CXXXVIa, Hodson 139, 140).

REFERENCE: HCL.

From *c.*1731
c. 1731 (Thomas Bakewell), *c.*1750 (Elizabeth Bakewell).
The atlas remained on sale for about twenty years and the county maps went through various states. The Herefordshire maps have been found (1) without a plate number, (2) with an erroneous plate number 17, (3) with the corrected plate number 18 and (4) with roads added. (Chubb CXXXVIb, CXXXVIc, Hodson 141).

1694?

John Seller (?1658-1697).
HEREFORD SHIRE. (Plate 6).

Published in *Anglia Contracta, or A Description of the Kingdom of England & Principality of Wales in Several New Mapps of all the Countyes ...* Undated (Chubb CXVIII, Skelton 115. See also Hodson 274-280).

Crude maps, reduced from Speed, poorly presented and engraved with a strange selection of place-names and misspellings. It is thought that some original surveying was carried out under Seller's supervision. As Hydrographer to the King he based his maps on the London (St. Paul's cathedral) meridian. Lists the hundreds. Lacks compass rose and graduation.

SCALE: 10 miles to 1⅜ inches. (1:460800).

SIZE: 123 x 145.

Other issues:

1696-7
(Samuel Clarke, John Gwillim, H. Newman). In *The History of England*. (Skelton 119- 120a).

1701
(Joseph Wild). In *Camden's Britannia abridg'd*. (Skelton 122).

1703
(J. Marshall) in *The History of England*. (Skelton 124).

1711
(Isaac Cleave). In *Camden's Britannia abridg'd*. (Skelton 124a).

1772-1809
Published in a Supplement to Francis Grose, *The Antiquities of England and Wales* 1787 and subsequent editions. Revisions of the original plates include the spelling of some place-names. (Hodson 274-280).

REFERENCE: HCL, (1787 Supplement).

1695

Robert Morden (d.1703).
HEREFORD SHIRE. By Robt Morden. (top right) Sold by Abel Swale Awnsham & John Churchil (bottom left). (Plate 16).

Published in *Camden's Britannia, Newly Translated into English ... Publish'd by Edmund Gibson ... Printed by F. Collins, for A. Swalle* (sic) *... and A. & J. Churchill*. 1695. (Chubb CXIII, Skelton 117).

Probably drafted by 1693. Although Gibson claimed that copies of the best existing surveys were purchased and sent to local gentry for comment and correction he does not name his correspondents, if any, in Herefordshire. Morden had to fall back on reworking Saxton and Speed and the result was criticised by contemporary antiquaries.

Longitude is measured along the bottom border (from the meridian of St. Paul's, London) and the difference in local time from London along the top border. The maps are neat and spacious. A few roads are marked on some maps, but not in Herefordshire. Hundreds, towns, villages, castles are shown, hills (by cleft molehills), woodland, river crossings. Shows the detached areas of the county at Litton (Cascob) in Radnorshire and The Fothok (Ffawyddog) in Monmouthshire. Place-name spellings mostly follow Saxton and Speed. Leominster has the alternative spelling of 'Lemster'. There is no text on the back.

In the 1695 edition Saxton's and Speed's spellings of Standysshe (Stanedge), Lidbury (Ledbury) and Estnor (Eastnor) were repeated (with Eastnor further misspelt as Esnor) and Ludford was placed outside the county. These mistakes were corrected in the 1722 edition. (Ludford was divided between Herefordshire and Shropshire until 1895, when it was placed entirely in Shropshire).

The unusual three-fold scales of great, middle and small miles are on the Herefordshire map equivalent to 49, 52 and 55 miles to one degree of latitude. This is a considerable difference from the more common contemporary measurement of 60 miles to one degree or, more accurately, the statutory mile of 1,760 yards which is about 69 miles to one degree. The scales on each of the county maps vary, reflecting the different length of a local mile. It is thought that Morden may have taken the threefold scale from Saxton's large map of England and Wales, which was reprinted a few years earlier. No other map-maker followed his example. Herefordshire local customary miles were one-third longer than the statutory mile of 1,760 yards.

SCALE: Three 5-mile scales, of great, middle and small miles. The last is the equivalent for the measured or statutory mile at 5 miles to 2¼ inches. (1:140800) (bottom right).

SIZE: 359 x 423.

REFERENCE: DCA, B.5/ 301. HCL.

Other issues:

1722
Coloured (Mary Matthews for Awnsham Churchill) in the two-volume edition of *Britannia*. Gibson sought further information from correspondents to correct misspellings occurring in the 1695 edition. (see above). (Chubb CXV, Hodson 169).

REFERENCE: HCL. HRO, K38/D8.

1730
(James and John Knapton and other booksellers). (Hodson 170).

1753
(R. Ware and others). (Hodson 171).

1772
(W. Bowyer and others). (Hodson 172).

(c.1700)

(Unidentified). Comté d'Hereford.[1] (Plate 14).

Undated and lacking imprint. A small, rather overcrowded map based on Speed (Bassett and Chiswell) 1676. Title in a rosette (top right), compass rose with E cross (right) and scale bar (bottom right). Gives latitude and longitude.

The place-names include 'Next Inne' (Nextend, Lyonshall), which appears first on Speed 1676. Other spellings appear to have been taken from Jansson, for example 'Halvern Hills' (Malvern Hills), perhaps after the dispersal of his plates in 1694. The unnamed French, or perhaps Dutch, engraver made additional misspellings, 'Ytokelaye' (Stoke Lacy), 'Great Anrnthall' (Llanrothal), 'Bransbourg' (Bromsberrow).

SCALE: 8 miles to 1 1/4 inches. (1:405500).

SIZE: 155 x 200.

REFERENCE: HCL

1. The copy in Hereford City Library is the only one known to English bibliographers. It does not appear in Chubb, Skelton or Hodson. Dr Donald Hodson, who had not previously seen a similar map, considers that it must have formed part of a complete atlas and suggests a date of c.1700. (Letter dated 17 February 2003).

1701

Robert Morden.

HEREFORD SHIRE By Rob.t Morden. (top right).

Published in *The New Description and State of England, containing the Mapps of the Counties of England and Wales ... Newly Design'd ... by the best Artists*. London 1701. (Chubb CXXIV, Skelton 123).

The 'Small Morden'. Of pocket book size and consequently popular for inclusion in other works. A simplified map showing hills (as molehills), rivers, market towns and villages, main roads (not too accurately). Lists the hundreds with letter references marked on the map. The places named were selected to fill spaces on the map rather than reflect their importance. The maps are plain on the back. As on Morden's large map of 1695 latitude, longitude and minutes of time west of London are given in the borders. Thought to have been drawn and perhaps engraved before 1693, perhaps originally intended for Gibson's edition of Camden's *Britannia*, but rejected as too small a scale. The engraver is not named.

SCALE: Three 5-mile scales of great, middle and small miles (to 1 3/8 to 1 inch) (1:230400 to 1:316800).

SIZE: 175 x 221.

REFERENCE: HCL.

Other issues:

1704
Two issues, the second in a new format. (Chubb CXXV, Hodson 125, 126).

1708
(Hermann Moll, *Begun by Mr. Morden: Perfected, Corrected and Enlarg'd by Mr. Moll*, who added a simple compass rose and, as the maps were intended to illustrate a road-book, more roads, copied in later editions. It is the earliest set of maps to indicate, by an asterisk, towns that were parliamentary boroughs.

REFERENCE: HRO, K38/D10; does not bear Moll's name. (Chubb CXXVI, Hodson 127).

1718
Published in *Magna Britannia et Hibernia...Or, a New Survey of Great Britain*, vol 2, (1720). Originally published in parts in the *Evening Post*, the map of Herefordshire in Part 19-20 in January and March 1717/18. (Chubb CXXVII, Hodson 128).

1738
In *Magna Britannia*. (Chubb CXXVIII, Hodson 129).

1720

John Owen and Emanuel Bowen. (fl.1720-1767)
Britannia Depicta or Ogilby Improv'd; Being a Correct Coppy of M.r Ogilby's Actual Survey of all y.e Direct and Principal Cross Roads in England and Wales ... with suitable Remarks on all places of Note drawn from the best Historians and Antiquaries. By I.no Owen of the Midd: Temple Gent ... London Printed for, & Sold by Tho: Bowles & Em: Bowen...1720.

The maps are printed on both sides of the paper, with the road maps in three or four strips to a page reduced from Ogilby 1675 and the sketchy county maps reduced from the small Morden 1701. Historical notes were contributed by the antiquary John Owen. There is an ornamental cartouche across the top. Bowen was the engraver. The volume was rushed out to forestall rival reprints and had to be amended in subsequent issues. (Chubb CXLVII, Hodson 149).

The three routes crossing Herefordshire, drawn at scales between 3/8 and 1 1/8 inches to 1 mile (1:168960 to 1:56320), are:

Plates 1-7. The Road from London to Aberistwith ... A Map of Cardiganshire. The eleven coats of arms include that of Leominster.

Plates 143-147. The Road from Bristol to Westchester ... Plate 143. A MAP of HEREFORD-SHIRE. Five coats of arms include those of Hereford and Tho. Coningsby. Scale: English miles 9 to 1 inch. (1:570240).

SIZE: c.115 x 115.

Plates 185-187. The Road from Glocester to Montgomery ... A Map of Montgomery. The three coats of arms include those of the bishopric of Hereford and the deanery of Hereford. Plate 186 has three columns containing the length of the road from Gloucester by Huntley to Ross and on to Hereford and towards Leominster as far as Moreton-on-Lugg.

SIZE: 175 x 90 maximum.

REFERENCE: HRO, BE60/55 (No indication of which issue, but the plate shows signs of wear); DCA, B.5/311. Plate 186 only.

Plates 188-190. The Road from Hereford to Leicester ... A Map of Leicester Shire. The three coats of arms include that of Price Devereux Viscount Hereford.

Other issues:

1720 (3), c.1723, 1724, 1730 (2), 1731, 1734, 1736, 1749, 1751, 1753, 1759, 1764, post-1764.
(Hodson 150-165).

1724

Herman Moll (d.1732).
HEREFORD SHIRE.

Published as plate 36 in *A New Description of England and Wales, With the Adjacent Islands.* London 1724. Coloured maps with the title 'By H. Moll Geographer'. (Chubb CLX, Hodson 173).

The historical narrative is based on Camden, the maps on Morden. Lists hundreds, marks roads and gives latitude and longitude. Outside both left and right borders are columns of drawings of six Romano-British coins.

SCALE: 10 miles to 2$^{1}/_{8}$ inches. (1:298160).

SIZE: 191 x 253, frame; 191 x 320, plate. (Hodson 191 x 314).

REFERENCE: DCA, B.5/310. HCL.

Other issues:

1724
Uncoloured.

1733
(Bowles, after Moll's death in 1732).

1739
(Hodson 174, 175, 176).

1747
In *The Geography of England and Wales,* plate 19. (Hodson 177).

1753
In *H. Moll's British Atlas* (Bowles), revised by Emanuel Bowen. Herefordshire remains plate 19 but without the side columns of coins. Road distances given in Ogilby's measured miles instead of computed miles. (Chubb CLXIII, Hodson 178).

*c.*1766
In *The Traveller's Companion ... Composed by Hermann Moll, Geographer. Revised and improved by Emanuel Bowen.* Only one example of the atlas is known. The county maps contain many revisions, mostly concerning roads. A plate number in brackets in the top right hand corner indicates the later issues of Moll's plates. (Hodson 178a).

1742

Thomas Badeslade (fl.1719-1745).
A MAP OF HEREFORD SHIRE North West from London. (outside the border).

Published in *Chorographia Britanniae or a Set of Maps of all the Counties of England and Wales ... By Thoms Badeslade, Surveyor and Engineer, and now neatly Engrav'd by Will: Henry Toms.* London 1742. (Chubb CLXX,).

The small simple maps show only the towns, hills, roads and rivers. The Herefordshire map marks the eight principal market towns, those returning M.P.s being distinguished by asterisks, and the villages of Stretford and Great Birch. The only roads marked are those from Presteigne to Bromyard and Worcester, from Ludlow to Leominster, and from Pembridge to Hereford and Ross. With a left-hand column of historical notes. The foot of each map is signed 'T. Badeslade delin. Publish'd by the Proprietors T. Badeslade & W.H. Toms Septr. 29th, 1741. W.H. Toms sculp^t.' (outside bottom border).

SCALE: 10 miles to 1$^{1}/_{4}$ inches. (1:506880).

SIZE: 142 x 143, of which the area of the map is 142 x 105. (Hodson; c.154 x 150 including the side panels).

REFERENCE: HCL.

Other issues:

1742 (2).

REFERENCE: HCL, with Toms named as sole proprietor. (Hodson 189, 190).

1745, c.1746, c.1749.
(Hodson 191, 192, 193).

1743

John Cowley (fl.1734-1744).
An Improved MAP of HEREFORD SHIRE ... by I. Cowley, Geographer to His Majesty.
   Published in *The Geography of England ... Printed for R. Dodsley*. London 1744. (Chubb CLXXXI, Hodson 194).
   Based on the small Morden it contains little detail, marking hills, rivers, roads and mileage. Compass rose, explanation of conventional signs for boroughs, market towns and castles, border with latitude and longitude. The spelling of the relatively few place-names is careless.

SCALE: 5 miles to 7/8 inch. (1:362060).

SIZE: 131 x 181.

REFERENCE: DCA, B.5/295. HCL.

Other issues:

1745 (2)
In *A New Sett of Pocket Mapps Of all the Counties of England and Wales*, printed for R. Dodsley and M. Cooper. (Chubb CLXXXIII, Hodson 195-6).

1744

(John Rocque) (c.1704-1762), and **Thomas Read**.
HEREFORD SHIRE. 'To the R^t: Hon:^{ble} the Lord Carpenter. This PLATE is hum.^{bly} Inscrib'd by....Thomas Read'.
   Published by Thomas Read in *The English Traveller: Containing ... a map of every county from the best and latest observations ... after the designs of Herman Moll. London, Printed for T(homas) Read*, vol. 2, 1746.

   This publication came out in parts, the map of Herefordshire first appearing in part 16 on 17 March 1743/4. The unnamed surveyor was probably the Huguenot immigrant John Rocque, although the quality of the draughtsmanship and accuracy of the spellings fall well below the standards of his great maps of London and its environs. He certainly published the revised issue of 1753. (Chubb CLXXXV, Hodson 197).
   Principal roads are shown and the boundaries of the hundreds. Hills, still drawn as molehills except for 'Robin Hoods Butts' (Butthouse Knap and Pyon Hill) which appear monstrous, are hatched to indicate their steepness by the closeness of the engraved lines. The spelling of place-names is erratic. With the coat of arms of Lord Carpenter (top right).

SCALE: 10 miles to 1 1/2 inches. (1:422400).

SIZE: 150 x 188. (Hodson 192 x 161).

Other issues:

c.1753, 1753.
HEREFORD SHIRE in *The Small British Atlas: Being a new Set of Maps of all the Counties in England and Wales Publish'd by John Rocque*. Only one copy is known, containing 53 maps, blank on the back, of which 33 were revised from *The English Traveller*. These revisions included the map of Herefordshire, (no. 22 partly erased, and with some other alterations presumably by Rocque). (Hodson 198, 199).

SCALE: 10 miles to 1 1/2 inches. (1:422400).

SIZE: 152 x 196.

REFERENCE: DCA, B.5/311. HCL.

1753
In *The Small British Atlas ... By John Rocque ... and Robert Sayer*. Chubb describes 54 maps. (Chubb CCVII, Hodson 200).

1762, 1764
In *The Small British Atlas ... By John Rocque*. Published by Roque's widow, Mary. Herefordshire was plate 14 in 1762 and 15 in 1764. (Chubb CCVIII, CCIX, Hodson 201, 202).

1769-70
In *England Displayed ... By P. Russell Esq., and ... Wales, By Mr Owen Price*. No plate numbers. (Chubb CCX, Hodson 203).

REFERENCE: DCA, B.5/264. HCL.

19, 20. *Isaac Taylor 1754 (1786 issue) The earliest map of Herefordshire at a scale of one inch to one mile*

The decorative features include (19) the fine title cartouche displaying local products, the public fountain at Ross immediately above his own house, and side panels containing the arms of his subscribers. A table (20) explains his conventional signs, based on Henry Beighton's 1728 map of Warwickshire, reissued in 1750.

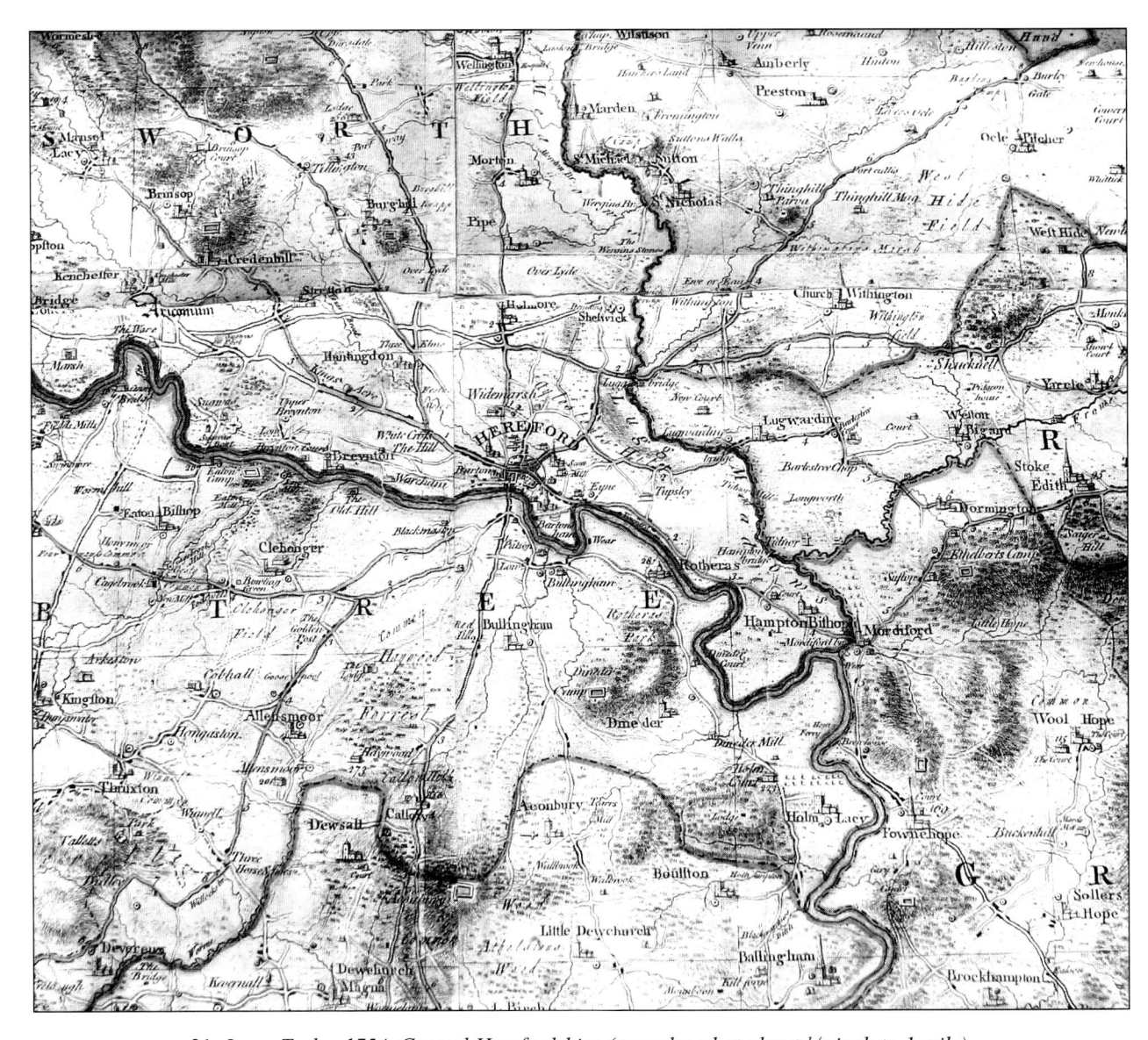

21. *Isaac Taylor 1754. Central Herefordshire (reproduced at about 1/2 inch to 1 mile).*

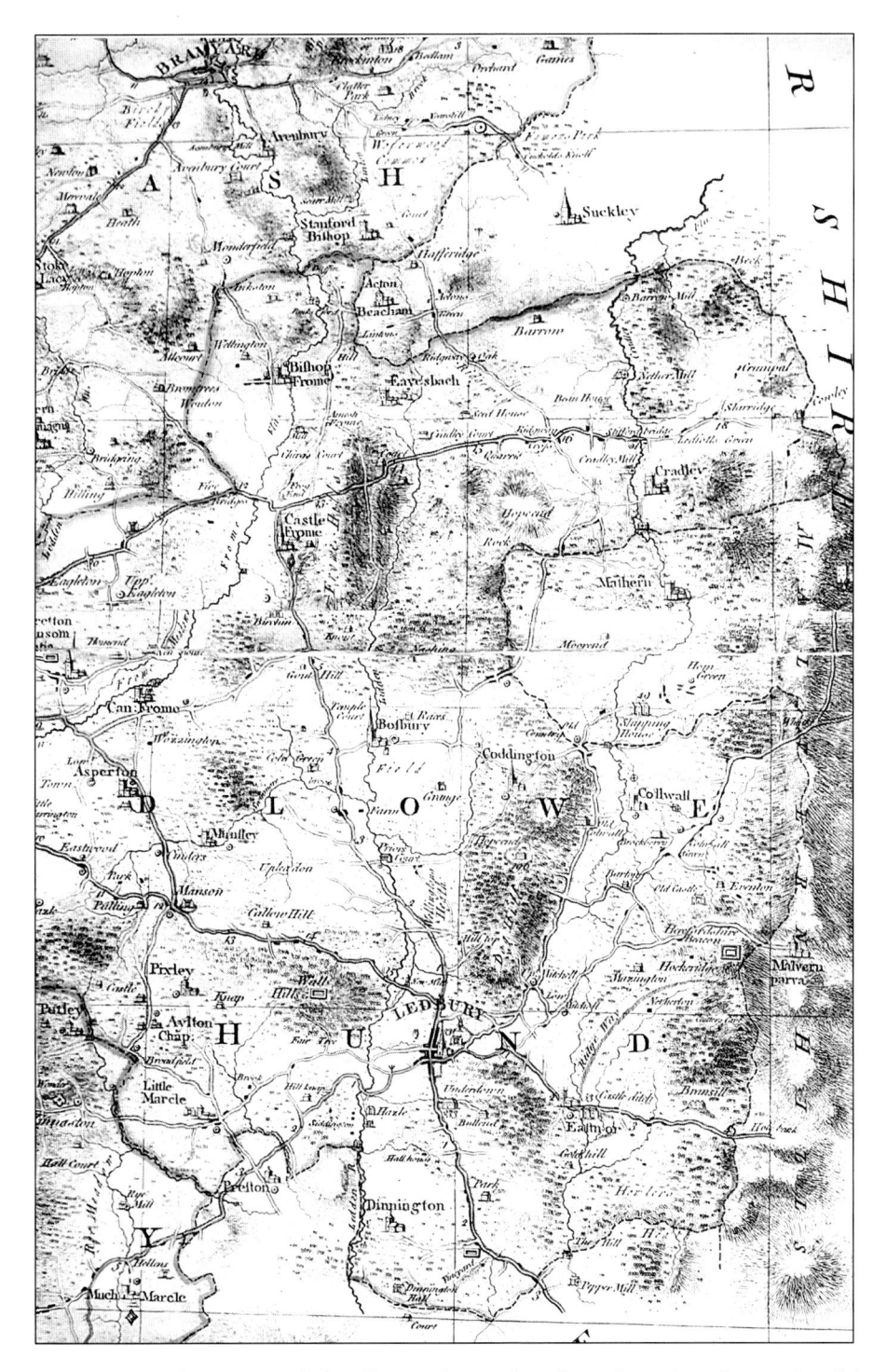

22. Isaac Taylor 1754. Colwall area.(reproduced at about 2 inches to 1 mile).

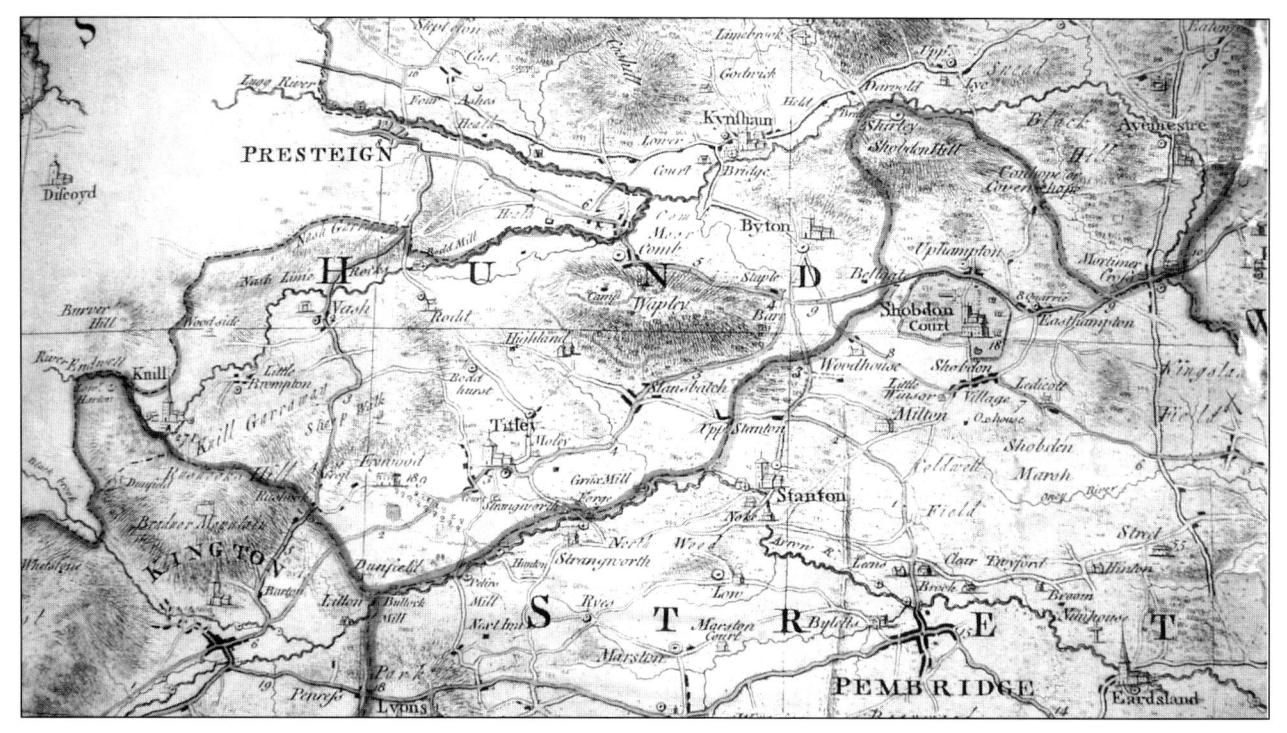

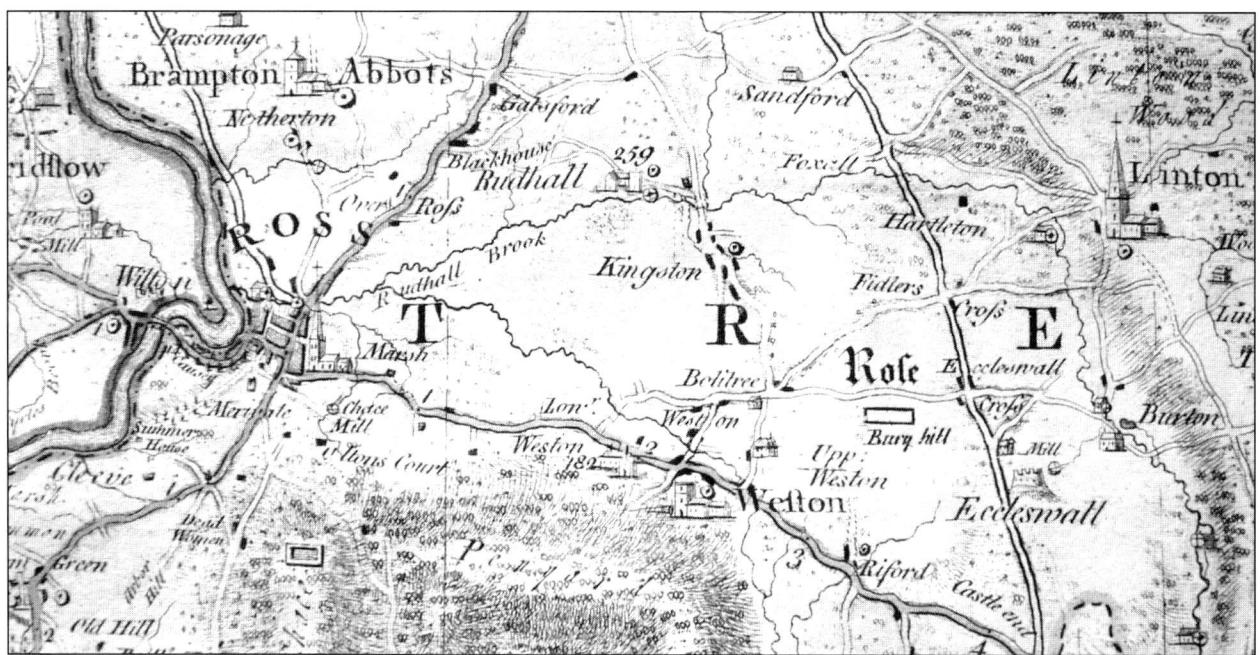

Isaac Taylor 1754.
23. Top : *Kington area (reproduced at about 1/2 inch to 1 mile).*
24. Bottom : *Ross area (reproduced at about 2 inches to 1 mile).*

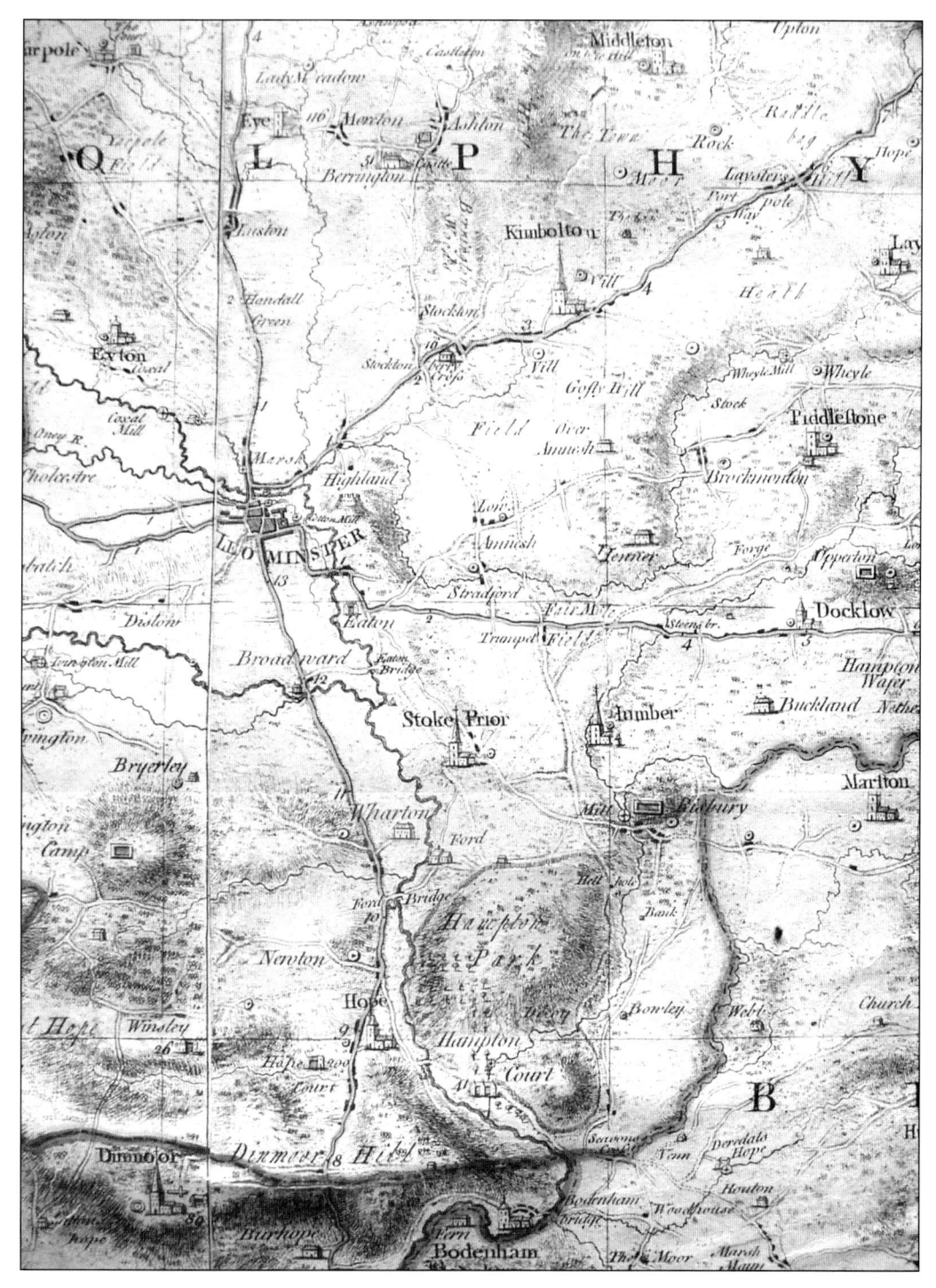

25. Isaac Taylor 1754. Leominster area (reproduced at about 1 inch to 1 mile).

26. *Isaac Taylor's house at Ross.*

27. *Isaac Taylor. Gloucestershire 1777, with a pair of dividers found in his house.*

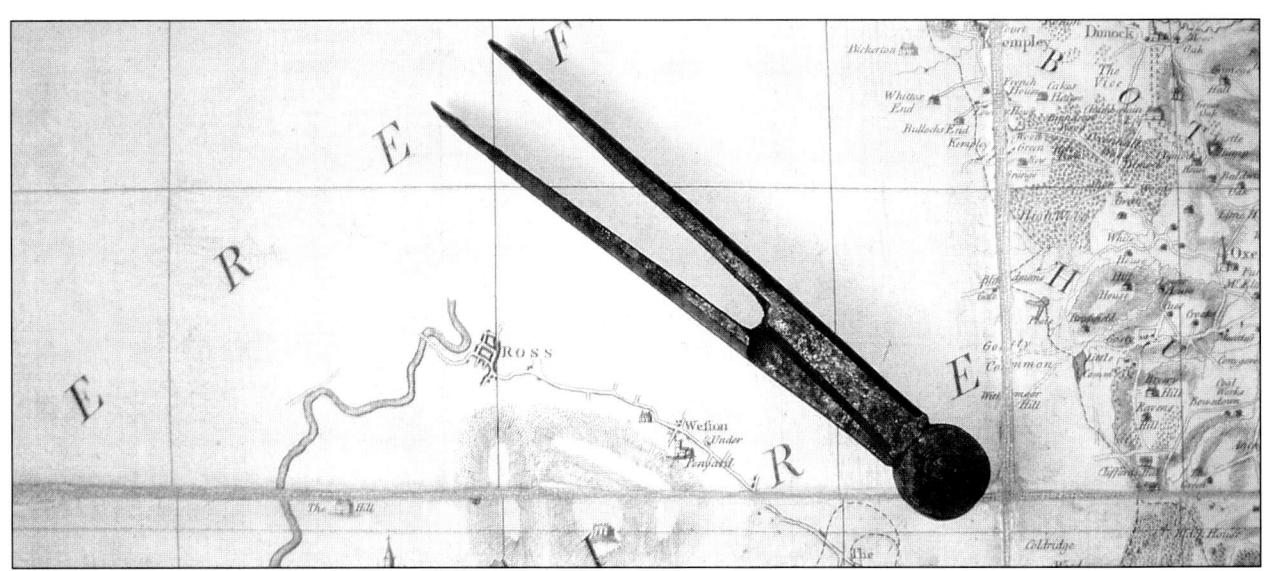

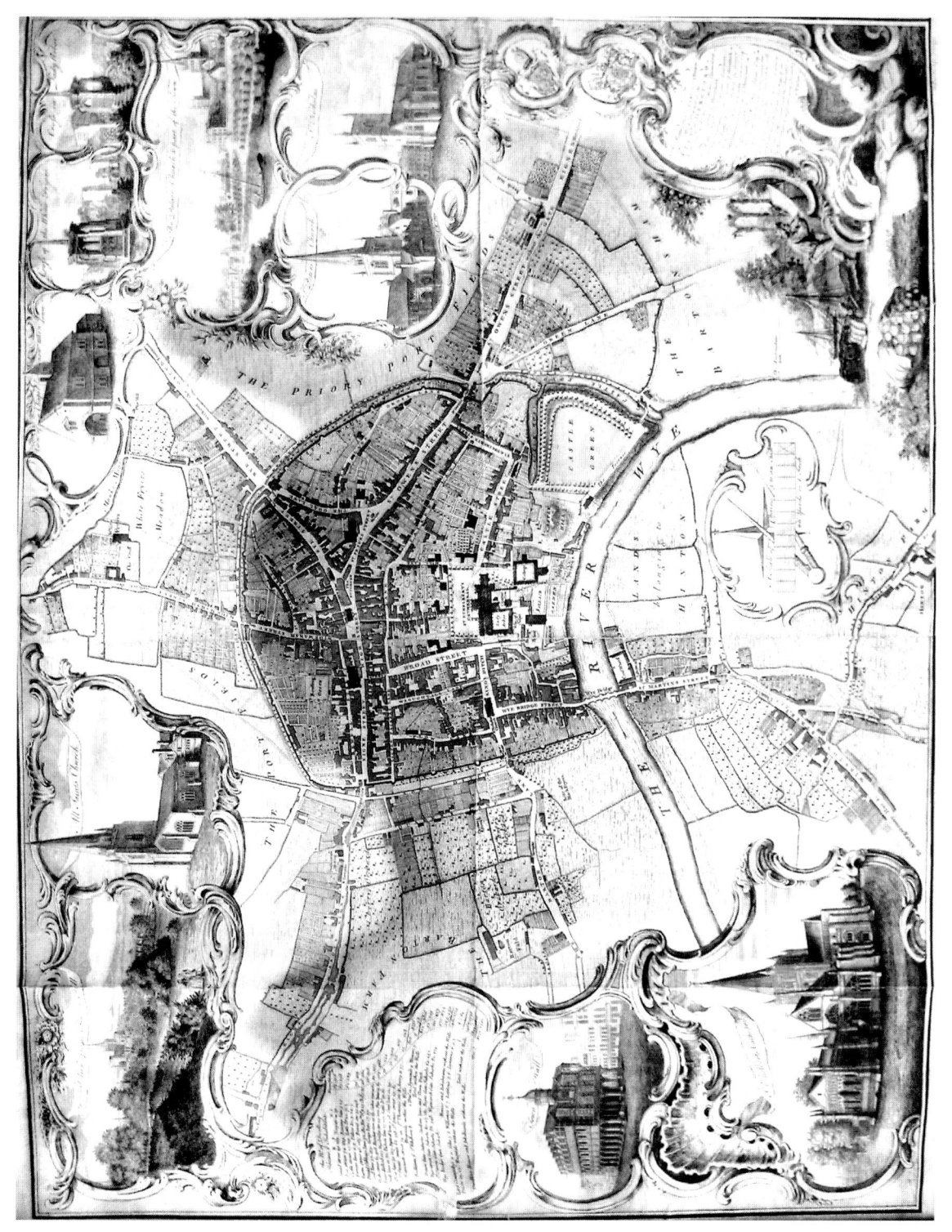

28. Isaac Taylor. Map of Hereford City 1757.

29. Isaac Taylor 1757. Title cartouche with a portrait of a draughtsman at work.

1746

Samuel Simpson.
HEREFORD SHIRE (top, outside frame).

Published in *The Agreeable Historian, or the Compleat English Traveller ... With a Map of every County ... after the Designs of Herman Moll, and others ... By Samuel Simpson, Gent. London: Printed by R. Wallker ...*, London 1746, vol. 2. (Chubb CLXXXIV, Hodson 204).

Very similar to Rocque 1744. The rather messy maps precede the histories of the counties but lack any reference to the publication or name of the engraver or publisher. There is no pagination or text on the back of the map. With the arms of Lord Hereford and the City of Hereford. But the arms printed are those of St. Albans; the city of Hereford's arms are printed on the map of Hertfordshire.

SCALE: 10 miles to 1³⁄₈ inches. (1:460800).

SIZE: 155 x 191. (Hodson 193 x 160).

REFERENCE: HCL.

1748

(Thomas Hutchinson).
A Correct MAP of HEREFORD SHIRE (top right, in a rectangular frame).

Published in *Geographia Magnae Britanniae. Or, Correct Maps of all the Counties ... Printed for T. Osborne, D. Browne, J. Hodges, A. Miller, J. Robinson, W. Johnston, P. Davey & B. Law.* (London), 1748. The Herefordshire map has the plate number 16 (top right, outside the border). Hutchinson engraved two of the maps and probably others that bear no name. (Chubb CXC, Hodson 205).

Based on Morden 1695 the maps are pleasingly simple and well engraved but comparatively rare. A limited number of place-names but reasonably accurately spelt. Hills are still drawn in Saxton's style, but less steeply portrayed than his molehills.

SCALE: 10 miles to 1³⁄₈ inches. (1:362060).

SIZE: 144 x 170. (Hodson *c.*147 x 167).

REFERENCE: HCL.

Other issues:

*c.*1756
(Chubb CLXXXIX, Hodson 206).

1748

Thomas Jefferys and Thomas Kitchin.
A MAP OF HEREFORD-SHIRE. (Plate 17).

Published in parts in *The Small English Atlas being a New and Accurate Sett of Maps of all the Counties...London 1749.* (Chubb CXCII, Hodson 209).

The maps do not bear an imprint with the name of the engraver or publisher and therefore appear anonymous and undated, but the map of Herefordshire was published in the first weekly part to be issued, announced in the *London Evening Post* on 19 November 1748. Jefferys is thought to have been the surveyor, engraver and publisher of at least some of the maps.

A limited selection of place-names. At the bottom of each plate and outside the border of the maps are notes about fair and market days and members of Parliament. The maps are blank at the back. Jefferys used the Greenwich meridian in place of St. Paul's, London, but though latitude and longitude is marked, the change is not indicated on the map.

SCALE: 10 miles to *c.*1 inch. (1:596330).

SIZE: 124 x 133.

Other issues:

1751
Plate number 20 added. (Chubb CXCIII, Hodson 210).

*c.*1775, *c.*1787.
Thomas Jefferys went bankrupt in 1766 and died 1771. The plates were revised for a re-issue of *The Small English Atlas c.*1775, when Kitchin sold his share to John Bowles and his son. Revisions include notes on the detached parts of Herefordshire in other counties and the addition of the road from Worcester to Hereford through Ledbury (Robert Sayer, John Bennett, John Bowles and Carington Bowles). (Hodson 211).
Published again by Sayer alone *c.*1787. (Hodson 213).

1776, 1787, *c.*1794, *c.*1796.
Plates used by Sayer and Bennett and by Sayer alone in *An English Atlas Or A Concise View of England And Wales.* (Chubb CCLIX, Hodson 212, 214, 215, 216).

REFERENCE: HCL, where it is dated 1780.

1750

(Emanuel Bowen).
HEREFORD SHIRE. (bottom right). Drawn from the best Authorities (bottom left). Printed for J. Hinton (outside bottom border).

Published in October 1750 in *The Universal Magazine of Knowledge and Pleasure*, 1747-66. (Chubb CLXXXVIII, Hodson 253).

The map of Herefordshire, with the arms of Hereford (top right), is based upon Morden with revisions. It does not bear a draughtsman's name but was probably by Bowen. The first four county maps in 1747 were signed by Kitchin. Between 1748 and 1750 the maps are anonymous until Bowen began signing them from 1751. The map appears somewhat overcrowded and unreliable. There is a key to the conventional signs.

Bowen and Kitchin were closely connected for thirty-five years. Bowen would appear to have been born *c.*1695 and was working in London as an engraver by 1720. Kitchin was apprenticed to him in 1732, married his daughter Sarah in 1739 and became associated with him in business and the Baptist church. The output of both was remarkable but Bowen remained an engraver, albeit an excellent one, whilst after Sarah's death and Kitchin's second marriage into a wealthy Baptist family in 1761 Kitchin flourished not only as an engraver but also as a successful printer and publisher. Emanuel Bowen died on 8 May 1767. His son Thomas, also an engraver, probably did rather more work in the partnership than he is credited with. Lacking premises and capital he appears to have sold the family interest to Kitchin shortly after his father's death. Thomas Kitchin moved from London in 1768, but his son Thomas Bowen Kitchin continued business there until 1776 when it was taken over by his former apprentice, William Hawkes.

SCALE: 8 miles to 1½ inches. (1:405500).

SIZE: 175 x 200. (Hodson 186 x 200).

REFERENCE: HCL.

1753 or 1754

George Bickham
(George Bickham senior 1684-1758 and his son George Bickham junior d.1771). A MAP of HEREFORDSHIRE, N. West from LONDON (1753 or 1754).

Published as plate 22 in *The British Monarchy: Or a New Chorographical Description Of all the Dominions Subject to the King of Great Britain*. London 1743, 1748, 1749-54. All the county plates are dated between 1750 and 1754; Herefordshire is one of the five not dated. (Chubb CLXXVIII, CLXXIX, Hodson 217).

An antiquarian curiosity. A bird's-eye view of the county as seen from the south but of little topographical value. Gives mileage from Hereford to London and the principal towns in the county and names the eleven hundreds. The viewpoint is a ruin above a river looking over the countryside N of Ross. The foreground features a spring issuing from a lion's head. The same spring appears in the dedicatory cartouche of Isaac Taylor's *Map of the County of Hereford* 1754 and would appear to be based on John Kyrle's spring at The Prospect, Ross (see below, Taylor 1754). All the text is outside the border. The Bickhams were noted for their calligraphy.

SIZE: 225 x 145 frame, 260 x 157 plate.

REFERENCE: DCA, B.5/296. HCL.

Other issues:

1796

Title reduced to HEREFORDSHIRE. No text but the plate number 15 added (top right) outside the border. (Chubb CLXXX, Hodson 218).

REFERENCE: DCA, B.5/296. HCL.

1754

Isaac Taylor (*c.*1725-1788).
New MAP of the COUNTY of HEREFORD. (Plates 19-25).

As the first one-inch-to-one-mile map of the county, at that time considered a large scale, it cannot be compared with any of the others in this listing. For a carto-bibliographical description see E.M. Rodger, 'Large scale English county maps and plans of cities not printed in atlases', part 14 Herefordshire', *The Map Collector*, Issue no. 53 (Winter 1990), pp. 22-25. Taylor's map was re-issued in 1786 but was not superseded until Henry Price published *A New Map of the County of Hereford; with part of The adjacent Counties* in 1817, based upon the Ordnance Survey's fieldwork in Herefordshire 1813-17. The Ordnance Survey did not publish its own revised one-inch to one mile *Old Series* until 1831-33.

There is no title. Taylor set out (1) a long dedication set within a pastoral scene, (2) 'A General Description of the MAP', (3) 'Explanation of the Characters', (4) an inset plan of the city of Hereford, (5) the scale bar and compass star, (6) the border and (7) two side panels (outside the frame of the map) of the coats of arms of the gentry. At the foot, beside (3) is 'T. Kitchin Sculp^t. 1754'.

1. Dedication (bottom right). 'To MY WORTHY SUBSCRIBERS in General but more particularly To those Noblemen Gentlemen &c who honour'd me with their Assistance This New MAP of the COUNTY of HEREFORD Engrav'd from the Original Drawing made from an Accurate Survey proved by Trigonometry Is Most Humbly Dedicated by ISAAC TAYLOR Ross Jany. 1st. 1754. Estates are survey'd & Mapp'd in a very Accurate & neat Manner at the Usual Prices; also Maps drawn in a New Taste'.

The dedication is written on a large monumental stone flanked by triple-shafted columns supporting a Decorated-style archway, standing in front of an apple tree with duck flying overhead. At its foot a spring gushes out of a lion's mouth. This appears to be a representation of the spring at The Prospect in Ross, constructed in 1705-6 on the site of the medieval bishops' palace by John Kyrle, the 'Man of Ross' (1637-1724) to provide the town with a reservoir and water supply. George Bickham also portrayed it in his 'map' of the county 1753-4 (see above). The reservoir contained over 550 hogsheads of water, which was conveyed by underground pipes to public standpipes in the streets. 'In the middle was a handsome spouting Image'. The fountain fell into disuse as the town's water supplies were improved and in 1794 it was filled in. (T. D. Fosbroke, *Companion to the Wye Tour. Ariconensia; or archaeological sketches of Ross*, Ross 1821, p. 99). In the foreground are baskets and sacks of apples, woolsacks, and a salmon net. Behind is a hopyard, an orchard and cider-press with a man at work, a cider-mill with a man leading a horse and cider barrels with a girl pouring in cider. In the background is a river with a trow.

2. 'A General Description of the MAP'. (On a scroll, bottom left)
Taylor describes his conventional signs. The boundary of the county is marked more strongly than those of its hundreds. Roads are distinguished between Roman roads (two thick lines), post roads (one thick and one thin line) and minor roads (two thin lines), with solid lines where crossing enclosed fields and pecked lines crossing open fields and waste. Streams grow wider into rivers. The importance and significance of places is expressed by the lettering of place-names. Distances along roads are given in computed and measured miles, triangulation being used for measurement. Latitude and longitude are given and a grid with letter and number references has been laid over the map to refer to the seats of the gentry whose arms are displayed in the side panels. Relief is indicated by shading, with steepness shown by the density of the hachuring, the first example of its use on maps of Herefordshire. Country houses are distinguished according to their size, their location being at the centre of the conventional sign. He concluded 'An Account of the Instrument and manner in which this Survey and Map are made will be given in a future Work' (but apparently not published) and noted that 'The Manor of Irchenfield is divided according to a Survey and Perambulation made in 1638'.

3. 'Explanation of the Characters'. (On a scroll, bottom left)
Taylor lists his conventional signs. They include churches with towers or spires, chapels, ruined chapels, old foundations, churches with fallen steeples, country seats with old or new buildings, Roman towns or depopulated places, camps, entrenchments, barrows, tumps and sites of battles, 'posts of direction' (signposts), parks, wind and water mills, forges, coal mines, river locks.

4. 'HEREFORD'. (On a scroll, top right)
A town plan with a key to the important landmarks and a note that it is intended only as a guide to strangers to find their way about the city but that for particulars 'he Refers to his large Plan on a Scale of twelve perches to one Inch' (see Taylor 1757 below).

5. Scale bar, compass rose (In a baroque frame, top left)
A scale bar with measuring poles, compass, chain, ruler, dividers, pen, roll of paper, drawing board. A compass star with four cardinal points and showing the magnetic variation of 17 degrees.

6. Border. The border, containing degrees of latitude and longitude and the grid references 1-10 horizontally and the letters a-n vertically.

7. 'A List of such Gentlemen's Arms, as were given agreable to the Proposals'.
Two side panels, each with four columns of 72 shields making a total of 288, of which 34 contain the coats of arms and a further 6 the names of county gentry who subscribed to the map, each with a grid cross-reference to their seat marked on the map. The remaining shields are blank. The number of subscribers' coats of arms may be used to determine the order in which copies of the map were printed.

The map covers the whole county, including the detached portions in Cascob (Radnorshire), Farlow (Shropshire), Ffawyddog (Monmouthshire). A small area (4a.) on Devauden Hill (Monmouthshire) is described but not mapped. The method of survey by triangulation, style of the map, conventional signs and side panels with coats of arms of the gentry owe much to Henry Beighton's map of Warwickshire 1728, of which a third issue was published in 1750. The large scale allowed Taylor to mark places and features not previously recorded on the small-scale county maps.

The map distinguishes enclosed lands from open fields and heath. Archaeological sites include *Ariconium* (but still identified with Kenchester and not Bromsash in Weston-under-Penyard), the Roman ford at The Weir in Kenchester, King Arthur's Table (Arthur's Stone, Dorstone), Bagdon Tump in Ocle Pychard (SO 590470), Clutter's Cave below the Herefordshire Beacon (SO 763394). Industrial sites include the forges at Bringewood in Downton, Docklow, Llancillo and Bishops Wood in Walford, water mills including paper mills on the right bank of the river Monnow opposite Llanrothal and at Pepper Mill in Bromsberrow (Glos.), Windmill Hill at Moreton Jeffries, quarries. The bridges over the river Wye at Hereford and Wilton are marked, and the other river crossings by ferries or fords, a weir at Bartonsham, Hereford and a riverside vineyard at Hampton Bishop. Other features include wells at Garnons in Mansell Gamage and Peterchurch, a bowling green at Clehonger, the White Cross at Hereford, a signpost W of Ledbury and 'David's Burial Place' S of Harewood Inn (Harewood End. The neighbouring churches of Pencoyd and Hentland were associated with St. Dyfrig; B. Coplestone-Crow, *Herefordshire place-names*, B.A.R. Publications, Oxford 1989, typescript, pp. 2-5).

Taylor's accuracy cannot be entirely relied on without supporting evidence. For example, he marks clearly a road across the open fields at Stretton Sugwas (SO 460420) which appears only in outline on Meredith Jones's estate map of 1757 (HRO, C99/III/216). There is no other evidence for the direct road to Brecon running straight across the hill ridges at Craswall from Wilmarsh Stone (Wilmaston in Peterchurch), where Taylor marks an otherwise unrecorded standing stone. Similarly in the same area, a direct road from Westbrook in Clifford to Bodcott above Dorstone would seem doubtful and minor roads elsewhere are questionable. On the other hand, his classification of country houses and outline of parks are accurate and the river Dore is correctly shown as rising from a number of minor sources, not the small lake, as previously copied unquestioningly from Saxton.[1]

SCALE: 69½ miles to one degree. *c.*1 inch to 1 mile (1:64360).

SIZE: 1175 x 962 frame, 1175 x 1122 including side panels.

REFERENCE: CL, with side panels. HRO, AP25/2 lacks side panels.

Other issues:

1754
(A copy of the 1754 edition in the British Library contains 78 complete coats of arms, two with arms or crest only and 21 with names only: E.M.Rodger, *op. cit.*).

1786
In the dedicatory title, the word 'Ross' and the advertisement about estate surveying were replaced by the imprint 'London. Printed for W^m. Faden, Geographer to the King, Corner of St Martin's Lane, Charing Cross Aug 21st.1786'. On the copy in the Herefordshire Record Office no changes appear to have been made to the map. For example, the bridges over the river Wye at Hay (built 1763) and Bredwardine (1769) are not marked.
The number of coats of arms illustrated in the side panels has increased.

REFERENCE: HCL, folded, lacking side panels. HRO, C99/III/232, with side panels, containing 79 complete coats of arms and 8 with names only. Private, with side panels containing 79 complete arms and 13 with names only.

***c*.1792**
(A copy of the 1786 edition in the British Library (*Rodger*) marks the Kington-Stourport and Gloucester-Hereford canals, with locks, road destinations and mileage at the county boundary and some other amendments in the area south of Tenbury. Robert Whitworth, the canal engineer, had proposed the two canal schemes in Herefordshire in 1777 without success. Fresh proposals were made in 1789, leading to Acts of Parliament in 1791 to enable construction. The only part of the Kington-Stourport canal to be constructed was opened between Leominster and Neen Sollers (Shropshire) in 1796. The Gloucester canal reached Ledbury in 1798 and Hereford in 1845. These amendments were published on H. Price's map of Herefordshire of 1817, which was based upon the Ordnance Survey's recently completed drawings, raising speculation that this copy of Taylor passed through the hands of either the Ordnance Survey or Price).

REFERENCE: BL, Maps 2755 (4).

1. I am grateful to the local historians Heather Hurley and David Whitehead for their comments on Isaac Taylor's accuracy.

1754

Thomas Kitchin.
HEREFORD SHIRE ... Drawn from the best Authorities. By T. Kitchin, Geog^r. (bottom left). Printed for R. Baldwin, junior (outside bottom border).

Published in *The London Magazine: or Gentleman's Monthly Intelligencer*, 1747-60, the undated Herefordshire map appeared in volume 23 in April 1754. (Chubb CLXXXVII, Hodson 229).

The map was based on Morden's map in Camden's *Britannia*, 1695 but lacks that map's uncluttered clarity. Very similar to Bowen 1750. Hills are shown as molehills, 'Principal or Direct Post Roads, Principal Cross Roads, Cross Roads' are distinguished but on the Herefordshire map the only post road shown is between Bromyard and Presteigne; some principal cross roads are marked but no minor ones. Towns returning M.P.s are indicated by asterisks and there is a key to the conventional signs. Its title (bottom left) is written on a monumental stone standing in front of a crenellated round tower with a riverside view (? of Ross) and in the foreground a cider barrel and apples. With the arms of Hereford. and (top right) a compass rose.

SCALE: 14 miles to 2³/₈ inches. (1:373490).

SIZE: 168 x 215. (Hodson, 175 x 212).

REFERENCE: DCA, B.5/310. HCL. HRO, K38/D11 does not have R. Baldwin's imprint.

Other issues:

1764
(HCL dates one of its copies to 1764).

c.1787-89
Published in Henry Boswell (alias Alexander Hogg) and Thomas Kitchin, *Historical Descriptions of New and Elegant Picturesque Views of the Antiquities of England and Wales [with] a complete Set of County-Maps*. (Chubb CCLVII, Hodson 281).

c.1793
In *Complete Historical Descriptions* (Hogg). (Hodson 282).

1795
In *The Antiquities Of England and Wales* (Hogg). (Hodson 283).

1798
In *A New and Complete Abridgement or Selection ... of The Antiquities of England and Wales* (H.D. Symonds and Hogg). (Hodson 284).

1755

Emanuel Bowen.
An ACCURATE MAP of HEREFORD SHIRE Divided into its HUNDREDS By Eman. Bowen (top right).

Published in Emanuel Bowen and Thomas Kitchin, *The Large English Atlas*, 1760, and described as by far the most important eighteenth-century English atlas before John Cary's of 1787. (Chubb CXCV, CXCVI, Hodson 221).

The county maps were carefully based on the latest available surveys and printed as a folio atlas, which breaks the mould of the atlases so repeatedly derived from Saxton's and Speed's maps, reproduced by Morden, Overton, Jefferys and others. It was correspondingly more expensive, the maps selling at 1*s*. 6*d*. each compared with Overton's reproductions of Speed at 6*d*. each. Publication of the first maps in the series started in 1749 but good progress was not made until John Tinney took over the publication from John Hinton in 1752-3 and about three years later brought in Thomas Bowles, John Bowles and Son and Robert Sayer as additional partners

Herefordshire was published in 1755, based on Morden 1695 with material from Isaac Taylor's map, which was being prepared and engraved by Kitchin at the same time. The title is written on a monumental stone surmounted by floral and harvest-time decoration, with woolsacks and cider barrel and two seated men with a cider flask and glass. Features taken from Isaac Taylor include hills indicated by shading in a transitional form of hachuring, roads with mileage, a more accurate rendering of the boundaries of the hundreds and the inclusion of many more place-names. Anglo-Welsh names like Michaelchurch Escley appear in their English form. Taylor's forms are followed, for example of 'Hatterel Hills or Black Mountain' rather than the earlier Hatterrall Hill only. But copying errors are made. In the Golden Valley area alone these include Bocko Hill for Bacho Hill, 'Claunston' for Chanston(e), Arthur Stoke for Arthur's Stone and the source of the river Dore in a lake (on Morden but not Taylor). Empty spaces around the map are filled with the arms of the Earls and Lords of Hereford (FitzOsborn, Miles, Bohun, Bullenbrook (Bolingbroke) and Dev(e)reux), copious notes about the county, its market towns and, in detail, about the landslip of Marcle Hill in 1575. Gives latitude and longitude.

SCALE: 69 miles to a degree. 12 miles to 5³/₈ inches. (1:141455). John Senex had introduced the measurement of 69¹/₂ miles to one degree in 1721 in place of the earlier 60 miles to one degree, as did Isaac Taylor. Kitchin, Bowles and Cary regularly gave this measurement.

SIZE: 517 x 693. (Hodson, 525 x 707).

Herefordshire Maps

Other issues:

*c.*1762
HEREFORD SHIRE (John Bowles & Son, T. Bowles, John Tinney, Rob[t]. Sayer). Dedicated to John Viscount Bateman, Lord Lieutenant. (Hodson 222).

REFERENCE: HCL. HRO, K38/D12.

*c.*1763
(John Bowles & Son, T. Bowles, Rob[t]. Sayer). Plate number 16. (Hodson 223).

REFERENCE: HCL.

*c.*1764
(John Bowles, Carington Bowles, Robert Sayer). Plate numbers 18 (top right) and 16 (bottom right). (Chubb CXCVII, Hodson 224).

REFERENCE: DCA, B.5/299.

1767
(Hodson 225).

*c.*1767
French edition. (Hodson 226).

*c.*1779
(Robert Wilkinson). (Chubb CXCIX, Hodson 227).

1787
(Robert Wilkinson, Carington Bowles, Robert Sayer). Despite the publication of Cary's atlas in 1787 *The Large English Atlas* remained on sale for a further forty years. (Chubb CC, Hodson 228).

See also below Emanuel Bowen *c*1764.

1757

Isaac Taylor.
CITY OF HEREFORD. (Plates 28, 29).

No title. Dedicated 'To The Right Hon.[ble] y[e] EARL of OXFORD High Steward of the CITY OF HEREFORD [and the Bishop, Lord Lieutenant of the county, the Members of Parliament, Mayor and others] this Plan of the aforesaid City is most humbly Dedicated by ... J. Taylor. Ross. 1757'.

A relatively early English town plan and the first large-scale printed map of Hereford, preceded only by the small maps of Speed 1610 and Taylor 1754. It shows the whole city within the city walls and the suburbs immediately outside the town gates on the radial approach roads. Buildings are shown in block plan including the cathedral, parish churches, nonconformist meeting houses, almshouses and hospitals, houses with gardens and yards, watermills, coal wharves and tenter grounds along the river bank, tanyards. Streets are named with the destination of the roads outside the city gates; streams, pools and causeways; cherry orchards to the W of the city; parish boundaries.

Individual buildings marked include the cathedral's chapter house and the site of the former chapel of SS. Katherine and Mary Magdalene, the sites of the former Black Friars and White Friars (*recte* Greyfriars, and note that their location has been mistakenly reversed); Castle Hill and Green with the county Bridewell, moat and pool, the county gaol; the Town Hall, Tolsey, Cooken Row and Butchery in High Street. Other landmarks include The (weighing) Machine outside St. Peter's church; turnpike gates in St. Owen's Street, Above Eign Street, Barton Street and St. Martin Street; the Wye Bridge; the Bowling Green, large private gardens at the Bishop's Palace, Deanery, College (of the Vicars Choral) and Coningsby almshouses.

The dedicatory cartouche has an ornate baroque frame containing the arms of the bishopric and the city, with a scene of the river with trows and in the foreground a sack of apples and a standing figure of a young ill-dressed man, standing to draw or write on a large sheet spread before him, perhaps a self-portrait.

Around the map are a series of framed views, entitled (clockwise, from top left) General View from Broomy Hill, View of a Chapel now taken down (SS. Katherine and Mary Magdalene), Cross of the White Fryers (*recte* Black Friars), Cross in Lady Arbor (at the Cathedral), The Palace, Church & part of the town (looking upstream towards the Wye bridge), St. Peter's Church, St. Nicholas, St. Ethelbert's Cathedral, Town Hall.

At its foot is a scale bar placed across a pair of dividers and both lying over a compass star. A table (top left) gives the number of dwelling houses and inhabitants street-by-street, totalling 1,279 houses and 5,592 inhabitants. Outside the border is the imprint 'Publish'd according to Act of Parliament March 21 1757' (bottom left) and 'R. Benning sculp.' (bottom right).

It is a fine map, frequently referred to by archaeologists and historians. The boundaries of properties and plans of buildings appear reasonably accurate. The five piers of the Wye bridge are drawn correctly, for example. On the other hand, the chapter house is drawn as an octagonal building whereas its remaining footings are decagonal. It would seem likely that some smaller buildings and the layout of small gardens are drawn more 'conventionally'.

SCALE: 3 chains to 1 inch, 66 yards to 1 inch (1:2376).

SIZE: 690 x 890.

REFERENCE: HCL. HRO, K38/D180 (facsimile).

Other issues:

?1757
Lacks 'Ross 1757' in cartouche and dated imprint at bottom.

REFERENCE: HRO, AP25/1.

1759

John Gibson.
HEREFORD SHIRE. No. 16 in *New and Accurate Maps of the Counties of England and Wales. Drawn from the latest Surveys By J. Gibson*. London 1759. (Chubb CCXIII, Hodson 219).

Derived from Morden 1695 and apparently aimed at the children's market.

SCALE: 10 miles to 5/8 inch. (1:1013760).

SIZE: (*c.*63 x 114, Hodson; 63 x 111, Chubb).

Other issues:

Post-1779
(Hodson 220).

1760

Emanuel Bowen.
HEREFORD SHIRE Divided into its HUNDREDS ... By Eman. Bowen. October 1760. Engraved for W. Owen 1760.

Published in *The Natural History of England ... Illustrated by a Map of each County ... By Benjamin Martin ... Printed and sold by W. Owen ... and by the Author* (Benjamin Martin 1705-82). Vol. 2, 1763. These volumes formed part of *The General Magazine of Arts and Sciences*, in which the map of Herefordshire map has the imprint 'Engrav'd for the General Magazine of Arts & Sciences for W. Owen at Temple Bar 1760' (outside the bottom border). (Chubb CCXV, Hodson 230).

The maps were based on the same sources as those in Bowen and Kitchin's *Large English Atlas*, 1749-60. Marks selected place-names, roads with post-stages but concentrates on naming hills, not always happily; they include Leysters Hill, Perriston (Perrystone) Hill, Bocko (Bacho) Hill, Snod Hill (Snodhill) and Arthur's Stoke (Stone) Mountain.

SCALE: 10 miles (to 1½ inches. 1:422400).

SIZE: 170 x 193. (Hodson 176 x 191).

REFERENCE: HCL.

Other issues:

1765
Closely copied as Het Graafschap HEREFORD 1765 in Pieter Meijer, *Algemeene Oefenschoole Van Konsten En Weetenschappen ... Behelzende De Natuurlyke Historie Des Aardryks ...,* Amsterdam, vol. 2, 1770. (Hodson 260).

*c.*1762-1764

Emanuel Bowen.
An ACCURATE MAP of the COUNTY of HEREFORD Drawn from Surveys; Illustrated with Historical Extracts ... with other improvements ... By Eman: Bowen (bottom right). Printed for H. Overton ... J. Bowles & H. Parker ... C.Bowles ... J. Ryall & R. Sayer ... & T. Kitchin (outside bottom border). (Plate 7).

Published *c.*1764 in *The Royal English Atlas ... By Emanuel Bowen ... Thomas Kitchin ... and Others*. (Chubb CCXVIII, Hodson 233).

Based on Isaac Taylor's map of 1754 and the similar but larger plates of *The Large English Atlas* of 1760, from which many of the minor names are taken. It has a smaller typeface than Taylor and may best be regarded as Taylor at a manageable size. Around the map are a drawing of Hereford cathedral (top right) and notes about the Herefordshire towns, Marcle Hill and the lords of Hereford. The plates are thought to have been engraved in 1762-3. The atlas was not a commercial success.

SCALE: 12 miles to 4½ inches. (1:168960). 69 miles to a degree.

SIZE: 408 x 510. (Hodson, 413 x 510).

REFERENCE: HCL.

Other issues:

***c.*1778**
(Carington Bowles). (Chubb CCXIX, Hodson 234).

1778
(R. Sayer and J. Bennett) (CCXIXa, Hodson 235). The Herefordshire map has 'Printed for John Bowles ... Carington Bowles ... Rob[t]. Sayer & John Bennett 1st. of Jan[y]. 1778'.

***c.*1780**
(R. Wilkinson). (Chubb CCIX, Hodson 236).

Herefordshire Maps

*c.*1828
(R. Martin) as *The English Atlas*. (Chubb CCXCIX, Hodson 237). Reproduced as J.B. Harley and D. Hodson, *The Royal English Atlas*, Newton Abbot 1971.

1763

Thomas Kitchin.
A New Map of HEREFORD SHIRE Drawn from the best Authorities By Thos. Kitchin Geogr. Engraver to H.R.H. the Duke of York (in a rococo frame, top right).

Published in *England Illustrated ... With maps of the several counties ... Printed for R. and J. Dodsley*, Vol. 1, London 1764. The volume actually came out late in 1763. (Chubb MDCCLXIV, Hodson 231).

The maps are based on the same sources as *The Large English Atlas* 1755. Main and cross roads are clearly shown, the hills as low pyramids. Conventional signs are used to distinguish boroughs with M.P.s, fairs and market towns, rectories and vicarages. English names are used for Anglo-Welsh place-names like Michaelchurch Escley, but misspellings are frequent.

SCALE: 9 miles to 1 3/4 inches. (1:325850).

SIZE: 254 x 195.

REFERENCE: HCL.

Other issues:

1764
Re-issued in parts through the year. (Hodson 231).

1765
The maps were republished, unaltered, in *Kitchen's English Atlas ... Printed for J. Dodsley*. (Chubb CCXXXVIII, Hodson 232).

1765

Joseph Ellis.
A Modern MAP of HEREFORD SHIRE Drawn from the latest surveys; Corrected & Improved by the best Authorities. W(illiam) Palmer sculp. Printed for Robt. Sayer in Fleet Street, & Carington Bowles in St. Pauls Church yard.

Published in *The New English Atlas ... Engrav'd in the best Manner, By J. Ellis, and Others*, London 1765. (Hodson 238).

A relatively small travelling atlas in a soft leather binding designed to be rolled up for travel. Only one copy of the first edition of the complete atlas is known but it sold well through Sayer and Bowles and was rapidly re-issued. The county maps are closely copied from Kitchin's *England Illustrated*. The maps are blank on the back.

SCALE: British Statute Miles 69 to a Degree. 9 miles to 1 3/4 inches. (1:325850).

SIZE: *c.*245 x 193.

Other issues (usually entitled *Ellis's English Atlas*):

1766
(5 issues, one being French). (Chubb CCXXVII, CCXXVIIa, Hodson 239-243).

1768
(2 issues). (Chubb CCXXVIII, Hodson 244-5).

1773
(Chubb CCXXIX, Hodson 246).

1777
(Chubb CCXXX, Hodson 247).

*c.*1782
(Hodson 248).

*c.*1786
(Hodson 249).

*c.*1790
(2 issues) (Hodson 250, 251).

*c.*1796
(Hodson 252). Still advertised for sale up to *c.*1825.

1767

Emanuel Bowen and **Thomas Bowen**.
HEREFORD SHIRE Divided into HUNDREDS, Exhibiting the County, Borough & Market Towns &c...By Eman: Bowen...& Thos. Bowen, (1767).

Published in *Atlas Anglicanus, or a Complete Sett of Maps of the Counties of South Britain ... Printed for T. Kitchin*.

Published in monthly parts from January 1767, the last of the rococo-style atlases of county maps, associated particularly with Bowen and Kitchin. Eleven maps were engraved by Emanuel Bowen, thirty-three by Emanuel and his son Thomas jointly and one by Thomas alone. In its first state a plate number is in the top right corner. In later issues the plate number was deleted and an imprint added outside the bottom border 'Printed for Tho. Kitchin Nº 59 Holborn Hill, London'. The map of Herefordshire has the plate number 5 (top left, outside the border). (Chubb CCXXXII, Hodson 254).

Also a copy entitled HEREFORD SHIRE Divided into HUNDREDS Drawn from Actual Surveys ... with Improvements not Inserted in any other Set of Half Sheet County Maps Extant By Eman. Bowen ... & Thos. Bowen. Lacks plate number and Kitchin imprint.

A messy design. Notes fill all the available space surrounding the map like Bowen *c.*1764.

SCALE: 12 miles to 2³/₈ inches. (1:320135).

SIZE: 275 x 320.(Hodson, *c.*225 x 325).

REFERENCE: HCL has copies in both these states.

Other issues:

*c.*1777
(Chubb CCXXXIII, Hodson 255).

1785
In *Bowles's New Medium English Atlas* with the title in a simple circular frame, as 'BOWLES's NEW MEDIUM MAP OF HEREFORDSHIRE DIVIDED into its HUNDREDS ... Printed for the Proprietor Carington Bowles'. A distinctive oval cartouche. (Chubb CCLV, Hodson 256).
SIZE: 215 x 312 .

Post 1793
(Hodson 257).

1769

Thomas Kitchin.
HEREFORD SHIRE.

Published in Kitchin's *Pocket Atlas of the Counties of South Britain ... Drawn to One Scale ... Being the first Set of Counties, ever Published on this Plan, 1769*. The experiment was a commercial failure, perhaps owing to the unavoidable result that the set of maps came in varying sizes, small counties being printed two to a page but large ones requiring a folded page. Based upon his four-sheet map of England and Wales 1752, he corrected county boundaries but is said not to have made use of maps, such as Isaac Taylor's of Herefordshire, which had appeared since 1752. (Chubb CCXXXV, Hodson 258).

SCALE: 10 miles to 1³/₈ inches. (1:460800).

SIZE: 155 x 172.

Other issues:

*c.*1778
In *Bowles's Pocket Atlas of the Counties of South Britain*. The map has a new title outside the top border BOWLES'S REDUCED MAP OF HEREFORDSHIRE with a plate number top right outside the border. (Hodson 259).

1784

Alexander Hogg.
A New MAP of HEREFORD SHIRE. Drawn from the Latest Authorities (top right). (Plate 18).

Printed on the right-hand part of the same page as a similar map of Monmouthshire in *The New British Traveller*, issued in parts, London 1784. It has a plain border, outside of which (at the bottom) is the name of the engraver, T. Conder, the imprint (Published by Alexr. Hogg at the K)ings Arms Nº 16 Paternoster Row and (at the top) 'Walpoole's New & Complete BRITISH TRAVELLER'. The Monmouthshire map has been described as a poor engraving, but the Herefordshire example in the Herefordshire RO is of a good quality. (Chubb CCLI, Hodson 269).

The map is simple, clear and uncluttered. It marks main roads, selective place-names including some castles, country houses and parks, but not the hundreds. Place names include the Black Mountain and Lug Mead (Lugg Meadow). The source of the river Dore in a lake persists. Arms of Hereford. Compass rose (top right).

SCALE: 15 miles to 2¹/₈ inches. (1:447250).

SIZE: 197 x 156. (Hodson, the earliest state, 216 x 319).

REFERENCE: HCL. HRO, K38/D9.

Other issues:

*c.*1792
(Hodson 270).

REFERENCE: HCL. (The map appears to have been trimmed and lacks an imprint).

*c.*1793
(Hodson 271).

*c.*1794
(Chubb CCLII, Hodson 272).

REFERENCE: HCL. T. Conder Sculpt. (bottom right).

Herefordshire Maps

1787

John Cary (1755-1835).
HEREFORDSHIRE.
Published in *Cary's New and Correct English Atlas* ... London 1787-89. (Chubb CCLX, Hodson 285).

Sometimes called 'the Small Cary'. The *New and Correct English Atlas* was the first genuinely new survey of the counties for a generation and a great improvement on its immediate predecessors. After apprenticeship as an engraver 1770-78 Cary was employed by a succession of unsuccessful publishers before starting his own business in 1783. He had worked on canal plans, mostly undated but one signed 1779. In 1784, the year of the introduction of the mail coach, he published the first in his series of popular and well reviewed road books. His *Atlas* was equally successful. The county maps were based on the latest available larger-scale surveys and were well designed and engraved. He concentrated on the needs of the growing number of travellers, and the canals, turnpike roads and post towns of late eighteenth-century England were prominent features. The destinations of the main roads to London and other major towns are indicated at the county boundaries, main and minor roads were distinguished by thick or thin lines and the mileage points along the main roads are marked. To avoid cluttering the maps with too much information he cut out all reference to the hundreds, although these ancient subdivisions of a county were still relevant for tax and administrative purposes. Longitude is measured from Greenwich. Relief is indicated by hachures. The atlas had a powerful influence on the nineteenth-century county maps and on the Ordnance Survey.

The *Atlas* was issued in twelve parts, with maps available coloured or uncoloured. Herefordshire appeared in part VIII about July 1788. The title of the map HEREFORDSHIRE By JOHN CARY, *Engraver* is on a simple label laid over a compass star (bottom left). All the maps were uniformly dated 1 September 1787. Cary, however, went on revising and issuing the plates. He also sold the maps loose. He amended the imprint on moving his business from one address to another in The Strand, London in 1792 but for the sake of consistency did not alter the date of the original engraving. The earliest imprint, therefore, reads 'London: *Published as the Act directs September 1st. 1787 by* JOHN CARY *Engraver, Map & Print seller, No. 188 the corner of Arundel Street, Strand.*'. Post-1792 imprints of the first edition are shortened, without altering the date, to 'London, *Publish'd Septr. 1st. 1787 by* J. Cary *Engraver & Map seller Strand.*'

The Herefordshire map is based on Isaac Taylor's 1754 map. It shows selected prehistoric hill forts such as Ivington, Wall Hills, Dinedor; country houses including Garnons, The Ware (Weir), Wigmore Grange, Gatley, roadside inns such as the Three Horseshoes at Allensmore, the Maidenhead at Orleton, and isolated oaks (Willey Oak, Ridgeway Oak at Cradley); hill ridges. Ferries and mills are marked but not woodland. Place-names, which are given in a recognisable modern form, include Michaelchurch Escle(y), Dewlas (Dulas), Hatterrall Hill or Black Mountain. The three detached parts of the county at Cascob (Radnorshire), Ffawyddog (Monmouthshire) and Farlow (Shropshire) are shown. Latitude and longitude is given within the ruled border.

Its size, clarity and detail made it a convenient choice for reproduction in other publications.

SCALE: 69$^{1/2}$ miles to a degree. 6 miles to *c.*1$^{1/4}$ inches. (!:320135).

SIZE: 262 x 210.

REFERENCE: HCL.

Other issues:

1793

Second edition. The imprints are dated 1 January 1793. (Chubb CCLXI-CCLXIII, Hodson 286). (Plate 8).

Herefordshire, like Hertfordshire, was reprinted from a re-engraved plate and redated with the imprint 'London: *published by* J. Cary *Engraver & Map seller N° 181 Strand. Jany. 1st. 1793*'. The title remained unchanged but the map was revised to correct errors in Taylor and brought up-to-date. Revisions include the alignment of roads between Bodenham and Amberley, the erasure of the words 'Road from Brecon' between Craswall and Peterchurch and downgrading it to a lane, the erasure of all but two mileage figures on the road from Hereford to Hay through Madley and Peterchurch , and the addition of the bridges over the river Wye at Hay, Whitney and Bredwardine, built between 1763 and 1774, but not marked on Taylor's re-issue of 1786.

In the last issues of this edition, the revised 1809 plate for the third edition of Herefordshire was used but, like those of Hertfordshire and Surrey, was dated 1793.

REFERENCE: HCL. HRO, K38/D14.

1809

Third edition, with the plates, worn out by the success of frequent re-issues, entirely replaced by new engravings, closely copied from the second edition.

The plates continued in use, revised further, until at least 1875.

1788

J(ohn) Lodge.
A MAP OF HEREFORDSHIRE FROM THE BEST AUTHORITIES. J. Lodge sc.(bottom right). London. Published ... Nov. 30 1788 by J. Murray Nº. 32 Fleet Street (bottom centre). Political Mag. Nov. 88 (top right). (Chubb CCXLIX).

Published in the *Political Magazine*, 1782-90, the Herefordshire map appearing on 30 November 1788.

The spellings of place-names follow Cary, based on Taylor. Hills are shown as low molehills.

SCALE: 9 miles to 2¼ inches. (1:253440).

SIZE: 322 x 261.

REFERENCE: HCL.

Other issues:

1795.

1789

E. Noble.
A Map of HEREFORDSHIRE from the latest Authorities. Engraved by J. Cary. (Plate 9).

Drawn by Noble, engraved by Cary and published in Richard Gough's 1789 translation of Camden's *Britannia* of 1607, vol. 2, plate no. 35. (Chubb CCLXXI).

Sometimes called the 'Large Cary'. Based on Isaac Taylor 1754 and Cary 1787. Because of its larger size it contains more features than Cary 1787, including minor roads, woods and minor place-names. It is, however, rather weakly drawn compared with the more attractive map by C. Smith, A NEW MAP of the COUNTY of HEREFORD Divided into Hundreds ... January 6th. 1802, which is about the same size and scale and contains most of the same features. Historians are advised to read the three maps together. The representation of minor roads is unreliable. The principal roads in some counties were revised in the second edition, but apparently not in Herefordshire. Latitude and longitude marked in the border.

It shows river Wye crossings by bridges, ferries and fords, but fails to update Taylor by not marking Hay and Bredwardine bridges built in 1763 and 1769 respectively. It marks Bringewood Forge. Hills are shown by hachuring. The map notes that the river Honddu is 'a very nice stream'.

SCALE: 69½ miles to a degree. 8 miles to 3⅝ inches. (1:139830).

SIZE: 523 x 427.

REFERENCE: HCL, HRO, K38/D18 (undated, uncoloured).

SIZE: 530 x 430).

Other issues:

1805

2nd. edition published in *The New British Atlas, London John Stockdale 1805* with the imprint A MAP OF HEREFORDSHIRE FROM THE BEST AUTHORITIES Published by John Stockdale Piccadilly 26th March 1805. Engraved by J. Cary. Anticipates the construction of the Kington-Leominster-Stourport canal, authorised by Act of Parliament 31 Geo. III, c.69 (1791), shown on the map but not completed.

SIZE : 530 x 430.

REFERENCE: HRO, K38/D16 (1805 coloured).

1789

(J.) Haywood.
A MAP OF HEREFORDSHIRE ENGRAVED FROM AN ACTUAL SURVEY, with Improvements (top right). Sudlow sc.(bottom left). Engraved for J. Harrison, 115, Newgate Street ... March 24th 1789 (bottom centre), Haywood del. (bottom right, all outside the border).

Published in *Maps of the English Counties with the Subdivisions of Hundreds ... London: Printed by and for John Harrison* 1791. (Chubb CCXCI).

A poor map. The place-names are roughly in the correct location but the many roads are unreliably drawn and relief is not indicated. A key to the hundreds (top left).

SCALE: 69½ miles to 1 degree. 12 miles to 3½ inches. (1:217230).

SIZE: 322 x 458.

REFERENCE: HCL.

Other issues:

1791.

1792.

1789

John Cary.
J. Cary HEREFORDSHIRE Engraver.
Imprint below the map 'London. Published Sepr 1789 by

J. Cary, Engraver N°. 188 Strand.'. Published in *Cary's Traveller's Companion*, 1790. (Chubb CCLXXIII).

A popular and clear lightweight pocket-size atlas on thin paper with an emphasis on the turnpike roads. These are coloured red and the mail roads blue, with distances marked along the roads. There is little decoration in order to concentrate attention on the map itself. The title (top) is in a panel of the border of the map with a compass star.

Cary was commissioned by the Postmaster General to survey some 9,000 miles of roads in England, on which he based his *New Itinerary* (1798) and other road-books.

SCALE: 8 miles to ³/₄ inch. (1:675840).

SIZE: 121 x 89.

Other issues:

1791.
At least eleven further editions 1806 to 1828.

1790

J. Aikin.
HEREFORDSHIRE.
Published in *England Delineated*, 2nd. edn. 1790. (The first edition, 1788, did not contain maps).

The name of the county is printed at the top of the map. Shows only an outline of the county, towns and rivers. Lacks scale and borders. (Chubb CCLXXXVI).

SIZE: (Chubb, of the volume, 190 x 113).

Other issues:

1795

1800, 1803, 1809.

c.1793

Benjamin Baker.
HEREFORDSHIRE. Engraved by B. Baker, Islington.
Published in *The Universal Magazine of Knowledge and Pleasure*, 1791-1797. (Chubb CCXCIII).

Roads and hills, shown by hachuring, are taken from Taylor, with selected place-names. The bridges at Hay and Bredwardine are marked.

SCALE: 8 miles to 1¹/₂ inches. (1:337920).

SIZE: 226 x 180.

REFERENCE: HCL.

Other issues:

1804
(Darton and Harvey, *Maps of the Several Counties and Shires in England*).

1807, 1816
(*Laurie and Whittle's New and Improved English Atlas*).

1796

John Price. Plan of the City of Hereford.
Published in J. Price, *An historical account of the city of Hereford*, Hereford 1796, between pp. 60-61.

Shows the city with the built-up area in block outline and major buildings marked individually. These include churches and chapels, 'The Music Room used for the Collegiate School', The Castle Walks, The Infirmary, The Theatre, The (Bowling) Green. There are key tables to these landmarks on pp.60-61 and to street names on pp. 59-60.

The compass rose incorporates a set square, rule and dividers. Scale bar. No borders.

SCALE: 377 yards to 1¹/₂ inches. (1:9048).

SIZE: 204 x 124.

REFERENCE: HCL.

1796

John Price.
Course of the River Wye from Brobery to Wilton.
Published in J. Price, *An historical account of the city of Hereford*, Hereford 1796, between pp. 176-177.

In simple outline gives the names of villages and principal country houses near the river. Marks prehistoric camps, *Ariconium* (as Kenchester), Wilton Castle, the Vineyard at Hereford and Monnington Walk.

Compass points, with an elaborate E point. Scale bar. Ruled border.

SCALE: 5 miles to 1¹/₂ inches. (1:211200).

SIZE: 162 x 95.

REFERENCE: HCL.

Catalogue of the Manuscript Maps of Herefordshire before 1800

The Catalogue of Manuscript Maps

The Catalogue

The *Catalogue of maps in the Essex Record Office,* 1947, edited by F.G. Emmison, was the pioneering model for imitators in other counties, who refined Emmison's approach without departing radically from it.[1] This Herefordshire catalogue broadly follows the same pattern, with some new features and, confessedly, some inconsistency. Over the period of its preparation refinements emerged which for practical reasons could not be applied to all the earlier entries.

The maps are arranged in chronological order. Square brackets indicate a date deduced from internal and associated evidence and round brackets indicate a later copy.

Each catalogue entry is headed by the name of the ancient pre-1851 parish, with the name of the modern parish added in parentheses in appropriate cases. The title of the map is given as written by its surveyor, but usually omits the repetitive words 'in the County of Hereford'. The description of the map begins with the size and location of the property, followed by:

(1) a note of the principal house or farmstead and other buildings (whether shown in bird's-eye view, perspective, profile or plan),
(2) the features of the main property (gardens, fields, orchards, woods and boundaries),
(3) the natural features (relief, rivers, ponds) and
(4) other landmarks outside the main property (roads, footpaths, bridges and the names of adjacent owners).

Areas have been given in statutory acres as measured by the surveyor, not in hectares. In a few cases surveyors also noted both statutory and local customary or computed acres. In Herefordshire three customary acres were equivalent to two statutory acres of arable land, one and a half acres of hops and five acres of woodland. Imperial and metric conversion tables are given on p. xv.

Specific landmarks and field-names of historical, topographical or etymological interest are noted selectively with Ordnance Survey grid references. The *Herefordshire field-name survey*, based on the tithe apportionments and maps *c.*1838-44 and compiled by the Woolhope Naturalists' Field Club in 1987-93, has been invaluable for identifying individual sites. This survey and Herefordshire Council's archaeological sites and monuments record became available on-line in 2003 (www.smr.herefordshire.gov.uk). The surveyors' spelling of place-names has been retained with the modern spelling, as recorded by the Ordnance Survey and other sources, added within square brackets. Decorative features (title cartouche, coats of arms, compass rose or north point, scale bar, borders) conclude the description of the working area of the map.

Separate notes give the scale of each map as expressed by the surveyor and (in brackets) converted to a fraction, the size, materials and condition of the map, and the name of the surveyor. The maps have been measured in millimetres to the frame or border where the drawing has one, or to the edge of the parchment or paper, height by width according to Anglo-American Cataloguing Rules. Because many parchment maps are buckled or irregular in size some measurements have necessarily been rounded to the nearest centimetre.

The location of the map by the name of its repository and call number complete the entry. With a few exceptions the maps in Herefordshire repositories were catalogued in 2001-02 and elsewhere in 2002-03. The location 'Private' indicates that the map remains in private hands but does not imply that it is accessible to the public for research. All enquiries about such maps should be addressed in the first place to the County Archivist (see pp. ix-x).

Whilst every attempt has been made to compile entries for all surviving maps, a few have been omitted at the wish of their owners. Others have inevitably been overlooked. Some of these may not have been entered on the national archival database (www.a2a.org.uk) at the time of my research or still lie in uncatalogued public collections. Some more surely remain safely but undiscovered in private hands. Information about all such omissions and errors would be much appreciated by the author (see p. 4 note 4).

---

1. An alternative model, published shortly after the completion of the groundwork for this catalogue, has been set by M.R. Ravenhill and M.M. Rowe, eds., *Devon maps and map-makers: manuscript maps before 1840*, Devon and Cornwall Record Society, new series, vols. 43, 45, (2002).

1577
BURRINGTON, PIPE ASTON, ELTON, RICHARDS CASTLE

Endorsed 'A Platte of Parte of the Chase of Bringewo[od] and of certayne groundes adioininge leased to Mr Walter of Ludlow Also a p[ar]ticular or Constat Und[er] the Audito[rs] hand howe the same groundes have bene holden by lease'[1]

Large area of woods and former woods of and adjoining Bringewood Chase, lying W of the road from Ludford bridge [SO 513743] to Richards Castle [SO 484703] and SE of the road from Ludford to Aston within the parishes of Burrington, Ludford [Shropshire], Pipe Aston, Elton and Richards Castle [Herefordshire and Shropshire].

Crudely drawn, showing buildings in perspective, woods and felled woodland, waste, clearings and enclosures, boundary ditches and hedges, gates, river Teme, roads. Marks Ludlow, Ludford bridge with three arches, a gallows at Whitclyffe [Whitcliffe], Overton, The Moore [Moor Park], 'The waste of Riccards Castell' [Hanway Common], the castle of 'Riccards Castle', Mary Knoll with Beckes Barn opposite 'the park yate' [Park gate at SO 482736], 'Hopkies the underkepers house and close' [at Gorsty, SO 479736], Fennalls [Vinalls], Shuttes alias Overies. 'The vallet woods of the Lordship of Riccards Castell' [Vallets] appear to be shown as woodland, most of which had recently been coppiced, and Mary Knoll, the Fennalls, Overies and Hopkies' close are shown as enclosed and partly or mostly felled and ploughed. A park ditch and hedge divides these lands from Bringewood Chase, in which deer, trees and tree stumps are drawn.

SCALE: Variable, [between *c.*1:15840 and 1:10560].

SIZE: 440 x 590. Ink and watercolour (pink, browns and blue) on paper. A few small pieces missing along the folds; repaired.

SURVEYOR: Not named.

REFERENCE: PRO, Kew, MPF 1/148 (extracted from SP 13/H/16). A high-quality facsimile monochrome photograph is in Shropshire RO, 2130/1.

1. The Crown owned Bringewood Chase from 1461 to 1636. The lessee in 1577 was Edmund Walter, a member of the Council for the Marches of Wales from *c.*1576 and Chief Justice of South Wales from 1581 until his death in 1594. The map is reproduced and analysed in H.T. Weyman, 'The Walters at Ludlow. An Elizabethan plan', *Trans. Shropshire Archaeological & Natural History Soc.*, 4th ser., vol.3 pt. 2 (1913), pp. 263-82 and plate. See also P. Cross, 'A 1577 plan "A platte of part of the Chase of Bringewood and of certayne groundes adjoining leased to Mr Walter of Ludlowe". A re-evaluation of the landscape', *Trans. WNFC*, vol. 48, pt.3, (1996), pp. 573-581, in which the map is compared with the map of Mocktree Forest and Bringewood Chase 1662; see 1662 DOWNTON below. I am grateful also to David Lovelace, who allowed me to read his paper on the history of Bringewood presented to a conference on ancient woodlands at Sheffield Hallam University in May 2003].

[c. 1580]
SOUTH HEREFORDSHIRE

No title. [The Forest of Dean and neighbouring area from Hereford to Newport and Gloucester to Bristol].

A pictorial map of the Forest of Dean, showing the river Wye, hills [as molehills], forest [as trees], principal roads with mileage. Marks Hereford, Ross and 'Goderich' [Goodrich] Castle [SO 577200] with buildings in perspective, and the roads from Hereford to Newnham and Monmouth, and Gloucester to Ross.

SCALE: [*c.*1:126720].

SIZE: 400 x 560. Ink and watercolour on parchment.

SURVEYOR: Not named.

REFERENCE: PRO, Kew, MPC 1/66 (extracted from DL 3176).

1582
MARDEN

No title.

55a. in Lawfield [Layfield, SO 515496] (coloured bright yellow), showing enclosures (coloured bright green) by Thomas Welford [12a.,] and John Camdyn [1½a.]. Marks Lafield Gate and field boundaries.

Submitted as evidence in case of Thomas Welford, steward of the manor of Marden *v.* Sir Thomas Coningsby relating to lands enclosed by Welford, 1595-6.

Cardinal points written at edge of map.

SCALE: [?*c.*1:10560].

SIZE: *c.*180 x 185. Ink and watercolour on parchment.

REFERENCE: PRO, Kew, E178/2002.

[c.1590]
BRILLEY

Endorsed 'The Common of Brillie in the Countie of Hereford'

Roughly rectangular area, measuring about 200 by 137 [perches], apparently in the E of the parish immediately W of

Pentre Coed Dingle [SO 285504]. Outline of boundary with measurements between each change of direction. Names of some adjoining owners.

No compass point or scale bar.

SCALE: [*c*.1:3960][1].

SIZE: 285 x 410. Ink on paper.

SURVEYOR: Not named.

REFERENCE: HRO, P22/20/4.

1. This scale is based on the measurements of the boundaries of the common as drawn on the map, assuming that these are on a scale of perches. But the area is not readily identifiable and, if its eastern boundary is Pentre Coed Dingle, a comparison with the Ordnance Survey 1:25000 map would suggest a smaller scale of about 1:5280.

[*c*.1600]
WALTERSTONE

No title. Endorsed 'Comes Sarum'. 'A Mapp of Alterennys'. (Plate 3).

10a. at Alltyrynys amid a larger surrounding area in the extreme SW of the parish [SO 335234]. Marks 'Allterenies howse' as a Tudor three-bay house with gables and ornate central porch and an additional one-bay W extension. Names a field called Clapper immediately W of Alltyrynys bridge and gives acreage and boundaries of riverside fields.

Shows other buildings crudely in perspective, road from Alltyrynys to Walterstone village, boundaries of disputed 'land in question', 'The River of Monnow', 'Nant Cravell' [Nant y Grafel, the stream flowing S from Walterstone Common to Alltyrynys] and 'a brooke devydinge the 2 p[ar]ishes' of Walterstone and 'Comeyoye in the com' of Hereford' [that part of Cwmyoy formerly in Herefordshire] flowing E to Alltyrynys from 'a mowntayne called Hatterhyll' [Hatterrall Hill]. Indicates villages of Trewalter [Walterstone] and 'Oulde castell in Monmouth' [Oldcastle] by out-of-scale drawings of churches and houses.[1]

Compass points of N and S written at the edge of the sheet. Scale bar. Ruled border at bottom, beneath which is written in a different and rather later hand 'The Mylle [Trewyn mill] of M[r]. Delahay is situate att the Upper end of the Clapper full west from the bridge & the old myll'.

SCALE: 'The Scale of perches according to the statute of wynchester'. 17 perches to 1 inch [1:3366].

SIZE: 380 x 330, irregular. Ink and watercolour on parchment.

SURVEYOR: Not named.

REFERENCE: Private. Hatfield House, Hertfordshire, CPM supp. 9. Published as 'Plan of the lands of William Cecil of Allt-yr-ynis' in R.A. Skelton and J. Summerson, *A description of maps and architectural drawings in the collection made by William Cecil first Baron Burghley now at Hatfield House*, The Roxburghe Club, Oxford 1971, no. 135 and p. 72. I am grateful to Lord Salisbury's Librarian and Archivist, Robin Harcourt Williams, for supplying me with this evidence about the context of the map.

1. Alltyrynnys comprised freehold and copyhold lands on both sides of the river Monnow, which marked the manorial, parish, hundred and county boundaries between Herefordshire and Monmouthshire. The Cecil family's freehold lay in Walterstone in Ewyas Lacy Hundred (Herefordshire) but copyhold lands were held of the Arnold family's adjacent manor of Oldcastle (Monmouthshire). A dispute about some of the copyhold lands went against the Cecils when Paul de la Hay lost a long-running action of ejectment in 1632 against Nicholas Arnold (Hatfield House, Legal 153/15 and General 72/4). The dispute went back a long time. This map, described in the context of the dispute as 'A Plott of Crageth, and other groundes for the inheritance whereof Mr. Arnold sued' (Hatfield House, CP 363, fol. 75v.), appears to be closely related to a sketch drawn in 1608 by Paul de la Hay (Hatfield House, General 40/6). The complexity of the land ownership was further complicated by flooding before *c*.1600, which had altered the course of the river, and the building about that year of a new house at Alltyrynnys. Some of the older building, left standing, stood in Monmouthshire.

1606
HEREFORD

'Hereford described at 50 paces to the inch September. 1. 1606'. (Plate 11). The earliest map of the city, drafted by John Speed for his map of Herefordshire printed in 1610 (see p. 67). Shows the principal buildings in bird's-eye view, streets, river Wye, streams. Marks the cathedral, parish churches, chapels, former friaries, standing crosses, almshouses and hospitals, castle, city walls and gates, town hall, gaol, mills and other landmarks. Names streets and buildings either on the map or in a reference table of symbols, letters and figures.

This manuscript draft differs from the printed version in many respects. It is clearly more accurate, defining the street pattern much better and showing the principal buildings in greater detail but omitting the conventional drawings of houses lining the streets. It provides significant new evidence about the castle and city walls before the Civil War to supplement the archaeological excavations from the late 1960s.[1]

Additional features not on the printed version include the walled circuit of the cathedral close with a standing cross NE of the cathedral (not beside the Chapter House as in the printed version), and a second walled area enclosing the Bishop's

Palace, Bishop's Chapel of SS Katherine and Mary Magdalene, and the College of the Vicars Choral; the Palace Gatehouse stands outside these walls. The complex ramparts and walls of the Castle indicate that the motte was defended by a small barbican between the water gate and a second gate leading from the bailey. The walls of the bailey joined the city wall on the E and were pierced on the N by the City Gate, adjoining an outer pool or enclosed swampy area. Within the bailey were a chapel and an isolated small tower. The city walls are shown with crenellation and bastions, of which only one of the three between Eign Gate and Friars' Gate has been omitted. Also shown are the market hall of *c.*1600 as a gabled building on pillars, the county gaol in St Peter's Square, a weighing machine in King's Ditch [King Street] and small buildings or street furniture in High Causey [High Town], Broad Street and Wyebridge Street. Outside the city gates are marked the 'pulpit' [preaching cross] at Blackfriars and St Giles's chapel; the extent of building development along the radial roads outside the walls is indicated by a few token houses at its limits. Gardens are marked between the College of the Vicars Choral and the river Wye and between Mill Street and Britoyns Street [Britons Street, now Green Street]. Names of minor streets are given and footpaths to the mill [in Mill Street] and Greyfriars are marked by pecked lines. The name Old Street on the printed map is here given as 'old scholstret' [Old School Street, now Union Street] and St Nicholas church is shown in the middle, not the side, of King's Ditch.

A few features not marked on the manuscript drawing were added to the printed map, suggesting that Speed also took notes of his fieldwork. The names of the cathedral as St Ethelbert's minster and of 'White Friers' [*recte* Greyfriars], and the road running N on the line of Edgar Street do not appear on the draft map. The prominent bastion marked on the printed map at the SW corner of the city walls by the Wye bridge is drawn more correctly at a reduced size on the draft. S of the river Wye the suburb beyond the former church of St Martin [in Wye Street] and the stream running from Hunderton beneath the Dry Bridge and along the Row Ditch are also not marked on the draft. The former St Guthlac's Priory, outside the walls on the NE of the city, is not marked on either map.

A graticule is laid over the map for copying purposes.

SCALE: 50 paces to 1 inch [but comparison with the OS map suggests *c.*1:3250].

SIZE: *c.*525 x 602 (maximum). Pen and ink on paper.

SURVEYOR: Not named [John Speed].[2]

REFERENCE: Merton College, Oxford, D.3.30, no. 7.[3]

1. For reports of excavations led by Frank Noble and Ron Shoesmith see *Trans. WNFC* from 1967 and R. Shoesmith, *Hereford city excavations, Vol. 1. Excavations at Castle Green*, Council for British Archaeology research report 36, 1980 and *Vol. 2. Excavations on and close to the defences*, CBA research report 46, 1982. I am grateful to Ron Shoesmith and Julian Reid, Merton College's archivist, for contributing to this entry.

2. One of the set of maps in the volume is signed 'John Sp'. See S. Bendall, 'Draft town maps for John Speed's *Theatre of the Empire of Great Britaine*, *Imago Mundi*, vol. 54 (2002), pp. 30-45.

3. A volume in the Pilley Collection in Hereford City Library of copies (1903) of drawings by S. Fisher of Hereford in the late eighteenth century also contains a simplified town plan of Hereford, based upon Speed and drawn by an anonymous Frenchman. This was taken from a series of British and European town plans *c.*1650 in the British Library. The copy marks the streets, city defences, cathedral and some of the churches at a scale of 100 paces to 1⁵/₈ inches. HCL, P.C. 2327.

n.d. [early 17th century]
PEMBRIDGE

Two maps of Strangworth Forge [SO 345594] and adjoining land relating to a legal dispute, apparently over the watercourses of the river Arrow and mill stream.

[1]
No title. Endorsed 'Map of the Lands about Strangward forge'

Shows the mill stream from its diversion at a weir [SO 338587] to Strangworth Forge [SO 345594] with the lands on the right bank of the river Arrow. Shows the upstream weir and river with the note 'The River in the Sum[m]er taken all to the Forge', the whole course of the mill stream with three double-floodgates; the first was at the weir, the second carried a footpath to Strangworth House on the right bank of the river, the third was above the Forge Pond and, together with gates just upstream of the Pond to allow waste water to flow back into the river, had been erected after the forge was built. Between the mill stream and the river is 'Arrow Meadow overflowed with the Forge pool & spoiled'. Below the forge is the note 'Forge stream diverted'.

Marks the ground plans of Strangworth House (with a central hall, stairway facing the entrance, four symmetrical rooms), 'The Mill' (with two water wheels) and 'Forge' (with one large and two smaller water wheels). Field-names include Millers Eight Acres, Mr Carrs Mill Meadow, Kitchen Meadow, Norton Woods.

Simple compass rose.

[2]
No title. Endorsed 'Penruddock & Lutterell Case'

A curious map, being a mirror image of [1]. It was apparently drawn shortly after and closely based on the earlier map. It copies the same features but some information is different and the writing is more legible. Only new information is noted here.

The mill is marked 'Corn Mill' with its 'Mill Streame time out of minde'. To N of the forge is 'Forge Mead Mr Carrs'. In the mill stream is 'Ford rendered not passable'. Opposite Strangworth House is a side stream letting water into the spoiled meadow marked 'this Sluce is to water this mead & is by them stopped w[i]th turves', the Mill Meadow to its N being bordered with a '[Ban]cke raised by them in the ten[an]ts time'.

Simple compass rose.

SCALE: Both maps [*c.*1:2112].

SIZE: [1] *c.*310 x 605. Ink and watercolour on parchment. Crumpled, with the original bright colouring faded. [2] 290 x 440 with an attached small extension 135 x 90. Ink and watercolour on paper. Badly torn.

SURVEYORS: Not named.

REFERENCE: Private.

1630
ACONBURY, HOLME LACY

Endorsed 'Juillet 2. 1630. Modelle des terres entre moy & caldicott'

A simple outline of nine adjoining plots of land [*c.*7½a.], apparently on both sides of a road or track at Caldecott [SO 526327]. Field-names include Plamsles meadowe, Peasefyeld, Knell Wood, Heathen Hall [3a.], all part of 'My Lo[rd's] land' of the Holme Lacy estate.

SCALE: [*c.*1:1530].

SIZE: *c.*165 x 100. Ink on paper.

SURVEYOR: Not named

REFERENCE: PRO, Kew, C115/109/8939. Among Chancery Masters' Exhibits relating to the Scudamore family estates.

n.d. [mid 17th century]
WIGMORE

'[PL]AN of a Survey of [WIG]MORE LODGE ...'
[Lodge Farm] in NW of the parish [SO 388692]. Shows buildings in bird's-eye view, enclosed fields, hedges, gates, woods, stream, ponds, roads. Marks Fish Pool, Darvell [Deerfold] Common, Park Gate, Rowley Gate [on boundary with Wigmore Rolls]. Field-names include The Park, Burnt Copy, Old Hopyard, Moor Bank. [N at top].

Title cartouche with drapes. Scale bar. Double ruled border.

SCALE: 4 chains to 1 inch [1:3168].

SIZE: *c.*392 x 495. Ink and watercolour on paper, mounted on a parchment deed. Map torn with both left and right edges partly missing.

SURVEYOR: Not named.

REFERENCE: Private.

1662
DOWNTON

'A SVRVEY OF MOCKTREE FORREST AND CHASE OF BRINGE=WOOD THE INHERITANCE OF THE RIGHT HO:ble WILLIAM LORD CRAVEN BARON OF HAMSTED MARSHALL All which landes are scituate in the County of HEREFORD' (Plates 30, 37)[1]

1,509a. in the chase of Bringewood 'as it is now enclosed' and 672a. in the former Mocktree Forest. Shows buildings in perspective, fields numbered and outlined in colour, woods, river Teme, Portman Lane leading into Ludlow and other roads. Names adjoining properties. Marks Burrington and Aston churches, Lodge House at Mocktree, Richard Moore's paper mill [SO 433735] and Bringewood forge and furnace [SO 454750]. Inset, Evenhay [2a.] adjoining Aston Common.

Reference table gives names of tenants, names and descriptions of fields, acreage. In Bringewood field-names include Park yate lessowes, Erles Arbor and Dog Hanging [SO 460740], Fire Piece, Bringewood Heyes. In Mocktree they include Rampart Louse [SO 453755], Gravelly Bank, Lodge lessow. Notes that 'Mr Frazer Walker Houldeth The Furnace, Forges & land thereto belonging'.

Cartouche with surveying instruments and two shelves of books within a baroque frame. Memorandum relating to three properties at Bringewood, including a paper mill. Table of lands at Bringewood Chase 'as now enclosed' and at Mocktree.. Two compass roses. Scale bar with dividers. Double-ruled border.

SCALE: *c.*48 perches to 1 inch (1:9500).

SIZE: 645 x 780. Ink and watercolour on parchment.

SURVEYOR: William Fowler.

REFERENCE: HRO, T87/1.

1. Map made for Lord Craven soon after its sale by the Crown. See D.G. Bayliss, 'The effect of Bringewood forge and furnace on the landscape of part of northern Herefordshire to the end of the seventeenth century', *Trans. WNFC*, vol. 45, pt. 3 (1987), pp.721-29; P. Cross, 'A 1577 plan "A platte of part of the Chase of Bringewood and of certayne groundes adjoining leased to Mr Walter of Ludlowe". A re-evaluation of the landscape', *ibid.*, vol. 48, pt. 3, [1996], pp. 573-81. See also 1577 BURRINGTON.

(1677)
HEREFORD

'The Ground plot of the Castle of Herefford'

Endorsed 'Hereford Castle designed by John Silvester Nov 17 77' [SO 512396]. Shows the castle precincts with the passages to Castle Street and the College [of the Vicars Choral]. Marks the circuit of the castle walls and within them, 'the Hill ... which the Great tower stood on', divided from the bailey [Castle Green] by a wall with 'a long building Mr Holland pulled down'. In the bailey is marked 'the Old Gate House', the dwelling house and 'that part of the house that was burnt' [Castle Cliffe House], the Mount and the Great Mount at the NE and SE corners of the bailey.

A proposal for moving the market from the city centre and building over part of the site of the castle. Running parallel inside the walls of the bailey are pecked lines marking a proposed street within the walls with houses and gardens on its inner side. The central area was intended for the market place with a granary and in the western part a new church [but not on the site of the former church of St. Guthlac].[1]

A reference table with the landmarks lettered A to L. At the end of the table, beneath the reference to the market in the central area, the word 'Granary' has been added in a different hand.

SCALE: 10 perches to 2½ inches [1:792].

SIZE: 450 x 295.

SURVEYOR: John Silvester.

REFERENCE: HCL, Pilley Collection 2326; copies, not identical and one dated 'G.L.O. 1910' at 914:244 (Large Print box).

1. F.G.Heys, 'Excavations at Castle Green, 1960: a lost Hereford church', *Trans. WNFC*, vol. 36 (1960), pp. 343-357.

[c.1680]
RICHARDS CASTLE

No title.[1]

The Byrry House [Bury] estate in the S of the parish [SO 497688]. 280a. enclosed arable land, measured to the nearest ten acres, and about 120a. meadow and pasture, not measured. Shows buildings in perspective, cottages, two with the names of occupants, orchard, hedges, fences, trees, streams, roads with destination. Marks The Byrry House prominently out-of-scale, with a barn and fenced yard with two gates, 'The fysh Pool', 'Ricards Castell' and church adjoining the castle, Dysbury Brook and the county boundary between Herefordshire and Shropshire on Clay Hills Brook. Field-names include Ullinghall Field, divided into four closes, Myddle Field, divided into two closes, Clay Hyll Field and Meadow, Ox and Cow Leasows.

Cardinal points written in the double-ruled border. Boldly coloured, now faded to pink and pale green. Crude draughtsmanship.

SCALE: [*c.*1:3730].

SIZE: 544 x 753. Ink and watercolour on paper, repaired.

SURVEYOR: Not named.

REFERENCE: Shropshire RO, 1141/190.

1. Dated by Shropshire RO. But the handwriting suggests a date generation earlier and the draughtsmanship is naïve. When compared with the maps of nearby Bringewood Chase it is more like 1577 BURRINGTON than 1662 DOWNTON.

1686
WELSH NEWTON

'Pembridge Castle & the Demesne thereof' (Plate 43)

302a. surrounding Pembridge Castle in N of the parish [SO 488192]. Marks the castle, outbuildings and house in bird's-eye and perspective views, trees, woods, a limekiln, roads. Field-names include Brick Field, Reaves Park, Warren Closes, Fishpool Closes.

Compass rose. Scale bar with dividers. Plain border.

Reference table on a separate sheet with the map, giving field-names, acreage and rents.

SCALE: 21 perches to 1 inch [1:4158].

SIZE: 420 x 534. Ink and watercolour on paper. Torn along folds. Formerly pp. 110-113 [in top margin] and pp.34-37 [in bottom margin] of a book.

SURVEYOR: W. Hill.

REFERENCE: HRO, C82/1-4.

1695
HOLME LACY ESTATES

Ten maps of the estates of the Scudamore family of Holme Lacy made in or about 1695. Some are signed by John Pye and the others may be attributed to him on the grounds of style or other internal evidence.[1]

The estates of the Scudamore family of Holme Lacy were surveyed in 1695 by John Pye, shortly after John 2nd. Viscount Scudamore made his will in 1694 but before his death in 1697 and in 1771-5 by Richard Frizell after the marriage of Frances Scudamore to Charles Howard, later 11th Duke of Norfolk. Other single maps were made of particular lands in 1723 by George Smyth, in 1729 by Robert Whittesley, in 1785 by John Harris and at the turn of the 18th century by anonymous surveyors. The archive of these 63 maps, together with 39 other miscellaneous maps of estates mostly in SE England, 1771-1884, were presented by Mr. D.H. Barry to the British Library in 1900 (Add. MS. 36307). Endorsements on some of the maps indicate that they have been numbered and renumbered at least twice and that at one period they were listed as part of a larger series of over 150 maps. A few stray documents of the Scudamore estates in Herefordshire, including two maps of lands in Much Cowarne 1824, are at Arundel Castle, Sussex (F.W. Steer ed., *Arundel Castle archives,* vols 2, 1972, and 3, 1976).

[1] 1695
BALLINGHAM, FOWNHOPE, KING'S CAPLE

'A SURVEY of Sr Barnabas Scudamore's land in BALLINGHAM CARY, and Severall other Lands and Woods on the other side Wie in the County of Hereford Surveyed Ano Dni 1695 by John Pye'[1]

E part of the parish of Ballingham [SO 5832] adjoining the river Wye with woods and meadows on the left bank of the river in King's Caple. Shows buildings in block plan, except for the church in perspective, enclosed fields outlined in green, orchards, woods, river Wye as 'Vaga Flu:', roads. Gives plot numbers. Field-names include Quarell Close and Quarell Field W of the church, The Sheep Summer Fatts Pasture, The Ox and Sheep Berevats Pasture, Werehead Meadow and Were Meadow [SO 580313], Mill Meadow [SO 572308], and on the left bank of the river Wye on the parish boundary between King's Caple and Fownhope is Forge Meadow [SO 571306] and Carey Wood.

Decorative oval title cartouche. Scale bar with dividers. Compass rose. Engraved decorative border pasted on.
[No reference book or table].

SCALE: 2 chains to 1 inch [1:1584].

SIZE: 882 x 1583. Ink and watercolour on parchment.

SURVEYOR: John Pye.

REFERENCE: BL, Add. MS. 36307, G3.

1. Sir Barnabas Scudamore had died in 1652.

[2] [1695]
BALLINGHAM, BROCKHAMPTON, FOWNHOPE, KING'S CAPLE

No title. Probably a draft for the preceding map [British Library, Add. MS. 36307, G3].

Most of the parish of Ballingham and parts of the parishes of Brockhampton, Fownhope and King's Caple on the left bank of the river Wye from Capler Wood to below Forge Meadow [SO 571306]. Shows all buildings, including the church, in perspective. Gives field-names and acreage. Names adjoining owners in red.

Small simple title cartouche, left blank. Compass rose. Ruled border.

SCALE: [1:1584].

SIZE: *c.*787 x 1575 maximum. Ink and watercolour on parchment of irregular shape, partly due to the shape of the parchment skin and partly because the corners have been trimmed. Stained.

SURVEYOR: Not named. [John Pye].

REFERENCE: BL, Add. MS. 36307, G4.

[3] [1695]
KING'S CAPLE

No title. Endorsed 'Map of the Estates late of Sir Barnabas Scudamore in King's Capell' and 'Pyes Map of Kings Caple 1740'[1]

Scattered lands, mostly enclosed, in the centre and W of the parish. Shows the church [SO 559288] and buildings in perspective, the houses coloured red, fields in crude outline with acreage and some field-names, river Wye, roads. Gives names of adjoining owners in red. Field-names include Cary [Carey] Wood, Castle Field [SO 562294], Court Ries Meadow.

SCALE: [*c.*1:5510].

SIZE: *c.*740 x 770, irregular. Ink and watercolour on parchment. Grubby and wrinkled.

SURVEYOR: [John Pye].

REFERENCE: BL, Add. MS. 36307, G49.

1. The first endorsement was probably contemporary. The second endorsement may have been added at the time Richard Frizell's surveys in 1771-5. It is clearly incorrect. No maps by Pye have been traced after 1704; he died in 1731.

[4] [c.1695]
LITTLE DEWCHURCH
No title. Endorsed in a later hand ' Pothether Wm Pyes Survey'

Area in SE of Little Dewchurch and the N of Hentland (Hoarwithy) [SO 536300]. Shows buildings in crude perspective, including mill at Tresseck, field boundaries in outline, streams, roads. Gives field-names, land-use and acreage. Names adjoining owners in red. Field-names include The Witch, Upper and Lower Weaven Meadow and Weaven Copse.

SCALE: [*c.*1:3960].

SIZE: *c.*475 x 530 maximum, irregular. Ink and watercolour on parchment.

SURVEYOR: [John Pye].[1]

REFERENCE: BL, Add. MS. 36307, G18.

1. The attribution to William Pye in the later endorsement is mistaken. The style of the map is similar to that of John Pye. No surveyor called William Pye has been recorded.

[5] [c.1695]
HENTLAND (HOARWITHY)
No title. Endorsed 'Treseck'

10a. in scattered closes including Ballows House adjoining 'Sir John Hoskins Coppice' [atTresseck in the N of the parish, SO 542296]. Shows buildings in crude perspective, field boundaries in outline with field-names, land-use, acreage, roads. Field-names include The Aspes, Plum-tree Field, The Red Combe.[1]

Ruled border.

SCALE: [*c.*1:5760].

SIZE: 275 x 440. Ink on parchment.

SURVEYOR: Not named. [John Pye].

REFERENCE: BL, Add. MS. 36307, G59.

1. Sir John Hoskyns of Harewood, the parish adjoining Hentland on the NW, succeeded his father as the second baronet in 1680. He died in 1705. The map is drawn in John Pye's style.

[6] [c.1695]
HOLME LACY, BALLINGHAM, BOLSTONE
No title. Reference table entitled 'A Copy of the Particular of the Lands contained in this Map'. Endorsed 'Xed [Examined] with new Survey'.

663a. in the S of Holme Lacy, N of Ballingham and part of Bolstone, including Hallanton [Hollington] Farm in Holme Lacy [SO 562338], Hancocks Mill Farm in Ballingham [SO 570325], Garolds Farm (49a.) [?Bolstone Court Farm] and Pritchards Farm. Shows buildings in block plan, fields outlined in colour with plot numbers and field-names, hopyard, trees (spaced out in Shepherds Rough), river Wye as 'Vaga Flu:', roads. Marks old (and new) farm houses and Church Close at Garolds Farm, decoy and Decoy House [SO 568340]. Field-names include Court Close at Hollington Farm, Blackwells Ditch along the river Wye N of Kidleys Hill, Upper, Middle and Lower Vineyard [SO 561335].

Two reference tables giving [1] plot numbers, field-names, acreage and [2] names of tenants.

Scale bar with dividers. Compass rose. Reference tables with ruled borders. Engraved decorative border pasted on.

SCALE: 3 chains to 1 inch [1:2376].

SIZE: 725 x 950 maximum, with rounded left-hand side. Ink and watercolour on parchment, formerly attached to a roller.

SURVEYOR: Not named. [In the style of John Pye].

REFERENCE: BL, Add. MS. 36307, G63.

[7] [c.1695]
LLANGARRON
No title.

[Lower Tre-Evan in the N of Llangarron, SO 522224]. Shows buildings in block plan, scattered closes (47a.) with boundaries in outline, field-names, land-use, acreage, roads. Gives names of adjoining owners in red. Field-names include Tredwarra [SO 530222], Llannock Hill [SO 527219].

N and S written in margins.

SCALE: [*c.*1:3960].

SIZE: 455 x 485. Ink and a little watercolour on parchment.

SURVEYOR: Not named. [In the style of John Pye].

REFERENCE: BL, Add. MS. 36307, G58.

[8] [c.1695]
LLANWARNE
No title. Endorsed 'Blewhenston'.

51a. in the NE of the parish of Llanwarne [SO 526298]. Shows buildings in perspecive, field boundaries in outline with field-names, land-use, acreage. Gives names of adjoining owners in red. Field-names include Sandy Way, White Wall Field, The Pike, Lawater.

SCALE: [c.1:3960].

SIZE: c.370 x 480 maximum. Ink and watercolour on an irregular scrap of parchment. Incomplete.

SURVEYOR: Not named. [In the style of John Pye].

REFERENCE: BL, Add. MS. 36307, G13.

[9] [c.1695]
WELSH NEWTON

No title. Endorsed 'Mapp of Newton' [contemporary], 'fo. 46 in Map book' [later], 'Welsh Newton' [19th century], 'No 150' [later still].

83a. N of the village of Welsh Newton. Shows buildings in block plan including [Newton Farm] immediately N of the church [SO 500180], with field-names and acreage. Marks the church. Field-names include Crossouk [? Crossoak SO 497183], The Meadow at the Well [SO 498181], New Ditch [SO 499183], Pulpenston to the NE of the church.

SCALE: [c.1:2880].

SIZE: 395 x 620. Ink on paper. Repaired and watermark not visible.

SURVEYOR: Not named. [Probably John Pye].

REFERENCE: BL. Add. MS. 36307, G60.

[10] 1695
TRETIRE, ST WEONARDS

'Parish of Tretire and allso the Lands at Llangunnock thereunto belonging…And allso the Lands at Trepenkennet Surveyed Ano Dni 1695 by John Pye'

[White House] in the centre of of the parish of Tretire [SO 519244] and two outlying plots at Llangunnocks Lower Mead and Trepenkennet in St Weonards. Shows buildings in block plan, fields outlined in colour with field-names and land-use, road.

Endorsed 'Folio 26 in the Map Book X[d] with New Survey'. 'No 152'.

Scale bar. Compass points in margins. Ruled border. Pencilled bearings and projections.

SCALE: 3 chains to 1 inch [1:2376].

SIZE: c.626 x 656. Ink and watercolour on parchment.

SURVEYOR: John Pye.

REFERENCE: BL, Add. MS. 36307, G56.

1698
GARWAY

'The Darrein, Hasell Field and other Lands Admeasured for George Lewis Gent'

70a. in S of the parish and W of river Monnow. Describes Daren [SO 468210] and Hazelfield farms [SO 475199] as the 'Total of Mr Bennets Land in the Map is 56 Acr: 1 Ro: 14 Perches'. Shows farm houses in perspective, fields outlined in colour. Marks St Michael's Well in Chapel Pleck and Our Ladye's Well. Gives names of adjoining owners. Field-names include Quarry Close.

Red and blue lettering. Compass rose. Scale bar. Coloured frame.

SCALE: 4 chains to 30/32th of 1 inch [1:2970].

SIZE: 220 x 500. Ink and watercolour on paper. Slightly stained and faded. Repaired by HRO 1972.

SURVEYOR: Thomas Croft.

REFERENCE: HRO, N87.

n.d. [late 17th century]
GOODRICH

'A Mapp of Thomas Fletchers land'

Pryors Field (14a.) on SE of road between Williams Cross [Goodrich Cross] and the road to Monmouth [SO 565186]. Outline of fields with a cottage and [?] a spring, including 1a. enclosed from roadside waste along road to Hunsom [Huntsham].

N and E points indicated in margins.

SCALE: [c.1:2534].

SIZE: 420 x 320. Ink on paper.

SURVEYOR: Not named.

REFERENCE: HRO, O68/Maps/1.

(1700)
STOKE PRIOR

'Copy MAP OF THE FORD ESTATE'[1]

216a. comprising Ford Farm [SO 512553] at Ford, formerly an extra-parochial area SW of Stoke Prior. Shows buildings in bird's-eye view, enclosed fields and some strips in

open fields, meadow, orchards, woods, coppice, gates. Marks dovecote at Ford Farm, river Lugg. Gives acreage and field-names, including Bridge Meadow, Mill Meadow, the Chair.
Reference table.
Compass rose. Scale bar. Plain border.

SCALE: 17 perches to 1 inch [1:3366].

SIZE: 525 x 590. Ink and pale watercolour on paper.

SURVEYOR: Thomas Cleer.

REFERENCE: HRO, A63/III/49.

1. Copy made by J. Yates c.1840 from a map lent to IA (?John Arkwright of Hampton Court).

n.d. [c.1700]
GARWAY

'An exact Mapp of the Estate of Sr WILLIAM COMPTON Bar^t call'd COMMADOCK In the Parish of Garway Com: Heref: Survey'd by Jo: Pye' (Plate 44)

Cwm Maddoc Farm in S of the parish [SO 479208]. Shows buildings in block plan, fields outlined in colour, orchards, roads. Gives some land-use. Marks avenue of trees NW of farm house. Field-names include Tuck Mill, Limekill Close, Nantawhaines [Nantywain] Wood.
Scale. Compass rose. Printed ornamental border pasted on.

SCALE: 3 chains to 1 inch [1:2376].

SIZE: 945 x 700. Ink and watercolour on parchment, attached to roller.

SURVEYOR. Jo[hn] Pye.

REFERENCE: HRO, AL40/7687.

n.d.[c.1700]
TARRINGTON

No title.
SE part of the parish [SO 6239]. Shows houses, fields, mostly enclosed, between Tarrington and Tarrington Common to the W and Durly [Darlow] Common and East Field to the E, roads. Gives field-names. Marks Worlds End House [Alders End], Tarrington Common and Radlow [hundredal meeting place] in Tarrington.
Compass rose.

SCALE [c.1:4526].

SIZE: 345 x 450. Ink and watercolour on paper. Poor condition.

SURVEYOR: Not named.

REFERENCE: HRO, E3/459.

1704
GOODRICH

'A Survey of the Priory Farme in the parish of Goodrich. Com Heref Taken Anno Domini 1704 By John Pye and plotted w.th a Scale of 3 Chaines or 12 Perches in an inch'

Priory Farm in NE of the parish [SO 579195]. Shows Goodrich Castle in perspective [as a crude drawing of the gate-house], Flanesford Priory in block plan, field boundaries in outline, orchard with pool at Flanesford Priory, river Wye, roads including a lane to a river crossing at the site of Kerne bridge. Gives acreage, land-use and field-names, including Sheppards Hill, Priory Meadow, Bannut Tree Close and Well Close.
Simple oval cartouche. Compass points written in ruled border.

SCALE: 3 chains to 1 inch [1:2376].

SIZE: 610 x 555. Ink and brown wash on parchment. Torn.

SURVEYOR: John Pye.

REFERENCE: HRO, O68/Maps 6 [LC 4429].

1705, 1708
OCLE PYCHARD, AYMESTREY, SHOBDON

A series of four maps of the estates which Sir James Bateman bought from Robert Chaplin of London in 1705, surveyed by William Whittell of Bodenham. The maps are similar in style but of different size and scale.

[1]

'A new Mapp and Survey of ye Mannor of LIVERS OCLE ... for ye Worshipfull Robert Chaplin Esq present owner Thereof' ' Measured by Mee William Whittell: A:D:1705 The estate Contains in Length One Mile and almost 3 quarters, in Breadth Half a mile half a quarter &c (?) Sevrall Content 457 Acres'

457a. comprising Livers Ocle estate [SO 577464] in W of Ocle Pychard on both sides of the Hereford-Bromyard road. Shows houses in perspective or bird's-eye view, fields outlined in colour with field-names and acreage, mostly enclosed but including a few strips in The Gore Common Field, orchards, ponds, brook on N boundary. Marks the road from Hereford to Bromyard before realignment and houses at crossroads at Harestone [SO 565458]. Field-names include New Inclosure, Coninger, Fish pooll Leasow.

Inset drawing within a decorative border incorporating a human face, 18a. in Lugg Meadow.

Coat of arms of Chaplin. Compass rose. Scale bar with dividers. Ruled border in yellow.

SCALE: 6 chains to 1 inch [1:4752].

SIZE: 410 x 620. Ink and watercolour on parchment.

SURVEYOR: William Whittell.

REFERENCE: BL, Egerton MS 2873/A.

[2]
'A Map of UPPER LYE in the Parish of ALMESTREY ... Belonging to the RIGHT WORSHIPFUL SIR JAMES BATEMAN, K.ᵗ and ALDERMAN of the CITY of LONDON 1708'

Lands at Upper Lye in NW part of Aymestrey [SO 395657]. Shows buildings at Upper Lye in bird's-eye view, coloured as if brick-built with slate roofs, formal garden, fields outlined in colour with field-names and acreage, some strips in open fields, river Lugg. Marks road bridge [Darley Bridge] W of Shirley Farm, ford near SO 390651 and footbridge on site of Lyepole Bridge [SO 398654], circular ditch and motte [SO 396654]. Gives names of adjoining owners. Field-names include Darleys Bridge Leasow, Meen Knowle [The Gamp]. Note explaining lettering of M[eadow], P[asture] and B for lands let to William Bedoes.

Coat of arms of Bateman and Searle impaled. Title cartouche in wreath. Compass rose. Scale bar with dividers. Yellow and black ruled border.

SCALE: 17 perches to 1 inch [1:3366]. [In later hand] 3 1/2 chains [1:2772].

SIZE: 673 x 815. Ink and watercolour on parchment.

SURVEYOR: Not named. [William Whittell].

REFERENCE: BL, Egerton MS 2873/B.

[3]
'A MAP of LOWER LYE in the Parish of ALMESTREE ... Belonging to the Right Worshipfull Sʳ JAMES BATEMAN, Kᵗ & ALDERMAN of the CITY of LONDON 1708' 'MEASURED by mee WILL: WHITTELL' (Plate 31)

Lands at Lower Lye [SO 406669] in NW of Aymestrey including the Snead Coppice (218 statutory acres or 135 customary acres). Shows houses in bird's-eye view, coloured as if brick-built with slate roofs, fields outlined in colour with some strips in open fields and some field-names. Gives names of adjoining owners. Marks river Lugg. Field-names include Oakley Common NW of Lower Lye, Speare Common Field, Dickindale Meadow. Note explains red lettering of M[eadow] and P[asture].

Coat of arms of Bateman and Searle impaled. Title cartouche, incorporating a scroll with peacocks and a human face. Compass rose. Scale bar with dividers. Ruled pink border.

SCALE: 16 perches to 1 inch [1:3168].

SIZE: 680 x 755. Ink and watercolour on parchment.

SURVEYOR: William Whittell.

REFERENCE: BL, Egerton MS 2873/C.

[4]
'A MAP of the WOODS in the Parish of SHOBDON, and other LANDS in COVENHOPE in the PARISH of ALMESTREE ... Belonging to the RIGHT WORSHIPFULL S.ʳ JAMES BATEMAN K.ᵗ 1708'

Shobdon Hill [SO 390644] and Caldoe woods (364 statutory acres or 225 customary acres), The Long Yards (127 statutory or 79 customary acres), Chaf Wood (28 statutory or 17 customary acres), Putnell Coppice (32 statutory or 20 customary acres) S of Covenhope and strips in open fields N of Covenhope. Shows houses in bird's-eye view, woods and some roads (as pecked lines). Marks Lye Ford [at site of Lyepole Bridge, SO 398654]. Gives field-names and acreage and names of adjoining owners. Field-names include The Rise [at Mortimer's Cross], Wasley Common, Covenhope Lawn N of Mortimer's Cross, Adeley Common NW of Covenhope, The Mear Hill.

Note 'SHOBDON HILL <inserted the little vallet> and ELBATCH VALLET lyeth in the Parish of SHOBDON. The rest of the WOODS and LANDS are Situate in the Parish of ALMESTREE

    The Characters distinguishing the Lands are:
        Red M   Signifies Meadow
        Red P   Signifies Pasture
                  Arable is distinguish'd by small Red lines in
                    the Form of Ridges
              Woods are Describ'd by small Trees
        Lands mark'd [A] are those late Hardway's
        Lands mark'd [B] are those in Possession of
           Rich: Cooke'

Coat of arms of Bateman and Searle impaled. Compass rose. Scale bar in frame with a human face and dividers. Explanatory note in yellow frame with a face at the top.

SCALE: 14 perches to 1 inch [1:2772].

SIZE: 690 x 820. Ink and watercolour on parchment.

SURVEYOR: William Whittell of Bodenham.

REFERENCE: BL, Egerton MS 2873/D.

1709
KINGSLAND

'A Map of KINGSLAND ... Belonging to the RIGHT WORSHIPFUL S.r IAMES BATEMAN, Knight And ALDERMAN of the CITY of LONDON'

Scattered lands spread across most of the parish. Shows church [SO 447613] in bird's-eye view, a few buildings including Kingsland mill, open and enclosed fields, river Lugg, Caniter [Kenwater] and Pinsley brooks, roads. Field-names include Shodgly, Kindon and Caseny Common Fields, Longate and Howgate at E end of village, Fair Field S of church, Sinders [SO 455605], The Mouse Snatch, Wegnall, Kingsland Great Common Meadow.

Title cartouche of a frame with a human face. Coat of arms of Bateman and Searle impaled. Scale bar in a frame with dividers, a human face and the surveyor's name and date. Compass rose. Red and yellow ruled frame.

SCALE: 34 perches to 1 inch [1:6732] or 4³/₄ inches to ½ mile [1:6670].¹

SIZE: 657 x 833. Ink and watercolour on parchment, a little crumpled.

SURVEYOR: William Whittell.

REFERENCE: HRO, G39/Maps. [Uncatalogued collection not available for research, January 2002].

1 These are the surveyor's figures. Comparison with the Ordnance Survey map indicates a scale close to 1:6733.

1709
WESTON UNDER PENYARD

'A Mapp of an Estate of Samuell Merrick Gentl' lately purchased from Ferdinando and Henry Georges Esq.s Scituate Lyeing & being in the Parish of Weston under Penyard in the County of Gloucester Measured and described in the yeare 1709 by Job Gilbert which s[ai]d Estate is called the upper & Lower Rudge'¹

66a. in the SE of the parish adjoining Lea Bailey. Shows buildings in profile, including the twin-gabled house at The Rudge [SO 643208] with farm buildings, with tiled or stone-slated roofs, old garden, enclosed fields with acreage and land-use, old and new orchards, isolated trees, gates, ponds, and the road from Hownhall and Ross to Lea Bailey. Field-names include Demcocks Hill, The Butts and Inedge. Describes boundaries of the estate with names of some adjoining owners, including Jonathan Ennis Lands [near Ennis Well SO 639205] and The Tumsses [SO 639206].

Title in a rectangular double-ruled frame. A similar frame with the note 'The whole tot[al] of this Mapp is sixty six Accres one Rood and fifteene Perches – Orchards and Gardens deducted it is but sixty Accres one Rood & thirty seven Perches'. Compass rose with a star for the cardinal points (S at the top) and quadrants marked out South West, Sout East [sic], North East, North West. Large scale bar of 'A Scale of Forty Statute Perches' with dividers. Ruled border of two pairs of double lines.

SCALE: 8 perches to 1 inch [1:1584].

SIZE: 695 x 525. Ink and watercolour on parchment.

SURVEYOR: Job Gilbert.

REFERENCE: Somerset RO, DD/GC 93.

1. Weston under Penyard has always been in Herefordshire, originally as a chapelry of Ross. It became a parish in 1671 and in 1883 and 1965 small neighbouring parts of Lea Bailey and Mitcheldean, Gloucestershire were transferred to it. The Rudge, however, lay in the ancient Herefordshire parish, adjoining the boundary with Gloucestershire.

1709
WIGMORE, BURRINGTON, LEINTHALL STARKES

'A MAP of certain Lands in the PARISHES of WIGMORE, BURRINGTON and LONG LENTHAL Belonging to the RIGHT WORSHIPFUL Sr JAMES BATEMAN K.t Alderman of the City of LONDON'

c.54a. of scattered pasture, meadow and wood in Wigmore, Burrington and Leinthall Starkes, including Bonds Vallet, part of Wigmore Common Wood [most probably Woodhampton Wood, SO 4067]. Shows woodland and hedgerow trees, gates, river Teme in Burrington.

Cartouche, incorporating a human face. Coat of arms of Bateman and Searle impaled. Scale bar with the surveyor's name and date within a frame. Compass rose. Yellow and red ruled border.

SCALE: *c.*8 perches to 1 inch [1:1584].

SIZE: 695 x 773. Ink and bright watercolour on parchment, backed with paper.

SURVEYOR: William Whittell.

REFERENCE: HRO, G39/Maps. [Uncatalogued collection not available for research, January 2002].

[1716]
BODENHAM, HOPE UNDER DINMORE

No title. Lands on the boundary between Bodenham and Hope under Dinmore [SO 523518]. Shows a pasture belonging to Lady Scudamore adjoining Great Bunhill Field and Kynetts Croft in Bodenham and a wood called The Councells in Hope under Dinmore belonging to Lord Coningsby. Marks a gated lane forming the parish boundary, a spring rising in the pasture [the spring at SO 522518], prominent oak and willow trees and pleached hedges, one of which was the source of the trespass in Lord Coningsby *v* Daniel Knight and others, tenants of Lady Scudamore.

No scale bar. No N point, but south is at the top. At the foot is a key to the landmarks noted by letter references. Plain border.

SCALE: [*c.*1:1056].

SIZE: 377 x 226. Ink and watercolour on paper.

SURVEYOR: Not named.

REFERENCE: PRO, Kew, C115/97/7084.

(1716)
HEREFORD

'An Exact Survey of the City of Hereford'

94a. Primarily a plan of the castle and city defences. Gives an outline map of the castle [SO 512396], city walls, bastions and gates, river Wye and Eign brook. There is a prominent rectangular wall bastion W of Bysters Gate, but the remainder, some of which survive, are semi-circular. Marks, but does not name, the cathedral, churches of All Saints, St. Nicholas and St Peter, indicating their arcades. The Wye bridge is shown with 5½ arches, and (only on the copy) Dogge Mill and Castle Mill.

No other buildings or streets are marked, but the highest part of the city is indicated about midway between All Saints' and St. Peter's churches. Within the castle are the Governor's House [not named], the Citadel or Castle, the Powder House 'as I suppose', 'The Great Gate of the Castle' leading to 'the Bridge of the Castle Yard' over the moat [Castle Pool], from which pecked lines are drawn to the city gates [named].

SCALE: No scale bar. *c.*21 inches to 1 mile. [1:3017].

SIZE: 324 x 267. Ink on paper.

SURVEYOR: [James Hill, 'The Antiquary'].

REFERENCE: Hereford City Library, L.C.942.44/AC, Hill MSS., Vol. 4. A copy in Hereford City Library, Pilley Collection 2326 refers to the Hill MS, 'The original plan from which this was copied is deposited in the Phillipp's Collection of MSS in the Library of St. Michael's Priory, Belmont, Hereford. W[alter] Pilley' [of The Barton, Hereford].

[1717]
GOODRICH

No title. An estate atlas of maps (pp. 1-29), mostly numbered and each with a separate title, with a terrier (pp. 31-67, some blank).

Lands throughout the whole parish of Goodrich [SO 5719], with small adjoining areas of Ross, Walford, Whitchurch, Ganarew, Llangarron, and Newland [Glos.]. Common features are buildings shown in crude perspective or profile, fields outlined in colour with field-names and reference numbers, hopyards, woods, roads. Most of the land is enclosed or named as common with a very few strips in open fields. Indicates cliffs and rocks. Marks the bounds of the Forest of Dean as a jagged line. Gives names of adjoining owners.

Each map has a bold title cartouche with zigzag decoration, unless indicated otherwise, and a compass rose.

1

'The Mannor of Goodrich ... beginning with the most eastern part of the Parish of Goodrich and the North West part of the Parish of Ross'

Shows scattered lands mostly on the right bank of the river Wye between Wilton Bridge and Goodrich Passage [the ferry at SO 576202]. Marks Wilton Bridge. Field-names include Old Wear Head below Wilton, Pencreak's [Pencraig] Green Common, Bannut Tree Close. Oval title cartouche. 2 pages.

2

[Out of order. See below].

[3]
'The Man'or of Goodrich Continued being in the North East & South East part of the Parish of Goodrich and the most Southern part of the Parish of Walford'

Lists names of 13 encroachments [8a.] at Copped [Coppet] Hill Common. Marks Byfield island in river Wye [SO 569177]. 2 pages and fold, 470 x 1460.

4
'The Mannor of Goodrich Continued being the North & by East & South and by West parts of the Parish of Goodrich'

Shows river Wye with Hunsons Passage and Boat House [the ferry at Huntsham bridge], Byfield island, riverside meadows including Lower Wear Homme, Priory Homme and Priors Field, Garron brook with Old Forge on the brook, New Mill Hill Common, Hangmans Acre and adjoining Warret-tree Field. 2 pages.

[5]
'The Mannor of Goodrich Continued being the Southern part of the Parish of Goodrich'

Shows Hunsons [Huntsham] Hill Common, with 2 encroachments [1a.] and river Wye with rocks.

Oval title cartouche.

[6]
The Mannor of Goodrich Continued being the most South & South west parts of the Parish of Goodrich'

Shows Coldwell Hill Common, Weare Hill Common and river Wye. Marks 'Rocks near Symons Yat', Hollow Rock, Long Stone Rock, and Minors [Miners] Path' to New Weir ,with a cross- reference to [a missing map on] p. 11.

7
'The Mannor of Goodrich Continued being in the eastern and Western parts of the Parish of Goodrich'

Area around the village. Marks the church, William's Cross [Goodrich Cross], a sheepcot by the roadside and a prison at crossroads [SO 574195]. Field-names include Sheppards Hill Common, [Marstow] Hill Field, Smith Close [site of a former smithy] at SO 573195.

2
'The Mannor of Goodrich being the most North East part of the Parish of Goodrich' (Plates 45, 46)[2]

Marks Goodrich Castle and pound opposite the castle entrance, Goodrich Passage, Flanesford Priory [not named] with river landing place.

8
'The Mannor of Goodrich Continued being the most Southern part of the Parish of Whitchurch and the most South East part of the Parish of Ganarew'

Marks Whitchurch Street and mill, The Criggals [SO 544170], Banisters Meese Place, Boson-holes [at about SO 530170], Goldsmiths Wood. 2 pages.

9
'The Mannor of Goodrich Continued being in the Southern part of the Parish of Langarran'[3]

Shows Longrove Common and Old Grove Common with 16 encroachments [6a.].

10
'Lead-brooks Mill with the Adjacent grounds in the Parish of NEWLAND in the County of GLOUCESTER'.

Shows a small area [at Lydbrook]. Ruled border. 210 x 265.

Terrier
Gives names of tenants in alphabetical order, abstracts of leases, with leases for lives dated up to 1715, description of properties, land-use and acreage, totalling 896a. Place-names include Old Priory House [Flanesford Priory], New Weir and Forge, Cutt Mill and The Furnace Piece, Whitchurch, [the site of a former furnace].

SCALE: 40 perches to $3^{1/4}$ inches [1:2137].

SIZE: Volume, calf-bound and tooled. Page size, 470 x 580. The maps are each drawn on one page unless indicated otherwise above. Ink and watercolour on paper.

SURVEYOR: Not named. [Edward Laurence].

REFERENCE: HRO, AW87.

1. Laurence's bill for making a survey of the Credenhill and Goodrich estates for Henry Grey, Duke of Kent in 1717 survives in Bedfordshire RO [L 26/1023]. See pp. 44-45.
2. Laurence published the second issue of *The duty and office of a land steward* in 1731, containing a folding frontispiece of a fictional estate map 'A SURVEY of DUN BOGGS Farm, in the Manor of HAVERSHAM, in ye County of HEREFORD by Edwd Laurence' with the note 'This Survey is according to the Author's Method of Drawing his fair Maps'. This fictional map of Dun Boggs Farm is based upon map 2 of the Goodrich atlas. It shows a castle and lands between a bend of the river Wye and a road to Monmouth with field-names including Priors Grove, Byfields Meadow, Tanners Close and Shepards Hill, all of which are taken from names near Goodrich castle. The representation of boundaries, and the precision of bird's-eye views of buildings, trees with shadows and the rise of Shepards Hill are, however, not found in his Goodrich survey. A pen and ink drawing of 'A SURVEY of Dun Boggs Farm in Hereford Shire' is in the Brtish Library [Maps.K.Top.15.106]. Unlike the printed version the

title cartouche is less ornate and the scale bar does not have the title 'A scale of Perches'. Some of the details of the map also differ from the printed version, including the addition in the bottom left-hand corner of the field-name Dun Boggs Marsh. It would appear to be Laurence's original drawing for the engraving.

3. This map of 'the Southern part of the Parish of Llangarron' [124a.], was copied at a scale of 12 perches to 1 inch [1:2376] in Laurence's style by an unnamed draughtsman, c.1813 [watermark]. HRO, O68/Maps/3.

1718

CLODOCK [LONGTOWN][1], EWYAS HAROLD, WALTERSTONE

A volume of fifteen maps of the estates of George Neville, 11th Baron Bergavenny [1659-1721] in Monmouthshire, Herefordshire and Worcestershire, surveyed June-August 1718. The atlas is now incomplete as fols. 20-43 are stubs, at least eight plans having been cut out, probably c.1800 when some of the estates were sold. Only the six described below relate to Herefordshire.[2]

All the maps show buildings in perspective, fields outlined in colour with field-names and acreage, orchards, woods, gates, rivers, roads, names of adjoining owners. Some maps contain notes about timber on the estates.

The maps are finely drawn with varied floral and other decoration and distinctive colouring, particularly in orange and pale green. The titles of the maps have some rubricated initial letters.

SCALE: 2 chains to 1 inch [1:1584] unless otherwise noted.

fol. 11

'A Draught of the Castle Yard and Green and the Closes joyning to the same here drawn round with a black and red line: with the close called Gwerlod Olphon Lying in Long Sown [sic] in the County of Hereford In the Occupation of John Watkins, also the five other closes joyning to Chrisr Prices Esq.r Estate called Llanfrank lying in ye Parish of Llanvehangel-Cucorney in ye County of Monmouth now in ye tenure or Occupation of Jno: Sanders Esq:r both belonging to ye Right Hono:ble George Lord Bergavenny' (Plate 40)

64a. mostly in Llanvihangel Crucorney and in the centre of Longtown village [SO 321292]. Inset plan of Longtown entitled 'Castle green mote and yards' contains perspective drawing of castle keep, approached between a 'shop' and a house, with garden and outer ditch. The surviving entrance to the north bailey is not shown.

Title cartouche in frame. Compass rose. Scale bar with dividers. Decorated leaf border.

DATED 19 July 1718.

[Davies 1805, No. 7. Field-names include Castle Green, Moat and Yard surrounded by a square ditch, on the S side of which were Great and Little Moat and Gwerlod y Grove. Scale: 1:3168. Size: 100 x 235].

fol. 12

'An Exact Map Of the Estate adjoyning to the Castle Yard Lying in ye P[ar]ish of Ewas:Herald ... belonging to ye Right Hono:ble George Lord Bergavenny now in the Tenure or Occupation of S.r Philip Jackson

82a. Marks Ewyas Harold Castle on a motte [SO 287385], Ditch and Park Piece to NW of the castle, mill on 'Trill flemen' [flumen], Dulas brook. Notes 'There is no timber on the Estate'.

Two decorative cartouches. Scale bar with dividers. Compass rose. Drawings of sixteen individual flowers.

DATED 8 July 1718.

[Davies 1805, No. 8. Marks a mill on the Dulas brook at SO 387287. Scale: 1:3960. Size 261 x 340].

fol. 13

'An Exact Map of the Demesnes & Long Meadow Lying Northerly from Pontrilas Bridge, also Lewis Crofts & Yns Water, in the Parish of Ewas:Herald ... belonging to ye Right Honourable Lord Bergavenny now in the Tenure or Occupation of S.r Phillip Jackson'

119a. Scattered lands in E of Ewyas Harold. Marks Colliers House, Gilberts Hill, Mount Hill Piece, two 'Quere' [quarries] in Great Callow Hill Wood, Cayya [Cayo] Lane, road to Helm [Elm] Bridge and road from Trap House [SO 393285] to Abbey Dore. Notes that the only timber trees were in Long Meadow.

Title cartouche decorated with flowers. Scale bar with dividers. Compass rose. Double ruled border with decorative leaves and berries.

DATED 30 June 1718.

SCALE: 2½ chains to 1 inch [1:1980].

[Davies 1805, No. 9. Marks Dog House and barn in field NE of Pontrilas bridge. Scale: 1:3960. Size 258 x 335].

fol.14

'An Exact Map of the Estate joyning to Colliers house the meadow lying Southerly to Oak Meadow Katherine Powell Wid: tenant, ye two Oak Meadows W.m Jones Gent. tenant ye two Spurr Meadows & two Lords pleck W.m Parry Gent . tenant, all ye other part of this Estate S:r Phillip Jackson is tenant lying in the P[ar]ish of Ewas-Herald belonging to ye Right Hon:ble George Lord Bergavenny' (Plate 32)

120a. in E of Ewyas Harold. Marks Collier House and roads to Helm [Elm] Green and Helm Bridge, and from Trap House to Abbey Dore.

Title cartouche and a cartouche for the surveyor's name and date, decorated with a colourful bird in a tree. Double ruled decorative border.

DATED 25 July 1718.

SCALE: 2½ chains to 1 inch [1:1980].

[Davies 1805, No. 10. Names include the Worm river Scale: 1:3960. 407 x 333].

fol. 15

'An Exact Map of the Estate called Bradleys al[ia]s Wallwyns homs & part of Wysomes land containing 52: 0: 23 those two Closes called Cathluin Containing 7: 2: 11 both in ye tenure or occupation of Sr Phillip Jackson: The Stoney furlong in ye occupation of John How Esq Containing 11:1:17 ye Close Joyning to Stoney furlong Notherly Called [blank] in ye occupation of Catherine Powell Containing 3: 1: 14 all Lying in ye Parish of Ewyasherold belonging to ye Right Honourable George Lord Bergavenny Containing in ye Whole 74:1:25'

74a. in E of Ewyas Harold, mostly between the rivers Monnow and Dore alongside the road from Llangua to Pontrilas. Marks Llangua bridge. Field-names include Pit Field. Notes that all trees except the orchard are timber.

Title cartouche and a cartouche for the surveyor's name framed with leaves. Scale bar with dividers. Compass rose. Double ruled decorative border.

DATED 5 July 1718.

[Davies 1805, No. 11. Scale: 1.3168. 360 x 328].

fol. 16

'An Exact Draught of the Estates called Cristfields lanes alias Crossfield lands lying betwixt helmebridge and ye Traphouse in ye parish of Ewas Herold in Com. Hereford. Jno How Esq:^r tenant to five closes Wm Jones Gent: tenant to footpath field, together wth Brockwalks, als. Brodwalls, Caefynnon & caeGrose, lying in ye p[ar]ish of Walterstone & County afores[ai]d Benja. Delahay Gent: tenant'

66a. in E of Ewyas Harold mostly N and W of Elm Bridge and scattered lands near Alltyrynys in Walterstone including two fields [21a.], on W side of the road from Alltyrynys to Walterstone church. Marks Helmen [Elm] Bridge, Trap House Corner, Footpath Field.

Cartouche framed with leaves. Scale bar with dividers. Compass rose. Ruled border with flowers at the four corners.

DATED 17 July 1718.

[Davies 1805, No. 12. Two of the fields in Walterstone lie along the road from Alltyrynys to Walterstone, in which the footpath in Footpath Field is marked running parallel with the road on the field side of the fence. The other two lie along the road from Alltyrynys to 'Coed y Gravel' [Grafel, at Walterstone Common] and the 'Nant y Gravel' brook. Scale: 1:3168. 260 x 335].

SIZE: 720 x 570. Ink and watercolour on parchment, leather-bound. Fols. 20-43 cut out. The ink titles within the cartouches have faded but the overall condition of the volume is excellent.

SURVEYOR: Benjamin Fallowes of Maldon, Essex.

REFERENCE: BL, Add. MS. 60746, fols. 11-16.

1. Longtown, an ancient chapelry of Clodock, became a separate civil parish in 1866.

2. David Davies of Llangattock, Brecon redrew Benjamin Fallowes's maps for Henry, Earl of Abergavenny at a reduced scale of 4 or 5 chains to the inch in 1805 and presented them in a red morocco-bound volume dated 1 January 1806, measuring 286 x 195 mm. The maps, nos. 7 to 12, are in ink and watercolour on parchment, mostly on a double-page spread, sometimes folded. His plain draughtsmanship, writing-master's script, simple north points and ruled borders are in stark contrast to Fallowes's exotic original maps. Apart from the titles, he made few changes, retaining the earlier names of fields and even of neighbouring landowners. Gwent RO, Man/A/2/0270.

1720
EARDISLAND

'An Exact Map of the estate called the Broome Farm lying in the Parish of Eardisland and County of Hereford belonging to The Hon;^ble William Bateman of Shobdon Court and County aforesaid; Esq. Now in ye occupacion of Hugh Gravner Containing A. R. P [blank]. Survey'd By Benjamin Fallow's Iuly 28th 1720' (Plate 38)

Broome Farm (277a.) in NW of the parish [SO 596341]. Marks farm house in perspective with hall and cross-wing. Shows fields with field-names and acreage, orchards, trees, gates, river Arrow, roads, footpaths. Field-names include Fludgate Field, Fishpool Field. Gives names of adjoining owners.

Note 'NB Round the Circumference the black dotts represent The ditch to shew who makes the fence; if they are on the outside the Hedge belongs to the Farm if on the inside the Fence belongs to the adjoyning Land'.

Two floral cartouches. Elaborate scale bar with dividers. Colourful compass rose. Coat of arms of Bateman and Searle impaled. Thirty four flowers of alternate design arranged all round the edge.

SCALE: *c*.3.6 poles to 1 inch [*c*.1:713].

SIZE: 820 x 640. Ink and watercolour on parchment, framed.

SURVEYOR: Benjamin Fallow's [Fallowes].

REFERENCE: Private.

1720, 1730
LEDBURY, DONNINGTON

An estate atlas containing four maps of an estate at Ledbury bought by Jacob Tonson in 1721.

'I. An Accurate Survey of HAZLE FARM …'. 1720. (Plate 33)

'II. An Accurate Survey of HALL HOUSE & HILLFIELD FARMS …'.1720

'III An Accurate Survey of HAZLE WOODS & WOOLPITS FARM Part of the HAZLE MAN^R For MR Jacob Tonson'. 1720

584a. S and E of Ledbury town. Shows buildings in block plan. Distinguishes between woodland and coppice. On Map I marks gates, windmill in the field E of Hazle Farm [SO 704364], avenue of trees leading to Hazle Farm. Field-names include Malmpool Mills and Hazle Mills on river Leadon, Perry Field, Cherry Orchard, Hopground. On Map III field-names include Hopground, Flax Close, Lime Pitts. Surveying lines are marked on Map III.

Reference tables: Tables [4 pp.] for Maps I and III, give names of tenants, field-names, land-use, computed acreage, measured acreage, value per acre, annual rent. The names of two tenants have been amended by parchment slips slotted into the reference tables.

Unnumbered, [4]. 'The VINEYARD And two other FARMS in HEREFORD-SHIRE … 1730'[1]

154a. in the E of the parish comprising Vineyard [9a.] [SO 718334], Lawne House Farm [32a.] and Camp Farm [113a.]. Shows strips in Hawfield open field, hopyard and lime kiln. Marks the Vineyard with orchard and vineyard, and Haffield iron-age camp. Field-names include Coney Berry Hill, Owlers, Fish Pool, Quarry Hill at Camp Farm and Brick Yard Mead, Burnt Land, Black House Orchard and Colepit Hill at Lawne House.

Reference table.
Colourful title cartouches with cherubs. Compass roses. Scale bars. Plain yellow borders. Map 4 is rather more delicately drawn.

SCALES: Map I, 23 perches to 1 inch [1:4554]; Map II 16½ perches to 1 inch [1:3267]; Maps III and 4, 16 perches to 1 inch [1:3168].

SIZES: In vellum-bound volume. Map I, 395 x 500; Map II, 390 x 495; Map III, 390 x 500; Map 4, 400 x 515. Ink and watercolour on parchment.

SURVEYOR: Charles Price.

REFERENCE: HRO, J95/1.

1. Map 4 is partly reproduced in J.G. Hillaby, *The book of Ledbury*, Buckingham 1982, p. 54. The Vineyard is recorded in production in the mid twelfth, late thirteenth and late seventeenth centuries [*ibid.*, pp. 77-78].

1721
BROCKHAMPTON (by ROSS) (formerly FOWNHOPE)

' A Map of that part of the Estate of Much Fawley in the Parish of Fown-hope in the County of Hereford which belongs to Francis Brydges of Tibberton in the said County Esq. By Ed: Moore Surveyor'[1]

Much Fawley farm [SO 590295] in the former detached part of the ancient parish of Fownhope with Fawley in a loop of the river Wye. Shows buildings in bird's-eye view, enclosed fields outlined in colour with plot numbers [but no reference table], orchards, roads with destinations. A very small part of one field lay in the parish of How Caple. Gives names of adjoining owners. Marks Much Fawley farm house and chapel [SO 590295], Fawley Cross as a wayside cross [SO 586304]. Field-names include Upper and Lower Pound Closes, Cross Field, Pitt Field [SO 584304], The Meend [SO 588308], the road to Hereford (running N and then NE from Fawley Cross), river Wye. Gives names of adjoining owners.

Title in a ruled frame. Simple compass rose. Scale bar.

SCALE: 16 perches to 1 inch. [1:3168].

SIZE: 765 x 550. Ink and watercolour on paper. Edges damaged and a little flaking, repaired 2001.

SURVEYOR: E[dwar]d Moore, surveyor.

REFERENCE: HRO, K11.

1. Transferred from Fownhope to Brockhampton 1884.

1721
HEREFORD

'A SURVEY of the MANNORS of BARTON and SHELWIC, Done for the Right Rever.d Father in GOD PHILIP Lord BISHOP OF HEREFORD. A.D. MDCCXXI'

*c.*2,250a. N of the river Wye and W of the river Lugg [SO 5140]. Shows buildings in perspective, streets and roads, the bishop's estates coloured green with field numbers but no field-names, orchards, especially at Holmer, streams, mill pools, ponds. Marks Hereford cathedral with W tower, the Bishop's Palace and formal gardens, bishop's chapel of SS. Katherine and Mary Magdalene [destroyed between 1737 and 1746], the churches of St Nicholas, All Saints and Holmer, White Cross, the Castle Mound, city walls and gates, Castle Mill, Wye bridge and Lugg bridge, ships on the rivers. Shows roadside encroachments E of White Cross, on the edges of the 'moors' and by road to Leominster.

Reference table gives plot numbers and acreage.

Baroque title cartouche with foliage and a mitre. Compass rose.

SCALE: [*c.*1:3520].

SIZE: 1670 x 1750. Ink and watercolour on parchment. When drawn this map and its companion (see 1722 HEREFORD) must have been the earliest and most informative large-scale maps of the city and its environs. It is badly stained by damp and much of the surface is obliterated or has flaked away. Grubby. Repaired unsympathetically with a cloth backing, probably *c.*1960.

SURVEYOR: Not named.

REFERENCE: HRO, AA59 (Church Commissioners, 12123).

1721
LEDBURY

No title.

Diagram of a small area at Eybridge [Highbridge, SO 355707] S of Ledbury town adjoining Hazle Farm, disputed between John Skipp and James Swift under the arbitration of Henry Collet. Describes the lands and gives names of adjoining owners. A legal document rather than an estate map.

SCALE: [*c.*1:2400].

SIZE: 305 x 190. Pen and ink on paper.

SURVEYOR: Henry Collet, [arbitrator].

REFERENCE: HRO, B38/251.

1722
BRAMPTON BRYAN

'A PLAN OF BRAMTON HOUSE. Castle. Gardens and lands contiguous: Belonging to the Right Hon.ble ROBERT EARL OF OXFORD MDCCXXII'

Brampton Bryan House [SO 370726], gardens and lands between the river Teme and the road from Knighton to Brampton. Shows buildings in perspective, formal gardens, canals and avenues, one of clipped shrubs or pyramids, orchard, trees with shadows, fencing and main gate. Marks river Teme with bridges, ruins of Brampton castle, church, dovecote. Field-names include the Bowling Green, Kitchen Garden.

Oval cartouche in baroque frame. Compass rose. Scale bar. Ruled border.

SCALE: 2 chains to 1 inch [1:1584].

SIZE: 411 x 539. Ink and watercolour on paper, rebacked.

SURVEYOR: Not named. [Charles Bridgeman].[1]

REFERENCE: Sir John Soane's Museum, Lincoln's Inn Fields, London, Volume 111, no. 43. Reproduced in P. Willis, *Charles Bridgeman and the English landscape garden*, Newcastle 2002, plate 78a and in D. Whitehead, 'Brampton Bryan park, Herefordshire: a documentary history' in D. Whitehead and J. Eisel eds., *A Herefordshire miscellany*, Hereford 2000, plate 6.

[1] Willis attributes the plan to Charles Bridgeman, one of whose leading patrons was Edward Harley, 2nd Earl of Oxford. The style, however, is quite different from Bridgeman's other plans and drawings.

1722
HEREFORD

'A SURVEY of the MANNORS of HAMPTON and TUPSLEY Done for the Right Rever.d Father in GOD PHILIP Lord BISHOP OF HEREFORD. A.D. MDCCXXII'

Hampton Bishop and a large area SE of the city centre [SO 5239]. Shows fields, with some strips in open fields, rivers Wye and Lugg, roads. Similar to the companion map, 1721 HEREFORD, (Manors of Barton and Shelwick), but this map is in even worse condition.

Similar cartouche, reference table and compass rose.

SCALE: [*c.*1:2880].

SIZE: *c.*1690 x 1620. Ink and watercolour on parchment. Badly stained by damp and parts flaked away. Almost entirely illegible. Repaired in an unsympathetic manner.

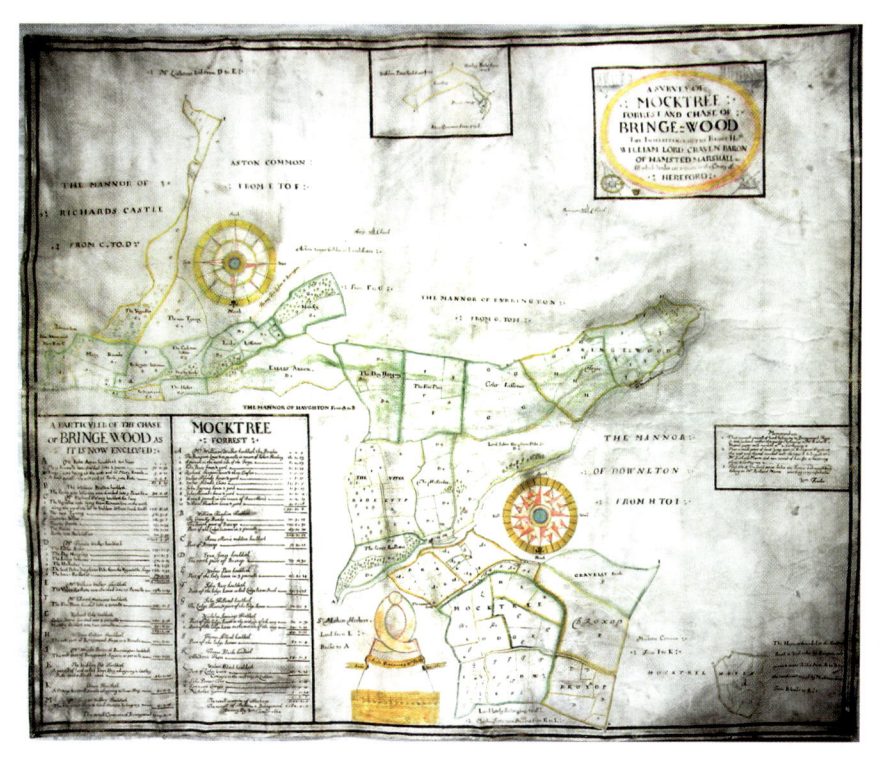

30. *Mocktree and Bringewood, Downton*
by William Fowler 1662.

The handwriting and decorative features, such as the title cartouche, framed tables, compass roses and scale bar, reflect Fowler's age and old-fashioned style.

31. *Lower Lye, Aymestrey*
by William Whittell 1708.

Though less experienced and younger than Fowler (above), William Whittell's decorative title cartouche, simple compass rose and dividers, and the ornate coat of arms of his client followed the fashion and text books of his time.

32. *Ewyas Harold by Benjamin Fallowes 1718.*

33. *The Hazle, Ledbury by Charles Price 1720.*

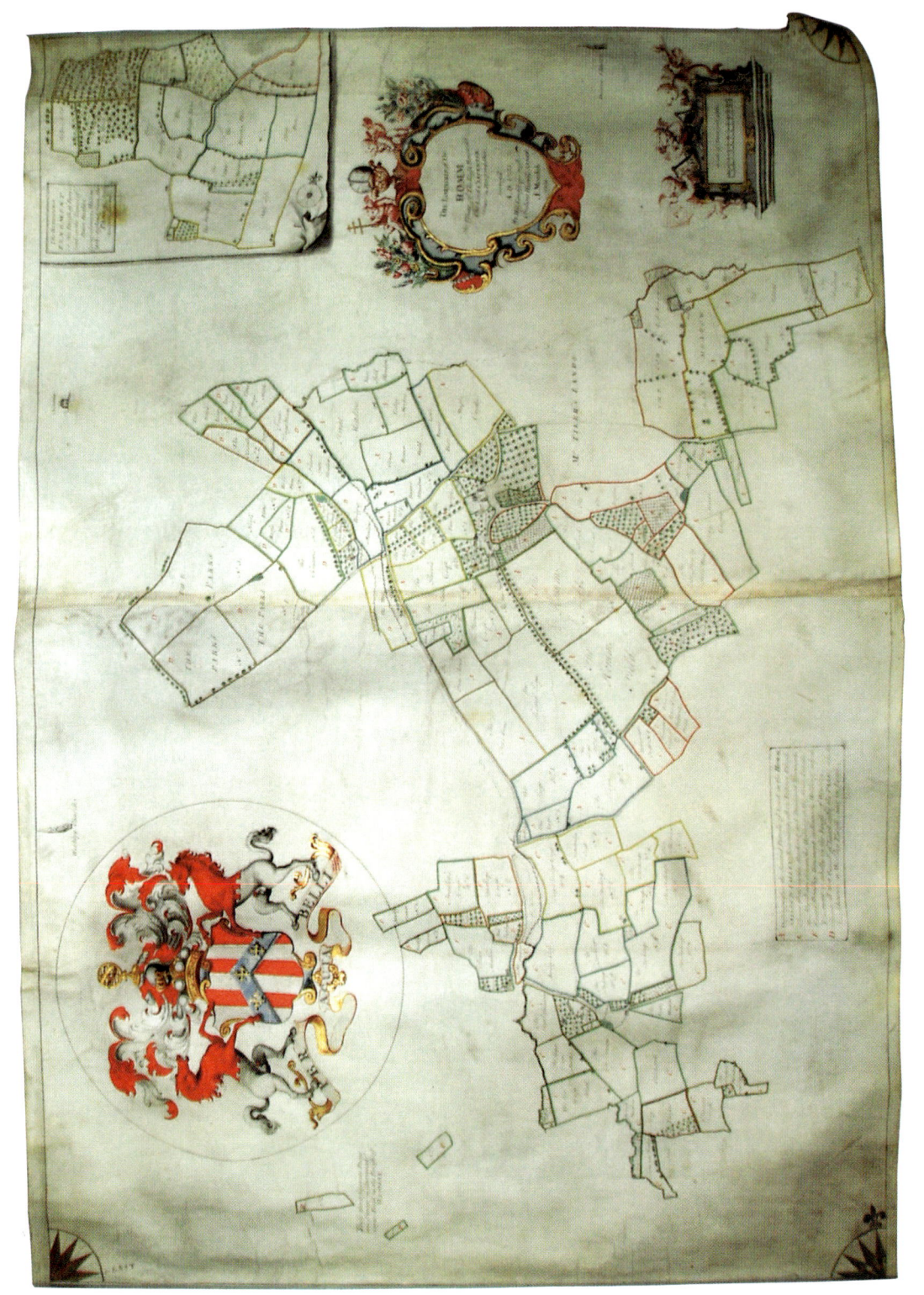

34. *The Homme, Weobley by J. Meredith 1733.*

35. *Richard's Castle by John Corbet 1743.*

36. Stretton Sugwas by Meredith Jones 1757 (see also Plate 63).

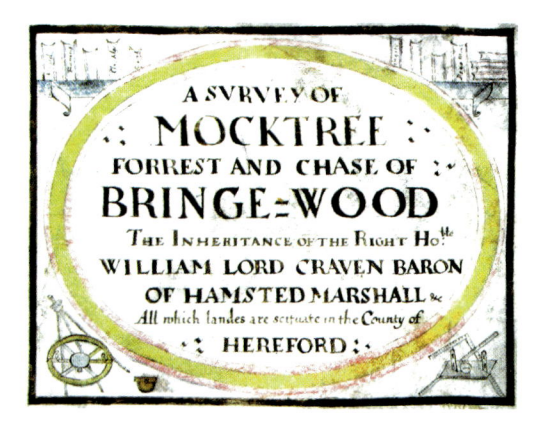

37. *Mocktree and Bringewood, Downton by William Fowler 1662. Cartouche.*

38. *Broome Farm, Eardisland by Benjamin Fallowes 1720. Signature.*

39. *The Homme, Weobley by J. Meredith 1733. Scale bar.*

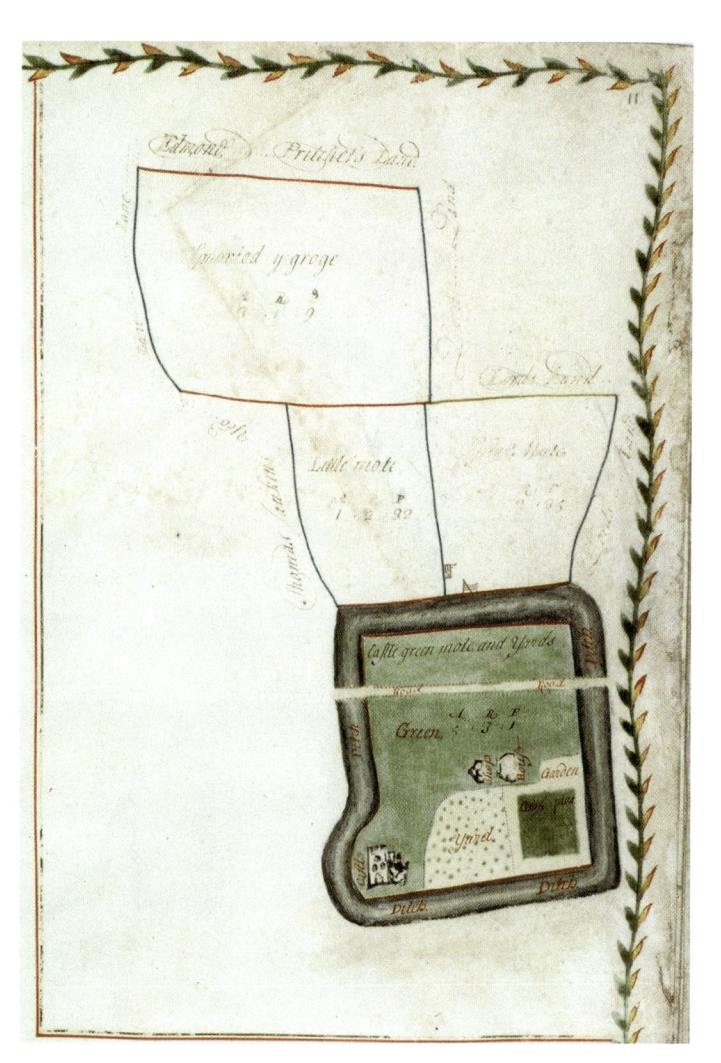

40. *Longtown castle by Benjamin Fallowes 1718. Detail.*

The Moccas estate atlas by John Lambe Davis 1772.

41. *Moccas Court and parish, showing a page of the reference table and the whole of the folding map.*

42. *The map of part of the neighbouring parish of Bredwardine marks the sites of Bredwardine castle and the manor house of The Radnor, the church and the fine brick bridge built only three years earlier.*

SURVEYOR: Not named.

REFERENCE: HRO, AA59 (Church Commissioners, 12122).

1723
HOLME LACY, BOLSTONE

'AN EXACT MAPP of Part of LADY SCUDAMORES WOODS in her LORDSHIP of HOMLACY ... Surveyed and described by George Smyth 1723'

Ramsdens Coppice and The Widow's Wood in Holme Lacy [SO 535348] and Trilloes Court [Wood, SO 547324] and Bowens Coppice in the S of Bolstone. Shows boundaries outlined in colour, trees [drawn in formal rows], roads, tracks.

Title within a frame. Scale bar with dividers. Compass rose. A printed border pasted on the map and to frame the title.

SCALE: 3 chains to 1 inch [1:2376].

SIZE: 695 x 800. Ink and watercolour on parchment.

SURVEYOR: George Smyth.

REFERENCE: BL, Add. MS. 36307, G22.

1724
MUCH DEWCHURCH

'A SURVEY of an Estate in the Parish of Much-Dewchurch, in Herefordshire; lately Purchased of Florice Gronow, by James Gunter of Saddlebow, Gent', taken by E: Moore. 1724.'

Enclosed lands N of Saddlebow Hill in SW of parish with a farm house and buildings [at Proberts Orchard, SO 454278]. Shows buildings in bird's-eye view, fields outlined in colour with reference numbers, orchard, isolated tree in Pear-tree Close, stream, bridleway towards Hereford. Adjoining landowners were the Saddlebow Estate and William Gronow.

Large oval baroque title cartouche. Compass rose. Scale bar of perches. Double-ruled border.

SCALE: 8½ perches to 1 inch [1:1683].

SIZE: 395 x 503. Ink and watercolour on parchment.

SURVEYOR: E. Moore.

REFERENCE: NLW, Mynde Park papers 1265.

1725
HOPE UNDER DINMORE

'Plan of the Gardens, Plantations &c of Hampton Court, in Herefordshire, the Seat of the R^t. Honourable The Lord Coningesby &c'[1]

Engraving. Shows the ground plan of Hampton Court and its formal gardens [SO 520525]. Marks the avenues, orchard, flower garden, fountains, parterre, rivers and summer house.

SCALE: 150 feet to 1 inch [1:1800].

SIZE: 375 x 240. Engraved.

SURVEYOR: Ca. Campbell Delin. H. Hulsbergh Sculpt.

REFERENCE: Published in Colin Campbell, *Vitruvius Britannicus or, The British Architect*, vol. 3, London 1725.

1. Thomas, Earl Coningsby remodelled the house, gardens and park in the 1690s. Its layout was painted by Leonard Knyff in 1699, by John Stevens c.1705-10 (both reproduced in J. Harris, *The artist and the country house*, London 1979) and engraved by Campbell in *Vitruvius Briatnnicus*, vol. 2, 1717 (the house) and vol. 3 (the gardens). Lord Coningsby died in 1729.

c.1725
MARDEN, SUTTON ST MICHAEL, SUTTON ST NICHOLAS[1]

'A MAP of the MANNOR of MARDEN ...'

Whole of the parishes of Sutton St Michael and Sutton St Nicholas [SO 5345], most of Marden and small parts of Wellington and Withington. Shows buildings in block plan, enclosed fields and strips in extensive open fields outlined in colour to distinguish tenants, hopyards, orchards, woods, river Lugg, streams, roads, bridges. Marks Marden church [SO 512471], vicarage and mill leet, dovecote at Wistaston [Wisteston] Court, chapel and dovecote at Amberley Court, Sutton Walls iron-age camp, Freens Court with six ponds in an orchard called The Moat; Sutton St Michael church, Sutton St Nicholas church, rectory and school, mill and Tan House Pond at S end of Sutton St Nicholas village, Mercer's Ford.

Names the principal farms, open fields and closes, including Hope Common, Keeper Knoll [Kipperknowle] Farm, Queens Wood, The Ferne [Vern], Holbach Cross, Berling [Burling] Gate, Ridgeway, Yew [Eau] Withington, The Wergans Stone Acre [at Wergins Stone], the Connygree [twice].

Explanatory note about ownership of a swilly [watercourse] in Lugg meadows. Table of tenants arranged alphabetically with key to colours of fields and strips.

Later note concerning examination of the map in Chancery proceedings Abiah Hill v Mary and Thomas Evans 1832.

Highly decorative. Ornate title cartouche with floral design, colours and gold leaf, two black cherubs, surmounted by earl's coronet but apparently not completed as a blank space has been left for the names of the owner of the estate and the surveyor and the date. Coat of arms of Thomas Coningsby, first Earl Coningsby with crest, supporters and motto. Compass rose. Scale bar with dividers, cherubs' heads and leaves. Broad border in yellow and red.

SCALE: 3 chains to 1 inch [1:2376].

SIZE: 2200 x 4320. Ink and watercolour on twenty parchment skins, lightly repaired and now separated for ease of storage and handling. The colours have faded.

SURVEYOR: Not named.

REFERENCE: HRO, J94/1.

1. Dated between 1719 when Thomas Coningsby was created Earl and before his death in 1729. HRO catalogue suggests c.1720 but the consistent use of block plans for all buildings, including the churches would indicate a later date.

n.d. [c. 1725]
YARPOLE

'A MAP OF BURCHER HALL'
149a., mostly enclosed and with some strips in open fields. Shows Bircher Hall [SO 476657] with dovecote and outbuildings in profile, fields coloured in pink or pale green, hopyards, orchards, gates, isolated trees, roads with destinations. Field-names include Red Oak Field [SO 470660], Great and Little Bym Field [Bine Field, SO 480650], Broad Heath, The Tump [SO 481662 near Birtley Knoll]. Gives names of adjoining owners.

Reference table within a ruled border, giving field-names and acreage.

Title in large lettering across the top of the map. South half of a compass rose. Scale bar. Patterned border, the same pattern being picked up in the top part of the border of the reference table. Rather crude draughtsmanship with pencilled ruled lines for the lettering.

SCALE: 4 chains to 1 inch [1:3168].

SIZE: 655 x 785. Ink and watercolour on parchment.

SURVEYOR: Not named.

REFERENCE: HRO, F76/III/18.

n.d. [early 18th century]
BLAKEMERE

'A MAP OF Blackmere farm the Parish of Blackmere ...'
314a. in the centre and W of the parish [SO 360410]. Shows a few buildings in perspective, enclosed fields outlined in colour, distinguishing arable, woods, roads.

Not completed. Reference table gives acreage only, leaving blank a column for field-names.

Title cartouche is also blank. Compass rose. Scale bar with dividers. Coloured border. Some later annotation.

SCALE: 5 chains to 1 inch [1:3960].

SIZE: 618 x 650. Ink and watercolour on parchment.

SURVEYOR: Not named.

REFERENCE: HRO, A81/III/34.

n.d. [early 18th century]
GOODRICH

Endorsed 'a rough Sketch of a Map of Huntsomehill Common'
Huntsham Hill [SO 563168], showing the Common (43a.) and its boundaries, including rocks on N boundary and the river Wye, with three cottages on the river bank [at about SO 565165].

N point, the map aligned with N at the bottom. Crudely drawn.

SCALE: [c.1:2755].

SIZE: 333 x 425. Ink on paper.

SURVEYOR: Not named.

REFERENCE: HRO, O68/Maps/2.

1726
EASTNOR, LEDBURY

Two maps of the estates of John Cocks in Eastnor and Ledbury.

[1]
'A Survey of The Mannor of Castle-Ditch in the Parish of Eastnor ... taken for John Cocks: Esqre by John George.'

W half of the parish [SO 7237]. Shows buildings carefully drawn in profile, mostly timber-framed, with cross-wings and large chimney stacks, open fields with strips and closes, hopyards, streams, roads, including Cannocks Lane (the road from Ledbury to Malvern N of Massington). Field tracks, including Rudgeway, marked by pecked lines. Marks Massington Farm with dovecote [SO 740395], Castle Ditch House [SO 735370] with adjoining Fish Pool, Old Court, 'Terras Walks', avenue from Castle Ditch through Cover Wood to 'Summer House Hill' with a summer house [SO 735367] and Park; Wain [Wayend] Street], Woodwardsend Farm [SO 743375], mill and flood-gates [SO 740366]. Field-names include Parks to E of Massington [SO 744395], Cock-shute Field, Rudgeway Common, Ashing [Ashen] Field, Gold-hill Green, Rowicks Green Common [about SO 750363], the Welch Ground, Owelers [Howlers] Heath, Joyners [now Hill] Farm [SO 740351]. Gives names of adjoining owners.

Two inset plans, framed:

'LANDS IN COLWALL' in and near Chanine Field and Malvern Hill. Marks the Mathon Brook, Mapleton [SO 745427] and the Crown Inn [at Colwall Stone].

'Lands in Teynton Parish call'd Bullocks Grounds'. A small area near Mr Holder's House [at Taynton, near Newent, Gloucestershire, SO 722222].

Within a square frame: Explanation of the lettering M, P, A, for meadow, pasture, arable, 'o:l:' for old leys and 'Red trees fruit trees', some trees being marked in faded red on the map. Other symbols, perhaps for different tenancies, are not explained.

Within a circular frame: 'The priked lines by the Perimeters represent where the Ditch is and Soe Shews to what Ground that hedg belongs'.

Ornamental circular cartouche. Patterned compass rose. Scale bar with dividers within a circle.
Firm clear lettering. Double-ruled and decorated border.

SCALE: 'A Scale of 60 perches', but with the scale bar marking only 16 perches at 3.2 perches to 1 inch. Comparison with the Ordnance Survey 1:25000 map suggests a scale of about 16 perches to 1 inch [1:3168].

SIZE: c.2030 x 880. Ink and a little watercolour on paper, linen-backed.

[2]
'A Survey of Some Lands in the Township of Mitchell in the Parish of Ledbury. And other Lands in the Parish of Eastnor ... taken for John Cocks: Esqre 1726 by John George'

Large area in N of parish [centred upon Upper Mitchell Farm at SO 723394]. Mostly open fields, with orchards, woodland, commons. Similar features to the first map. Marks Pettye France, Mitchell Little [Lower Mitchell] Farm, Old Houses [White House Farm], windmill [on Kilbury Camp, SO 723388], Eastnor church. Field-names include Upper and Lower Wake Field, Bannut- tree Banck, Burnt Land and Wormwood Acre [at Upper Mitchell], Upper and Lower Windmill Field, Broadley [Bradlow] Common, Gatleys End and Gatleys Hill [Eastnor Hill] and Gatleys Hopyard, Dead-mans-thorn Copse [Dead Womans Thorn, SO 723377].

Title in top left corner as an incomplete oval. Compass rose of complex pattern of circles and arcs. Double-ruled and chequered border with the letters G H I K in the vertical borders for reference. Framed explanations as in the first map.

SCALE: 'The Same Scale as the other Mapp' [in frame].

SIZE: 1145 x 620. Ink and a little watercolour on paper, linen-backed.

SURVEYOR: John George.

REFERENCE: Private.

1726
EASTNOR

'A SURVEY of the MANNOR of Bromeshill, alias Bromehill, in the Parish of Eastnor...taken for Tho. Read Esq: LORD thereof; by Edw. Moore Surveyor, 1726'[1]

Bronsil estate, containing enclosed lands (c.237a.), rough ground and woods [not measured]. Shows Bronsil Castle [SO 749373] in bird's-eye view and other buildings in perspective, fields outlined in colour, woods, coppices and groves, streams, pools, roads. Marks the moat and ruined Bronsil castle, mill [SO 747369], the road from Ledbury to Tewkesbury, a road to Gloucester running from Bronsil to the Whiteleaved Oak and 'A Team Way' running northward W of the castle. Field-names include Gullet, Rough Leasow and Malvern Hill, The Park [W of the castle], High Park and Shepheards Park.

Cartouche as a baroque monument. Compass rose. Scale bar. Decorated border.

SCALE: 16 perches to 1 inch [1:3168].

SIZE: 380 x 450. Ink and watercolour on parchment.

SURVEYOR: Edward Moore, surveyor.

REFERENCE: Private.

1. With a later copy in 19th-century style by W. Roberts, 1832.

n.d. [early 18th. century]
WIGMORE

'Plan of Woodwood'

Woodhampton Wood (*c.*72a) S of village [SO 405675]. Outline drawing of woodland subdivided into small plots of 1/2a. to 10a. Gives names of landholders and adjoining owners.

No scale bar. No compass point.

SCALE: [*c.*1:3960].

SIZE: 390 x 260. Ink on paper.

SURVEYOR: Not named.

REFERENCE: HRO, G87/47/141.

1727
WILLERSLEY, EARDISLEY

'A SURVEY of PARTON in the Parish's of WILLERSLEY and EARDESLEY … the Estate of IOHN BRASSEY Esq[r] I: GREEN Surveyer 1727'

Parton in Willersley and the S of Eardisley [SO 316483]. Shows fields outlined in green, red, blue and yellow with four fields coloured in purple, woods, orchards and isolated trees with shadows, roads and footpaths, but no buildings. Marks river Wye and Willersley's Brook. Reference table gives plot numbers, field-names, acreage and land-use.

Title cartouche with decorative leaves and shield [blank]. Compass rose, large and decorative. Scale bar with dividers. Decorative border with leaves and flowers.

SCALE: 4 chains to 1 inch [1:3168].

SIZE: 740 x 970. Ink and watercolour on parchment.

SURVEYOR: J. Green.

REFERENCE: Leicestershire RO, DG9/Ma/M/39.[1]

1. I am grateful to Dr Margaret Bonney and Kim Goodwin of the Leicestershire Record Office for this catalogue entry.

1728
GOODRICH

'A Map of New Weare Weare Hill &c'

Huntsham Hill area, surveyed to determine the boundaries of Weir Hill and Coldwalls [Coldwell] Hill with Mailscot Wood, with notes concerning these. Shows the river Wye and tributary streams, roads, merestones and boundary landmarks. Marks road to Symons Yatt [Symonds Yat], Jute Weir [SO 567158], Old Weir [SO 567169] and The New Weir [SO 560160]. Reference table of merestones and landmarks, including a yew tree, Long Stone and 'An Ancient Ditch' [? a ditch of the late iron-age promontory fort].

Compass rose. Ruled border.

SCALE: [*c.*1:14080].

SIZE: 306 x 402. Ink and brown wash on paper.

SURVEYOR: George Smyth.

REFERENCE: HRO, O68/Maps/7.

1729
HOLME LACY

'A SURVEY of part of the Mannor & Parish of HOM LACY … The Estate of their Graces The Duke and Dutchess of Beaufort Taken in July 1729 by Robert Whittesley'

383a. in the E of the parish and adjoining the river Wye. Buildings shown in bird's-eye view include the church [SO 568348] and a summer house or pavilion at the entrance to an octagonal enclosure with an avenue leading to the river. Shows enclosed fields outlined in colour, an orchard between the church and the river Wye, trees, ponds, roads. Marks one close and a strip in the riverside meadow in Fownhope called 'A piece by the Ferry' [SO 564355], wooded hill [West Wood] in Fownhope, island in the river Wye and the mouth of the river Lugg.

Reference table gives plot letters, field-names, acreage. Field-names include Lugg Mouth, Church Rise [W of the church].

Baroque cartouche incorporating the Beaufort arms with an escutcheon of pretence for Scudamore. Scale bar. Compass rose. Ruled border.

SCALE: 20 perches to 1 inch [1:3960].

SIZE: 688 x 599. Ink and watercolour on parchment. Stained and repaired.

SURVEYOR: Robert Whittesley.

REFERENCE: BL, Add. MS, 36307, G23.

1729
LEOMINSTER

'A SURVEY of the Manours of the Berry of IVINGTON and the Park...the estate of Edmund Pitts Esquire taken NOVEMBER 1729 by Jon:ª Green'

Ivington Farm (358a.) and Park Farm (313a.) SW of Leominster town. Marks The House [Ivingtonbury], mill and Park Farm buildings in perspective, enclosed fields, orchards, woods, roadside and boundary trees, river Arrow, road from Leominster to Ivington. Field-names include Great and Little Camp [SO 484547]. Names adjoining owners.

Reference table headed 'Ivington Estate and The Park Farm' gives field-names and acreage.

Title cartouche with foliage. Reference table in ruled frame. Two 16-point compass roses. Scale bar with dividers, but lacks figures. Decorative border of foliage with flowers in each corner.

SCALE: 5/16 of an inch to 1 chain [ie. a little less than 3 chains to 1 inch] [1:2227].

SIZE: c.1365 x 940. Ink and watercolour on parchment, consisting of five unequal pieces of parchment with traces of an earlier paper backing. Edges damaged and two corners missing.

SURVEYOR: Jonª [?Jonah] Green.

REFERENCE: Worcestershire RO, s705:255 [BA 1545/74].

1. I am grateful to Robin Whittaker for this catalogue entry.

1732
PENCOMBE WITH GRENDON WARREN

'A Map of the Mannor of Grendon Warren Being the Hereditary Possession of THOMAS CONYNGSBY of Grendon Court ... Drawn by Joseph Dougharty of Worcestr Land Surveyor AD 1732' (Plate 47)

804a. comprising the whole of Grendon Warren. Shows buildings in bird's-eye view with shadows, mill, fields, all enclosed, orchards and tracks. Marks Grendon Court [SO 600548] as a three-storey jettied house with four gables and smoking chimneys, chapel and dovecote.

Reference table gives field-names, including Piddle Wharf by stream and Kiln house Orchard, acreage and land-use, including 15½a. of hopyards.

Title cartouche with shell ornament. Two compass roses. Scale bar with pen and ink drawing of two doves and a man's face.

SCALE: [c.1:2880].

SIZE: 650 x 1080. Ink and watercolour on parchment.

SURVEYOR: Joseph Dougharty, land surveyor.

REFERENCE: HRO, AA59/8245.

(1732)
LEDBURY [WELLINGTON HEATH]

No title, being a memorandum about the Swallow estate in the manor of Ledbury 1792.

Small area at Loxter [SO 719409] adjoining the road from Wellington Heath to Hope End 'Taken from a Map of Hope End Estate made in 1732'. Marks buildings crudely in perspective, fields with names, roads.

N and S points.

Includes extracts from deeds 1726-8 and survey 1792. [The copy was possibly drawn in connection with the sale of the Hope End estate to E.M. Barrett in 1814].

SCALE: [c.1:3520].

SIZE: 160 x 320. Ink on paper, crudely drawn.

SURVEYOR. Not named.

REFERENCE: DCA, 4874.

1733
LLANGARRON

'AN Accurate Survey of ye Glascomb part of ye Estate of Robert Gwillym Esqr in ye Parish of Langarron'

A small area of fields and woodland [SO 520193]. The fields are outlined in colour and trees are marked. Field-names are illegible.

SCALE: [Probably 1:3168].

SIZE: 232 x 367. Ink and watercolour on parchment. Almost illegible from the effects of damp.

SURVEYOR: George Smyth.

REFERENCE: HRO, R54/1.

1733
WEOBLEY, DILWYN, EARDISLAND

'THE LORDSHIP of The HOMME the estate of the Right Honourable The Lord CARPENTER, Situate in Herefordshire' (Plates 34, 39)

The Homme estate [SO 474530] in NE of the parish of Weobley and lying between Weobley and Dilwyn villages. Shows buildings, including The Homme, in rather crude bird's-eye view, fields outlined in colour with reference letters to distinguish tenancies, hopyards, orchards, trees, [lime] pits, roads and lanes, some gated. Marks Weobley and Dilwyn churches, two avenues of trees leading to The Homme and its gardens, a small piece of common at Dunwood and a moat [SO 419538]. Field-names include The Parks, N of Weobley, and Gatebridge [Gad Bridge]. Also shows three separate small pieces of land in Sherwoods Cross Field in Weobley.

'The RUDDOCKS TENEMENT in the Parish of Earsland in the tenure of Peter Bathos lying about a Mile & 1/8th distant from the PARKS North Westward'

A small area of land, not identified, in the adjoining parish of Eardisland.

Elaborate title cartouche with cherubs and flowers. Coat of arms, with crest, supporters and motto of Carpenter.[1] One-quarter segments of a compass rose in three corners of the map. Scale bar on a pediment with three cherubs holding dividers. Ruled border. The map of Ruddocks Tenement is drawn as a *trompe d'oeil* piece of parchment lying on the main map.

SCALE: 4 chains to 1 inch [1:3168].

SIZE: 820 x 1205. Ink and watercolour on parchment.

REFERENCE BOOK: Missing.

SURVEYOR: J. Meredith, Salop.

REFERENCE: HRO, W89/1.

1. George Carpenter, 2nd. Baron Carpenter, had succeeded his father in 1731.

n.d. [c. 1735][1]
LEINTHALL EARLS

'A MAP of Gatley Park ... the Property of Thomas Dunne Esq' Gatley Park estate. Shows enclosed fields, outlined in colour, and coppices with plot numbers and field-names. Marks Gatley Park house in profile, of three storeys with a central chimney [?or cupola], dovecote. Field-names include New Park Hopyard W of the house, Upper and Lower Windmill Bank [about SO 447686], Limekiln Leasow, Upper Holloway, Camp Piece [SO 444684].[2]

Rectangular floral cartouche. Compass points with N arrow and E cross. Scale bar.

SCALE: 4 chains to I inch [1:3168].

SIZE: c.630 x 1040, framed. Ink and watercolour on parchment. Faded.

SURVEYOR: Not named.

REFERENCE: Private.

1. Thomas Dunne I of Gatley died in 1734 and his successor, Thomas Dunne II may have commissioned the map. A contemporary landscape painting of the estate (Royal Comm. Hist. Mons., *Herefs*., vol. 3 (1934), p. 12) remains at Gatley.

2. Named as the viewpoint to Croft Ambrey hill-fort. Archaeological investigation of Camp Piece before the extension of a quarry in the late twentieth century confirmed that it contained no prehistoric features (*ex inf.* the owner).

1737
BROCKHAMPTON [by BROMYARD]

'The Manor of Brockhampton Survey'd by Jnº. Perkins 1737'

Whole of the manor. Shows houses in perspective but other buildings in block plan, fields, woods, streams, roads and tracks. Marks Lower Brockhampton [SO 687560] and a house near the site of Brockhampton House [built *c.*1765]. Fields with reference numbers, which are duplicated for each holding, and letters to distinguish arable, pasture, meadow, hopyards.

Scale bar with dividers. Compass rose.

Separate reference table, undated and framed. Arranged by holdings, with the same letter and number references, field-names and acreage. Holdings named The Hill, The Grove, Hill House, Lower Brockhampton, The Mill, Howley, Warren or Hillbarns, The Farm, Crumble Berry. Field-names include Mill Hopyard, Forbury, The Avenue (all at The Hill), Clover Piece, Berrow Field (Hill House), Churchyard Meadow (Lower Brockhampton), Old Brockhamton (at The Warren), Limepitts Field, Paradise, Over Spittle Field, Potters Court. Total acreage, [?] 1,243a. [The writing is faint and partly illegible].

SCALE: [*c.*1:6336]

SIZE: 490 x 695. Ink and watercolour on parchment, framed.

SURVEYOR : John Perkins.

REFERENCE: National Trust, Lower Brockhampton Court.

1741
MUCH MARCLE

No title.

821a. comprising the manor of Marcle Audleys or Hellens [SO 661332]. Shows houses crudely in perspective, roads, enclosed fields and strips in open fields, orchards. Gives names of adjoining owners. Marks Much Marcle church, two avenues leading to Hellens.

Simple, almost a draft, pen and ink outline of field boundaries and features.

SCALE: 4 chains to 1 inch [1:3168].

SIZE: 10 sheets in a volume, marbled paper on card, half-leather bound. 385 x 525. Pen and ink on paper.

REFERENCE BOOK: 'References to yᵉ Map of yᵉ Manour of Marcle Audleys or Hellens Survey'd &c by George Smyth 1741'. Contains explanatory note for the reader. Gives field numbers, statutory acreage and estimated [customary] acreage, bounds. Field-names include Stone Barrow, Stoneing House [Stone House], Cares Wall [Farm], Streets End, Brick Kiln Close, Monkfield Orchard, The Middle Parks, The Lawns, The Park, Holly's Green [Hallwood Green], Halwood.

SURVEYOR: George Smyth.

REFERENCE: HRO, F35/RC/M III/1a [map], 1b [reference book].

1742
BROMYARD (LINTON)

'An EXACT MAPP of part of the *Township* of *Linton* being the ESTATE of Yerrow Arrowsmith Genⁿ. in the PARISH of *Bromyard*'

256a. E of Bromyard town adjoining Bromyard Down and Overwood Common [SO 5367]. Shows a few houses in bird's-eye view, fields coloured or coloured in outline, orchards, woods, streams, roads. Gives names of adjoining owners and field-names including Meerditch and Hopyard.

Cartouche. Compass rose. Scale bar with dividers. Decorated border. All in ink or ink and wash. Delicate but faint penmanship.

SCALE: 4 chains to 1 inch [1:3168].

SIZE: 920 x 780. Ink and watercolour on parchment, attached to a roller.

SURVEYOR: G. Smyth.

REFERENCE: HRO, B9/5.

1742
MADLEY

'AN EXACT MAPP of LOW.ᴿ CHILSTON and other Estates and Tenements belonging to EDWARD KINGDON Esq.ʳ in the PARISH of MADLEY ... Survey'd and Map'd by George Smyth MDCCXLII'

Lower Chilstone estate [SO 409398] NW of Madley village. Shows buildings in bird's-eye view, fields in outline colour with field-names and reference numbers, orchards. Madley church shown in perspective, village streets and roads with destinations. Marks Lower Chilstone within a circular enclosure [? a moat] and with a dovecote. Names adjoining owners. Field-names include Old Hopyards, the Gravel Pitts.

Title cartouche. Compass rose. Scale bar. Patterned border. No reference table or book.

SCALE: *c.*4 chains to 1 inch [1:3168].

SIZE: 890 x 730. Ink and watercolour on parchment, the ink affected by damp and very faint. Formerly attached to a roller.

SURVEYOR: George Smyth.

REFERENCE: HRO, AS80/4.

1743
RICHARDS CASTLE

'Survey of Mʳ. Henry Jordan Estate at Richards Castle. Part in Herefordshire and part in Shropshire' (Plate 35)

Shows houses in perspective including the principal house with a dovecote adjoining Mill Green and a fishpool on the Shropshire boundary [SO 504704], scattered lands mostly in open fields but some closes, distinguished by solid or outline colouring, gates, stiles, roads, bridleways and footpaths. Field-names include Castle Wicket on parish boundary with Orleton, Hop Yard, Burnt House, Park Lane, Tumpy Plock, Cynder Meadow, Mine Field [open field] Richards Castle Common.

Two explanatory notes:

[1] 'Note. The round Red spot by the House is the Ground where the Dovehouse stands. The Black squares the Ground where the

Barns stand and the Green line that crosses the Road by Mill green Denotes a Gutter that parts Herefordshire and Shropshire. Note. all lands in Common Fields that belong to the Estate and are not Inclosed are expres'd by Letters. Number'd and wash'd over with several Colours and all pieces Inclosed have their names wrote in the middle Number'd and washed only round the Boundary line. Note. wherever any Person has land adjoyning. There their names are wrote which shows the Bounding of every piece or parcel of Land whether it be Inclosed or not. Note. The House near Linehalls Eye. Commonly call'd by the name of Cams House. And adjoyning to that House in the middle of a Field. is about three parts of an acre belonging to the Estate and Ioyn'd on the South side thereof with Cams Land. And on the North with Land of Sq.re Salwey. Note. The Black line draw'd from the Corner of the Fishpool by Mill Green. Denotes a small stream of Water that runs there. And here Note. That the Black line at the bottom of the Orlly Ground. Denotes a Brook that parts Orltuns [Orleton] Parish and Richards Castle. &c'.

[2] 'Note. all Bridle ways are double spac'd
Foot path single
Gates expres'd thus   I=I
And Stiles thus   H'.

Simple title cartouche. Large perspective drawing of dovecote and a house. Two crude pen and ink drawings of cherubs. Two framed explanatory notes [see above]. Two compass roses. Elaborate scale bar with dividers. Yellow border.

SCALE: 4 chains to 1 inch [1:3168].

SIZE: 855 x 1205. Ink and watercolour on paper. Repaired.

SURVEYOR: John Corbet.

REFERENCE: HRO, F76/III/22.

1747
LEOMINSTER

'A Survey of the Manour of Bradward in the Parish of Leominster HEREFORDSHIRE belonging to Brazen Nose College Oxon, taken A.D. 1747 By R.F. Bursar'[1]

[Broadward Hall estate, SO 497572]. 151a. on both sides of the road from Leominster to Hereford, S of the town of Leominster. Shows buildings in block plan, red for dwellings and black for outbuildings, fields mostly enclosed but also a few strips, arable (68a., coloured pale yellow), meadow and pasture (83a., coloured pale green) with field-names and land-use, hedges and fences, gates, orchards, isolated trees, roads with destination. Gives names of adjoining owners. Field-names include Upper, Middle and Lower Bradward Fields, Warton Field, Great and Little Bielt, Six Acres and Pond [SO 510567, but no pond is marked], The Pools [SO 505585; no pools are marked].

Reference table, giving field-names, land-use, acreage.

Title written with fine decorative penmanship for the capital letters. Compass rose, small. Scale bar in chains, furlongs and poles. Triple ruled border.

SCALE: 4 chains to 1 inch [1:3168].

SIZE: 578 x 630. Ink and watercolour on parchment.

SURVEYOR: R[ichard] F[orster], bursar.

REFERENCE: Brasenose College, Oxford, B14. 1/39.

1. Richard Forster, Fellow 1732-49, vice-principal 1744-5 and senior bursar 1746-8.

n.d. [mid 18th century]
BISHOP'S FROME

'PLAN of ESTATES in the parish of BISHOPS FROOME ... belonging to Archibald Duncan Esq.r and John Hyde'

109a., comprising Wellington Farm [SO 656487] and Green Dragon Lane. Shows buildings in block plan, scattered closes and a few strips in open fields outlined in colour, hopyards, woods, streams and roads. Gives names of adjoining owners. Field-names include Forked bridge, Quarry Meadow, Philley Brook.

Title cartouche with ears of wheat. Drawing of the parish church. Simple compass rose and scale bar. Plain border.

SCALE: 2.4 chains to 1 inch [1:1901].

SIZE: 395 x 930. Ink and watercolour on paper, frayed.

SURVEYOR: Not named.

REFERENCE: HRO, B9/4.

n.d. [mid 18th century]
CROFT

'A Map of Croft Castle Demesne'

N part of parish. Shows Croft Castle [SO 449655] in block plan, field boundaries, commons, woods, trees, roads with destinations, gates. Marks avenues in the Park, limekiln, conduit and

'Top of the Ambury' [Croft Ambrey iron-age fort] (272 a.). Gives letter and number references to fields and distinguishes glebe lands.

A rough and simple sketch.

SCALE: [c.1:3520].

SIZE: 943 x 770. Ink on 4 sheets of thin paper, once pasted together. Some flaking from damp and small parts missing.

SURVEYOR: Not named.

REFERENCE: HRO, O98 [LC 5506].

n.d. [mid 18th century]
PEMBRIDGE

'A MAP OF the Low in Parish of Pembridge An Estate in the County of Hereford belonging to the Honourable Robert Harley Esq.r Survey'd by John Smith'

[Lowe Farm, SO 370584]. 149a. in NW of the parish. Shows buildings in outline block plan, enclosed fields and strips in open fields, orchards, hedges, gates, roads and tracks over fields. Marks Pembridge church in profile with a slender W tower.[1] Field-names include Tin Cock, Brick Close.

Reference table in a classical frame of two columns gives plot numbers, field-names, acreage, land-use.

Framed title. Large compass rose. Scale bar with dividers. Ruled and painted border. Crude draughtsmanship.

SCALE: 3¼ [chains] to 1 inch [1:2574].

SIZE: 715 x 885. Ink and watercolour on parchment, faded, especially the ink.

SURVEYOR: John Smith.

REFERENCE: Private.

1. The surveyor can never have seen Pembridge church with its distinctive late 14th-century bell- tower, a detached, squat, octagonal stone-walled and timber-framed building

n.d. [mid 18th century]
PENCOMBE

'A SURVEY of a Farm belonging to M.rs Foley at Pankham ...'
180a. in W of parish [Churchyard Farm, SO 575538]. Shows buildings in block plan, roads, enclosed fields outlined in colour. Distinguishes arable. Gives names of adjoining owners.

Reference table gives acreage and field-names, including Dumbfield [Dumpfield], Limekiln Field.

Title cartouche in delicate pen and ink. Compass rose. Scale bar. Plain border.

SCALE: 6 chains to 1 inch [1:4752].

SIZE: 356 x 480. Ink and watercolour on parchment, rather faint.

SURVEYOR: Not named.

REFERENCE: HRO, B9/8.

n.d. [mid 18th century]
PIXLEY

'Manstone Farm in the Parish of PIXLEY Belong'g to JOHN COCKS ESQ:'[1]

[Mainstone Court, SO 656399]. 64a. in scattered lands, mostly enclosed at The Trumpet [SO 656395] and The Pullen [Poolend Farm, SO 648391]. Shows houses in bird's-eye view, fields coloured in outline, gates. Marks The Trumpet inn.

Reference table, giving field-names including 'Four Lands & 2 piques [pikes] in Manston Field'.

Title, ruled, across the top of the map. Framed reference table. Compass rose. Double-ruled border.

SCALE: [c.1:7920].

SIZE: 493 x 377. Ink and watercolour on paper.

SURVEYOR: Not named [? John Doharty, Worcester].

REFERENCE: Private.

1. John Cocks married his cousin Mary c.1723 and died 1771. The draughtsmanship of the map is in the style of John Doharty of Worcester, who was active 1731-55.

n.d. [mid 18th century]
STRETTON GRANDISON

No title. Endorsed 'Plan of Herefordsheir Estate The Huming in the Parish of Streton Gansom'. [In a later hand] 'Small lots in Stretton'

[Homend Farm, SO 641441]. 34a of scattered closes and strips in open fields. Shows the farm house in perspective. Fields drawn in outline.

Reference table gives acreage and field-names, including Ash Field and Sallings Field.

Ruled border.

Herefordshire Maps

SCALE: *c*.3 chains to 1 inch [*c*.1:2376].

SIZE: 575 x 655. Ink and watercolour on parchment.

SURVEYOR: Not named

REFERENCE: HRO, T51 [LC 4475].

1751
BYTON

'Part of Mons Land in Byton'

Outline sketch of land adjoining the road from Byton to Kinsham bridge [SO 363643, part of Rylands Wood adjoining Maunds Wood]. Marks boundaries, measurements in yards, prominent trees, road.

With two witnesses' depositions relating to a dispute about the land.

SCALE: 32 yards to 1 inch [1:1152].

SIZE: 187 x 300. Ink on paper.

SURVEYOR: Not named.

REFERENCE: Private.

1754
LYONSHALL

'AN EXACT PLAN OF PENROSE FARM Being a tenement belonging to [blank] Scituate in the Parish of Lyonshall'

117a. (or 175 customary acres) in W of parish, comprising Penrhos Farm [SO 317562], mostly N of the Kington-Leominster road. Shows Penrhos Farm in bird's-eye view, fields in colour, both enclosed and strips in open fields, roads. Gives names of adjoining owners. Marks river Arrow. Field-names include Spittal Oval, Red Hill.

Reference table giving land use, statutory and customary acreage.

Scale with dividers. Grey border.

SCALE: 3 chains to 1 inch [1:2376].

SIZE: 590 x 610. Ink and watercolour on parchment.

SURVEYOR: Meredith Jones.

REFERENCE: HRO, B33/209.

1755
WEOBLEY

Endorsed 'Mr. Thomas Survey of Petty Furlong field'.[1]

Sketch plan of Petty Furlong Field to the E of the village [SO 4151], March 1755. Shows Kate Harper's house and plock, the road to Weobley and the road from Weobley to [King's] Pyon, the horse road to Dunwood, the foot road to [Weobley] Marsh and a footway into Sherrard Cross Field, New Tinings and Goblett Hedge. Gives names of adjoining owners or occupiers.

Found loose in the Herefordshire rental for 1754-56.

SCALE: [*c*.1:1440].

SIZE: 185 x 310. Ink on paper.

SURVEYOR: [Philip] Thomas, [bailiff to Thomas Thynne, 3rd Viscount Weymouth and 1st Marquess of Bath].

REFERENCE: Longleat House muniments, Weobley 49 01/03/1755. Open to established scholars by appointment only on payment of a fee. All enquiries should be directed in the first place to the Curator, Longleat Historical Collections, The Estate Office, Longleat House, Warminster, Wiltshire BA12 7NW.

1. I am grateful to Dr Kate Harris for this catalogue entry.

1755
BRIDSTOW, PETERSTOW

'A SURVEY of Lands in the MANOR of WILTON & County of HEREFORD, belonging to GUYS Hospital' (Plate 48)

Most of the parish of Bridstow [SO 5824] and SE part of Peterstow. Shows most buildings in block plan, arable, meadow and woods, isolated trees, the river Wye, streams, roads and paths. No field-names except Buckcastle. Marks Bridstow and Pitstow [Peterstow] churches in perspective, and Mynd [Wilton Castle], Ash and Pye [The Broome] farm houses, Wilton bridge. Small marginal plan of Ross town.

Title cartouche with plough. Inset view of Wilton Castle with a boat on the river in the foreground. A grid superimposed on the map. Compass rose. Scale bar with dividers. Floral border.

SCALE: 8 chains to 1 inch [1:6336].

SIZE: 695 x 705. Ink and grey wash on parchment.

SURVEYOR: J. Green.

REFERENCE: HRO, C59/6.

[c.1755-60]
CANON PYON, DILWYN

'A SURVEY and MAP, Of Several ESTATES Belonging to ANT:NY SAWYER Esqr In the PARISHES of CANON PYON DILWYN, and WELLINGTON ... In which the different Tenures and all other particulars are carefully distinguish'd. Map'd by ISAAC TAYLOR Author of the MAP of Herefordshire'

Most of the parish of Canon Pyon [SO 4549], the manor of Alton in Dilwyn [SO 427534] and a small area in the W of Wellington. Shows buildings in block plan, gardens, enclosed fields, with boundaries outlined in faded colour to indicate tenant holdings and strips in open fields, distinguishing land-use and furlongs, orchards, woods, trees with shadows, streams and pools, roads. Marks Alton Court in Dilwyn, the Great House with formal gardens, Home Farm and Court House [Farm], Canon Pyon church and pound, all in the centre of Canon Pyon, Pyons Hill (wooded, but without the crenellated summer house tower on its summit being marked on the map), New Inn, tanyard and oven [in the hamlet of New End] by the Roman road from Kenchester and Hereford to Leintwardine, Foulbridge [Fullbridge] House and Kinford mill in the E of the parish of Canon Pyon. Field-names include Tumpy Croft in Dilwyn [SO 429535], The Bricks and Fish Ponds at Little Pyon Farm [The Brick House], Stroud Field, Croxhill [Crockshill], Millers Orchard [SO 449491, but no mill is marked on the stream], Coal Pitts and Upper Cole Pitts [SO 445484].

Large cartouche incorporating a stone tablet with an urn above and drinking trough below. Compass rose. Overall grid of squares with letter and figure references from a1 to u26, the references placed within a double ruled border.

Eight framed sketches of buildings on the estate arranged around the upper part of the map:

[1] 'The MANOR of ALTON In the PARISH of DILWYN...' [a blank space above for the drawing]
[2] 'LITTLE PYON FARM or THE BRICK HOUSE'
[3] View looking towards a walled garden with Canon Pyon church tower, Pyon Hill and its crenellated summer house tower in the background
[4] View of the classical Great House with Pyon Hill and its summer house with crenellated tower in the background.
[5] View of Canon Pyon church
[6] View of a large L-shaped timber-framed house of late 16th-century style with gables and diagonal shaft chimneys, barns and a farmyard containing a circular haystack [Court Farm]
[7] Another view of the same house or a smaller timber-framed house with a diagonal shaft chimney
[8] View of a farmyard containing a wooden crane.

SCALE: [c.1:3250].

SIZE: 1465 x 1590. Ink and wash with a little watercolour for tenants' boundaries. Overall brownish discolouration with parts flaked and missing. Repaired 2002.

SURVEYOR: Isaac Taylor, 'Author of the MAP of Herefordshire'.

REFERENCE: HRO, BN81.

1756
BRIDSTOW, PETERSTOW

'A Survey of the land in the Mannour of Wilton...belonging to Mr Guys Hospital Taken by In Green. 1756.'

Scattered lands in most of the parish of Bridstow [SO 5824] and the SE part of Peterstow. Shows buildings on the estate in block plan, painted pink or in perspective, and other buildings in block plan in black, fields in brown or green and a few signs of former strips in open fields, all with letter and number plot references, orchards, common land, woods, isolated trees, gates, the river Wye, streams and ponds, roads with destinations and pointing hands, field tracks and footpaths. Marks the churches of Bridstow, Peterstow and Ross in perspective, Wilton castle in block plan, a rough street plan of Ross, an avenue to the W of The Flann, and stock watering places or fishing access on the right bank of the river Wye at Wilton and, with a spring, at Weirend [SO 575236]. Names adjoining owners.

Floral title cartouche. Compass points [eight points]. Scale bar with dividers. A grid over the blank areas of the map with marginal grid references within the grey wash floral border, a-z [horizontal] and A-W [vertical].

SCALE: 4 chains to 1 inch [1:3168].

SIZE: 1200 x 1390. Ink and watercolour on parchment. Repaired and mounted on paper. The edges are a little frayed with one tear across part of the map, but nothing is missing.

REFERENCE BOOK, 'Quantity of land contained in the Map of the Manor of Wilton taken by J. Green 1756'.

SURVEYOR: John Green.

REFERENCE: Map, Private. Reference book, HRO, C99/III/243.

1757
ACONBURY

'AN EXACT PLAN OF ACONBURY COURT with Several other adjacent Tenements and Woods Scituate in the Parish of Aconbury ... being part of the Estate belonging to the PRESIDENT & GOVERNORS of the Hospital founded at the Sole Costs and Charges of THO^S: GUY ESQ:' (Plate 49)

706 statutory acres or 883 customary acres in centre of parish [SO 5133].[1] Shows buildings crudely in bird's-eye view, enclosed fields in solid colours, arable, pasture, orchards, woods, streams, roads, gates. Indicates relief by shading. Marks church, Aconbury Court, Pyes Nest, Aconbury Cott. Field-names include The Wallbrookes [Wood], Pigeon House Meadow at Aconbury Court, Mill Meadow on Tar's Brook and house called Nether Wood Tenement at Tar's Mill, Old Hopyard, The Bewgates, Quarry Field, Kings Pitt Close. Gives names of adjoining owners.

Reference table gives field-names, acreage.

Coloured floral title cartouche. Inset drawing of 'A SOUTH WEST PROSPECT OF ACONBURY COURT p[er] Mered: Jones', which includes the church. Compass rose. Scale bar. Ruled border.

SCALE: 3 chains to 1 inch [1:2376].

SIZE: 832 x 1200 on two skins. Ink and watercolour on parchment.

SURVEYOR: Meredith Jones of Brecon, land surveyor.

REFERENCE: HRO, C99/III/217.

1 The map refers to a second map, untraced, of lands to the west.

1757
LITTLE MARCLE

'A SURVEY of Little Marcle Estate ... belonging to Mrs Elizabeth Neale, Mrs Frances Barrett, Mrs Mary Annesley and Mrs Catharine Cornwall'

[Little Marcle Court estate, SO 665363]. Shows buildings in perspective, enclosed fields, orchards, woods, streams, roads. Fields coloured yellow or green [to indicate land-use] with letter and number references. Marks the former church and Little Marcle Court with an avenue of trees running E as far as Ludstock Brook. Gives names of adjoining owners.

Reference book not traced.

Cartouche with foliage, rococo. Scale bar with dividers. Graticule over blank parts of the map. Yellow ruled border.

SCALE: A little under 4 chains to 1 inch [c.1:3136].

SIZE: 1008 x 1265. Ink and watercolour on parchment.

SURVEYOR: I: Green [John Green].

REFERENCE. Private.

1757
STRETTON SUGWAS

'AN EXACT SURVEY OF STRETTON COURT and all other Messuages Tenements and Lands in the parish of Sutton ... belonging to THE PRESIDENT AND GOVERNORS of the Hospital founded at the Sole Costs and Charges of THOMAS GUY ESQUIRE' (Plate 36)

640a. comprising the N part of the parish [SO 4642] with small parts in S of Credenhill and Burghill. Shows Sutton Court in bird's-eye view with gables and timber-framed barns, other buildings in block plan, extensive open fields and some enclosed lands, coloured to distinguish tenancies, orchards, woods, streams, roads. Marks Three Mile Stone on road from Hereford to Brecon, Lime Cross at crossroads on the Roman road [SO 462426], church near Sutton Court [SO 468429] with a rectangular tree-lined walk in Church Field to its SE. The road from the Lime Cross SW to the Hereford-Brecon road is outlined as running across the open fields. Field-names include Kytes Nest, Brick Close, Swines Hill, Body Croft [N of Sutton Court], Brockall Coppice E of Kites Nest. Gives names of adjoining owners.

Reference table, arranged by names of tenants, gives statutory [640a.] and customary [c.935a.] acreage and field-names.

Title cartouche in oblong frame. Compass rose. Scale bar. Ruled border. Colourful but rather crude.

SCALE: 3 chains to 1 inch [1:2376].

SIZE: 800 x 1360. Ink and watercolour on parchment.

SURVEYOR: Meredith Jones.

REFERENCE: HRO, C99/III/216.

1758
COLWALL

'A Map of the Manor of Brockbury Lying in the Parishes of Colwall and Coddington ... Belonging to Robert Bright Esqr'

Centre [SO 745418] and W of parish with a small area in Coddington. Shows buildings at Brockbury in bird's-eye view with gardens, fields, all enclosed but with some indication of recent removal of hedges, hopyards, orchards, ponds, roads, paths. Marks traces of moat [SO 745419].

Inset, a drawing, framed as if hanging, with inconsistent perspective of an enlarged bird's-eye view of Brockbury, half-timbered farm buildings, stables and dovecote, gardens.

Title cartouche, framed. Scale bar. Compass rose. Ruled border.

SCALE: 4 chains to 1 inch [1:3168].

SIZE: 355 x 345 [of photograph]. [Ink and watercolour on parchment].

SURVEYOR: Samuel Addams.

REFERENCE: Private. Photograph, HRO, AT51.

1758
LEDBURY, COLWALL

'A Map of Two Farms call'd Ockridge and Netherton in the parishes of Ledbury and Colwall … the property of John Hager Esq.r'

Ockeridge [SO 750339] and Netherton farms in N of Ledbury with adjoining S part of Colwall. Shows buildings in block plan, lands, mostly enclosed but some strips in open fields, hopyards, woods [News Wood], gates. Fields outlined in colour with letter and number references. Marks 'The Hill Ground', The Rudgeway, Ledbury-Colwall parish boundary. Gives names of adjoining owners.

Reference book not traced.

Map amended by T. Davies in 1807 by adding Thomas's Place and Dugmoor Farm in Ledbury and Eastnor 'for the Right Honourable Lord Som[m]ers the proprietor of the whole'.

Compass rose. Scale bar. Single-ruled border.

SCALE: 'Scale of chains and links', 4 chains to 1 inch [1:3168].

SIZE: 960 x 738. Ink and watercolour on parchment.

SURVEYOR: John Broome, Worcester.

REFERENCE: Private.

1758
GOODRICH

'An Exact Map of the New Wear Works and Lands Allotted thereto for Building Workmen's Houses being part of the Estate of the Hon.ble Thomas Griffin Esquire within the Manor or Lordship of Goodrich … and Part of Wear Hill'.

11a. mostly on the right bank of the river Wye [SO 561156]. Shows houses in crude perspective, roads with destinations, gates. Marks Lock House, lock and weir, Forge pond, Slitting Mill, Fishing pool. Gives names of adjoining owners or tenants.

Reference table.

Oval decorated title cartouche. Contains drawings of a fish [salmon] and a sailing boat on the river being hauled upstream by seven men. Compass point, incorrectly aligned. The scale bar with dividers was on a missing portion of the map.

SCALE: [c.1:745].

SIZE: Photocopy. c.480 x 660. Paper with ragged edges, the bottom left-hand part torn off.

SURVEYOR: Daniel Williams.

REFERENCE: HRO, BR33.

1758
WHITCHURCH

'Survey of Doward Wood Distinguished by the name of the Inclosed Doward Belonging to The Honble Thomas Griffin Esqre Admiral of the Blue Squadron in his Majestys [Navy] in the Manor or Lordship of [Whitchurch in the] County of Hereford Surveyed by Danl Williams'

Doward [Lords] Wood (329a.) [SO 555150] and river Wye between New Weir and Seven Sisters rocks. Shows the woods, cliffs, rock outcrops and river Wye. Marks a house in perspective and limekilns [SO 543155] with a track running along the right bank of the river SE. to riverside cottages, also marked in perspective [SO 556147]. Other place-names include King Arthurs Hall [Cave] and [Dr]opping Well.

Circular 'mirror frame' cartouche with foliage. Compass rose. Scale bar with large dividers.

SCALE: 4 chains to 1 inch [1:3168].

SIZE: c.440 x 640. Ink and watercolour on paper, with damaged edges and some holes.

SURVEYOR: Daniel Williams.

REFERENCE: PRO, Kew, F 17/106. See also [1758] WHITCHURCH.

[1758]
WHITCHURCH

'Map of Great Dorward Lords Wood Little Dorward' (Plate 50)

Duplicate of WHITCHURCH 1758. The style and features are replicated, but with a little difference in scale and size. The house and limekilns are marked at Green Rocks [Seven Sisters] and a road is shown from 'Minepitt Gate [SO 555156] to New Wear'.

No compass point. No scale bar.

SCALE: [c.1:3335].

SIZE: 480 x 670. Ink and watercolour on paper. Repaired.

SURVEYOR: Not named.

REFERENCE: HRO, AS40.

1759
HEREFORD, ALL SAINTS and HOLMER

'To M.r Tho.s Church, This Plan of the Port Fields and other Parcels of Meadow Lands in the Parishes of All Saints and Holmer near the City of Hereford, is humbly Inscrib'd by Your most Obed.t humble Servant to Command Mered: Jones Jan: 1759'

20a. adjoining Old Wide Marsh [Widemarsh] Common, N of the city centre. Shows fields, trees (some with shadows), streams, roads, gates. Marks Wide Marsh mill and mill pond [SO 509408].

Reference table with acreage.

Simple compass rose. Ruled border. Unsophisticated draughtsmanship.

SCALE: [c.1:2545].

SIZE: 330 x 425. Ink and watercolour on paper, linen-backed.

SURVEYOR: Meredith Jones.

REFERENCE: DCA, 4549.

(1759)
HEREFORD

'Plan of the Canon Moor Farm Surveyed in 1759 by Meredith Jones'

Copy made c.1900 of an area N and W of the city centre [SO 510405], showing buildings in block plan, outline field boundaries, including some strips, roads, including Fryers Lane [Coningsby Street]. Gives names of adjoining owners and some field-names.

Inset is an area about 1/4 mile NW of White Cross.

Compass points. Scale bar.

Original map not traced.

SCALE: 3 chains to 1 inch [1:2376].

SIZE: 820 x 450. Ink on oiled paper.

SURVEYOR: Meredith Jones.

REFERENCE: DCA, 3916/58.

1759
MONKLAND

'The MANOR of MONKLAND ... Survey'd May 1759 by J. Ward'

Whole parish. Shows buildings in block plan with an accurate bird's-eye view of the church [SO 460576], open and enclosed fields, hopyards, river Arrow and Stretford river, roads. Marks Almond Bridge and Banister Bridge, smithy and pound. Field-names include Monkland or Shoredale Common, Quarry Piece.

Baroque 'silverware' cartouches for title and scale bar. Overall grid with reference letters and numbers in the margins. Compass rose. Scale bar. Yellow border.

SCALE: 7 1/2 chains to 1 inch [1:5940].

SIZE: 510 x 620. Ink and watercolour on paper, attached to a small roller.

SURVEYOR: J. Ward.

REFERENCE: Private.

1760
CODDINGTON

'AN EXACT PLAN of CODDINGTON FARM, being part of the Estate of Thomas Williams Esq.r in the Parish of Coddington ... Taken and delin.d in the Year of our Lord 1760 by Mered: Jones of Brecon Land Surveyor'

Coddington Farm [Old Farm, SO 718423] (111a.) in the centre and S of the parish. Shows buildings in bird's-eye view, enclosed fields and strips in open fields with plot numbers and field-names, land-use distinguished by pale brown or green colour, orchards, coppices, gates, roads, tracks across the open fields. Marks the church by an accurate drawing. Gives names of adjoining owners. Field-names include Bishops-law and Veldyate (open) Fields, Rakem [Raycomb] and Rakem Coppice [now Berrington Wood].

Note, 'N:B. All these several Scattering Parcels of Lands on this Farm, are taken and delineated here in their exact Proportion Bearings and Distances'.

Cartouche, circular with foliage and flowers. Compass rose with the four cardinal points and four mid points ruled and eight intermediary points with reversed-S wavy lines. Scale bar. Ruled border with diagonal shading (barber's pole effect). The Note is in a ruled frame. Rather crude and thick penmanship.

SCALE: 4 chains to 1 inch [1:3168].

SIZE: 435 x 547. Ink and watercolour on paper. Repaired by the Worcestershire RO.

REFERENCE BOOK: 'A Book of Acres 1760' (on cover). Title page 'A book of Reference to the Plans of Coddington and Eight Oaken Farms'; Wherein are Exhibited the Exact N[o]. of Acres Roods and perches p[er] Statute and Custom contain'd in each and every Piece and Parcel of Lands on the said Farms, as discovered by an Exact Survey taken in the Month of April 1760 p[er] Mered: Jones'.[1] Also contains (p. 4) an explanation of the statutory measurements of perches, roods and acres and the note 'The Statute bears the same Proportion to the Customary Acre as 2 to 3 Particularly in all Arable Lands and Pasture but Meadows most commonly are computed by the days Math which comes near the Statute'. Gives plot numbers, field-names, land-use, statutory and customary acreage. 215 x 123. Paper.

SURVEYOR: Meredith Jones, Brecon.

REFERENCE: Worcestershire RO, (BA 869) 705:24/1365 (map) and 705:24/1366 (book of reference).

1. Meredith Jones's map of Eight Oaks Farm in Castlemorton, Worcestershire, is also in the same archive (705:24/409). Both farms were owned by Thomas Williams (died 1766), who had married Elizabeth Berington in 1749.

(1761)
GOODRICH, BRIDSTOW (MARSTOW)

'A SURVEY of Little Ash farm with other Tenements and Lands in the Parishes of Goodrich, Hentland and Bridstow...the Property of George Roberts Esq.[r] Survey'd by Walter Green 1761'[1]

Ashe Farm [SO 564216] with lands mostly between Luke Brook at Glewstone [SO 557222] and the river Wye, now in the parish of Marstow, with outlying fields in Goodrich and Bridstow.[2] Shows buildings in block plan, enclosed fields with plot numbers, orchards, woods, river Wye with direction of flow, streams, roads and tracks. Marks Pencrecks [Pencraig] Green [SO 563212]. Names adjoining owners.

Oval ornamental title cartouche. Compass points, aligned incorrectly with W [not N] at the top. Scale bar with dividers. Double-ruled frame.

TERRIER: 191a. statutory and 288a. customary, divided by land-use as arable, meadow and pasture, orchards, gardens, folds and waste lands. Contains comments about lands that might beneficially be exchanged with Guy's Hospital. Field-names include Brincourt, the Vargates, Coldwell, Lower Mill Close, Hollywell Pasture, Kedistone Orchard. [Typescript].

SCALE: Of the photocopy, c.13 chains to 1 inch [c.1:10300]; of the original, [4 chains to 1 inch, 1:3168].

SIZE: Of the photocopy 200 x 224; of the original, c.675 x 745. [Ink and watercolour on parchment].

SURVEYOR: Walter Green.

REFERENCE: Private.

1. The map described is a photocopy, reduced in size, of a modern copy of the original in a private collection.
2. The ancient parishes of Bridstow, Goodrich, Hentland, Marstow and Peterstow were much intermingled. Marstow, anciently a chapelry of Sellack, became a separate civil parish in 1777. Boundary alterations in 1884 brought most of the property shown on the map within an enlarged civil parish of Marstow.

1763
BRILLEY

'An EXACT PLAN OF CUMME AND FERNHALL BEING PART OF THE ESTATE OF Thomas Legge Esq[r] Situate in the Parish of Brilley ... Taken and delin[d] in the Year of our Lord 1763 By Meredith Jones of Brecon Land Surveyor'

Endorsed 'An ancient Plan of the Cumma & Fernhall Farm's In the Parish of Brilley and County of Hereford 1763'.

Cwmmau [SO 277512] and Fernhall farms in the NE of the parish. Shows buildings in perspective and profile, enclosed fields in green or yellow with plot numbers, marshy or rough land, woods, coppice and isolated trees, gates, roads and footpaths. Gives names of adjoining lamdowners or tenants. Marks Brilley Mountain. Field-names include Quarry Field, Chymney

Meadow, Cae-myn, Cae-Mair, Kiln Meadow and Wood, Moor Coppices.

Reference table gives plot numbers, field-names, acreage. In the top right hand corner is the note 'Explanation Sixteen Feet and an half Square make one Perch. Forty Perches one Rood, and four Roods one Acre p[er] Statute Vid[elicet] Stat[ute] 33ᵈ 1ˢᵗ.[1]

Title cartouche with decorative border of leaves. Compass rose, coloured. Scale bar. Double ruled border.

Additional notes on some fields with their acreage have been written in a later hand. Ink on paper, pinned to the map.

SCALE: 3 chains to 1 inch [1:2376].

SIZE: 740 x 940. Ink and watercolour on paper, cloth backed. Apparently previously attached to a roller.

SURVEYOR: Meredith Jones of Brecon, land surveyor.

REFERENCE: Leicestershire RO, DG9/Ma/M/27. The catalogue reference is DG9/2726 [previously DG9/1766], a bundle of deeds of the estate 1668-1811.[2]

1. The statute of 33 Edward I (1304-5) c.6 laid down the size of the acre.
2. I am grateful to Dr Margaret Bonney of the Leicestershire Record Office for this catalogue entry.

1763
MATHON[1]

[1]
'THE PARTICULARS, PLAN AND CONDITIONS of SALE of TWO FREEHOLD and TWO COPYHOLD ESTATES, well-tenanted and in good Repair, Situated in the Parish of Mathon, near *Malvern Hills*, (Remarkable for growing the finest Hops) In the Counties of *Hereford* and *Worcester*'[2]

190 statute acres or 285 customary acres comprising (1) Little South Hyde or Baches, 40a. (Herefs.) [SO 739442], freehold; (2) South-End, Wall-house and Berkley's Lands and smith's shop, 43a. (Worcs.) [SO 738445], copyhold; (3) Burford's or Nash's, 23a. (Worcs.) [SO 747446], copyhold; (4) The Braise, 83a. (Herefs.) [SO 733442]. Gives descriptions of buildings and properties, field-names, acreage, annual rents and outgoings. Field-names include hopyards and orchards, Moat Close, The Stone, Sunshall Croft and Mead, Rem's Croft and Mead, Ryles Field, Grudge Moor.

To be sold by Samuel Paterson at The Bell, Worcester on 4 June 1763.

[2]
'Land lying in the Parish of Mathon'

Printed map, hand-painted in four colours to distinguish the lands of the four farms. Shows a compact group of enclosed lands at Ham Green, Smith's Green and South End. Field boundaries and plot references, gates, roads with destination, streams. Marks Shippend [South Hyde] House in profile.

Compass rose. Ruled border.

SCALE: [*c.*1:8600].

SIZE: 170 x 350. Printed and watercolour on paper.

SURVEYOR: N. Darley sculp.

REFERENCE: Bodleian Library, Oxford, Gough Worcs. 14(8).

1. Sale particulars rarely included maps before 1800 and this *Catalogue* does not include other printed sale particulars in its coverage of the manuscript maps of Herefordshire before 1800.
   An exception has been made in this case as these particulars are, and were designed to be, hand-painted. It also serves as a reminder of this source of cartographical evidence. Sale particulars illustrated by maps became common from the 1830s and 1840s when they could be cheaply based upon parish rating and tithe maps.
2. The rural part of Mathon, including this area, was transferred from Worcestershire to Herefordshire in 1897; the built-up part on the Malvern Hills remains in Worcestershire.

1763
RIVER WYE

'A SURVEY of the RIVER WYE from the CITY of HEREFORD to Bixwear Shewing the different Falls where Lock's should be erected to improve the Navigation'

Shows the course of the river Wye from Breinton to Bigsweir, Gloucestershire, riverside towns, villages and landmarks, bridges and ferries. Marks 22 proposed locks to be built at an estimated cost of £20,900.

Scale bar. Compass rose.

SCALE: about 1½ miles to 1 inch (of reproduction) [1:95040].

SIZE: 420 x 140 (of reproduction). Paper, printed.

SURVEYOR: Isaac Taylor.

REFERENCE: Reproduced (indistinctly) and transcribed in *Trans. WNFC*, 1905-7, at end of volume.

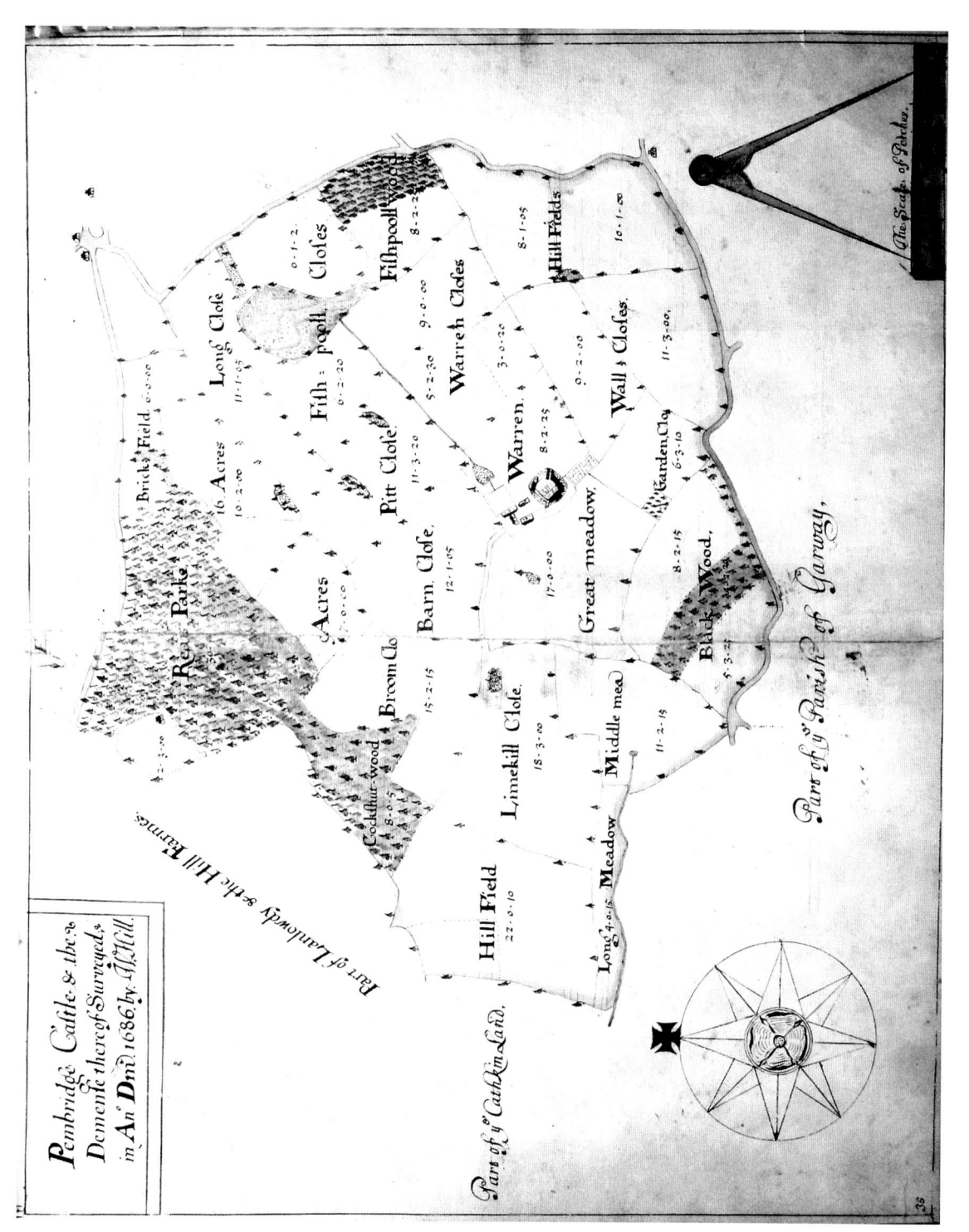

43. Pembridge Castle, Welsh Newton by W. Hill 1686.

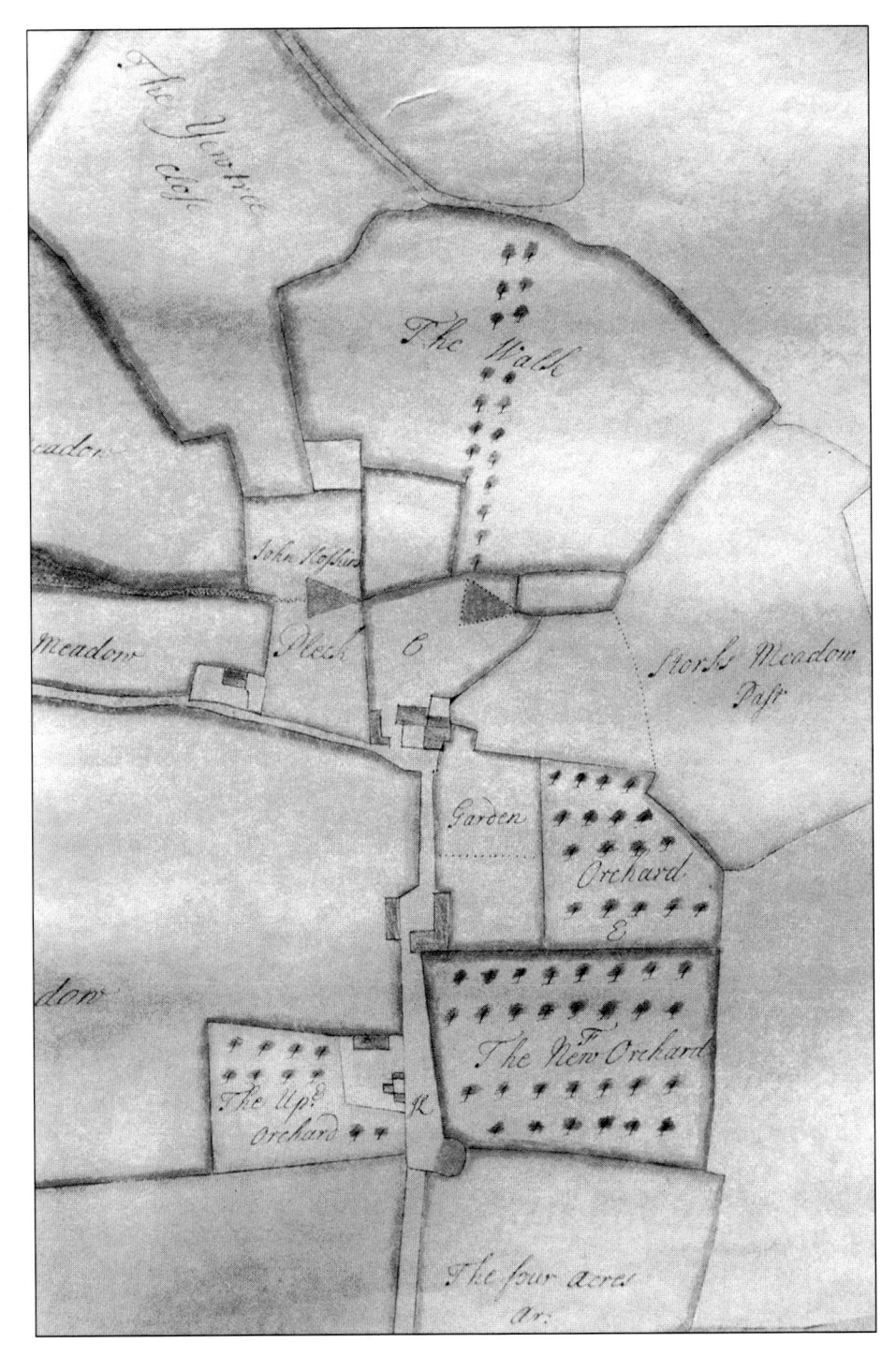

44. *Cwm Maddoc, Garway by John Pye* c.1700.

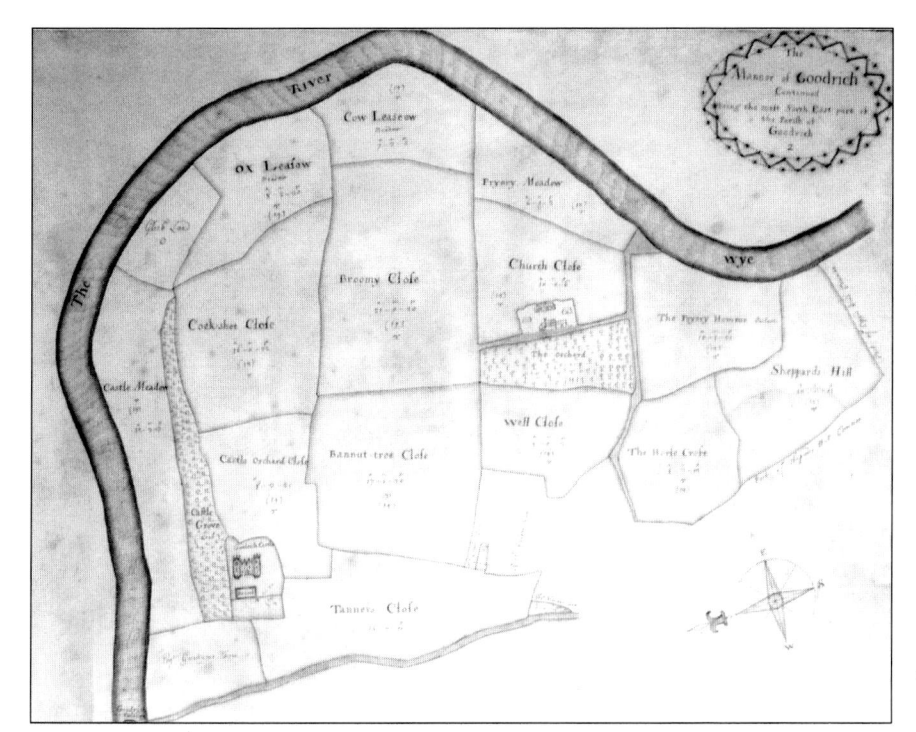

45. *Goodrich Castle by William Laurence 1717.*

46. *'Haversham' by William Laurence 1731.*

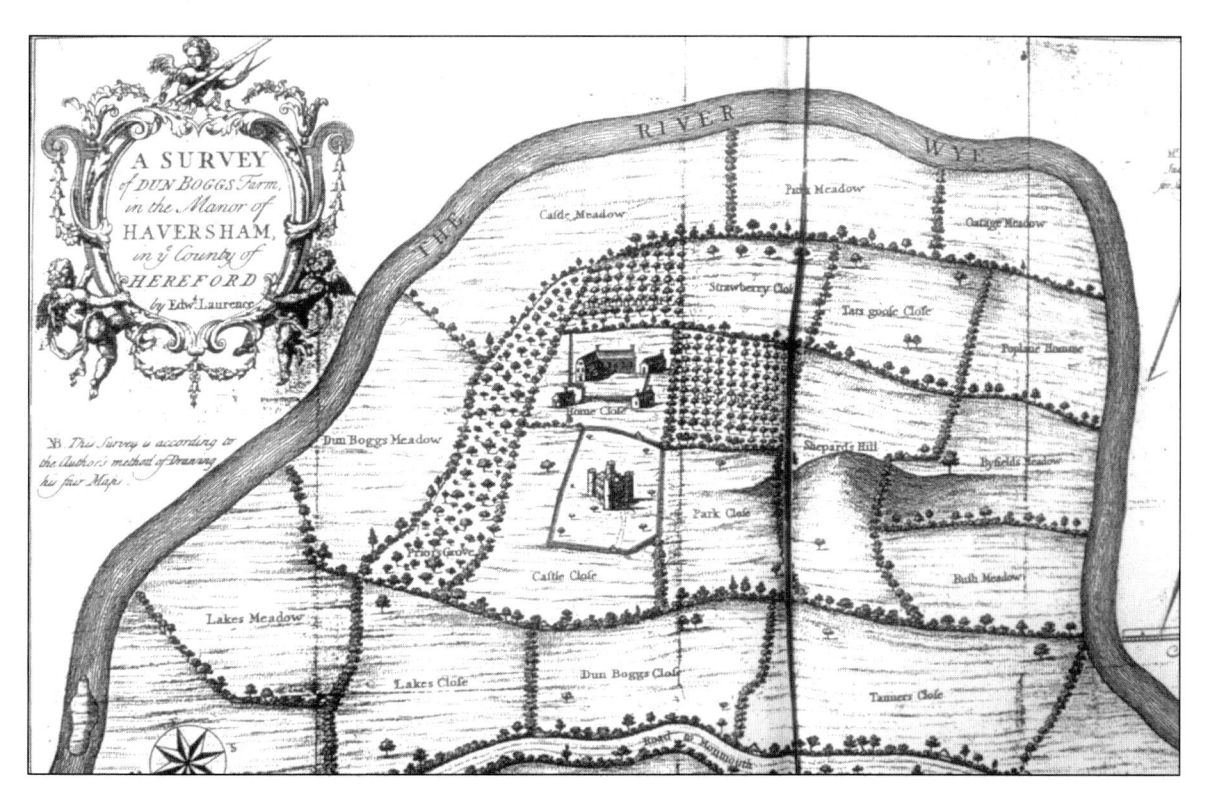

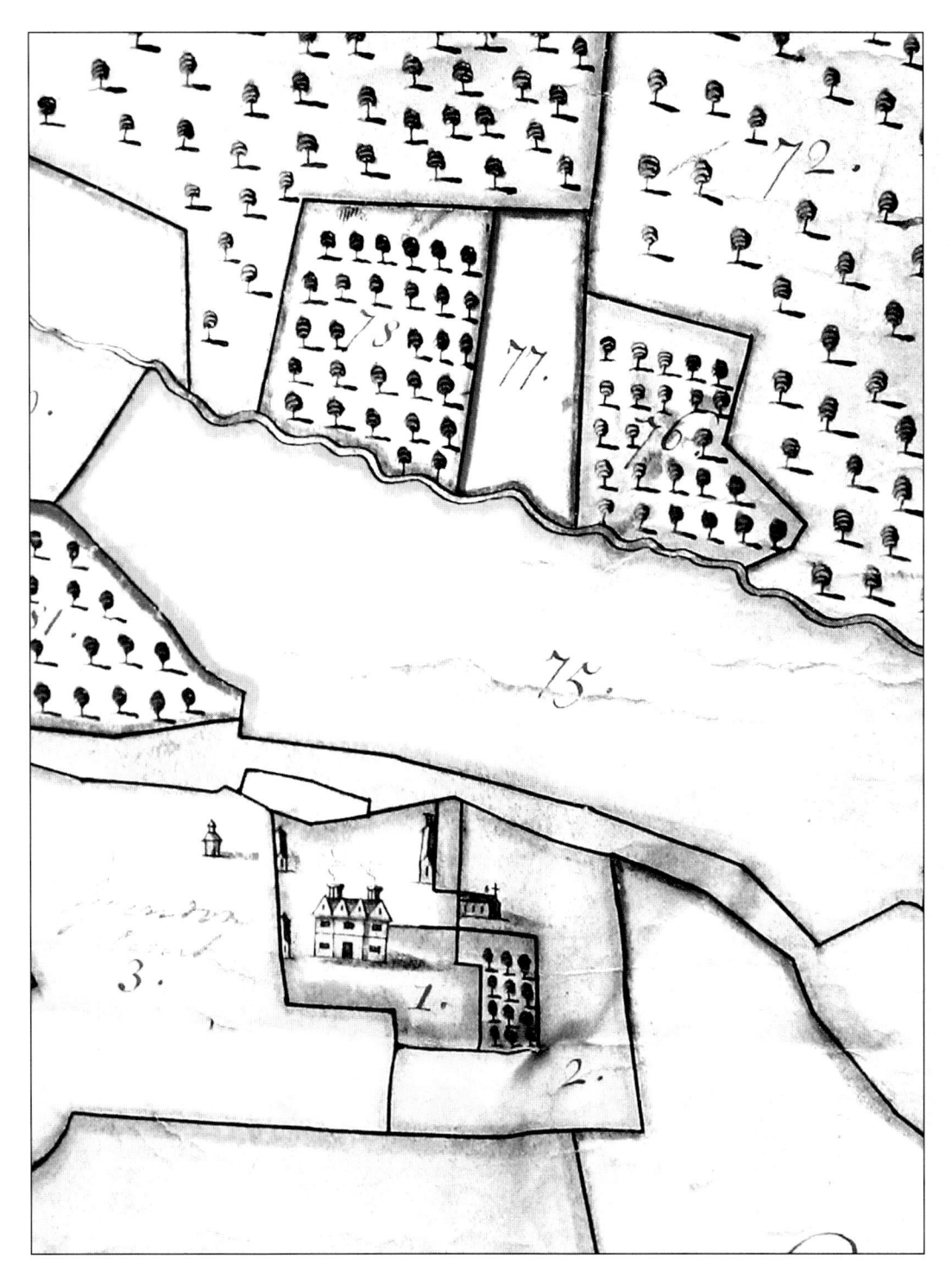

47. *Grendon Court, Grendon Warren detail by Joseph Dougharty 1732*

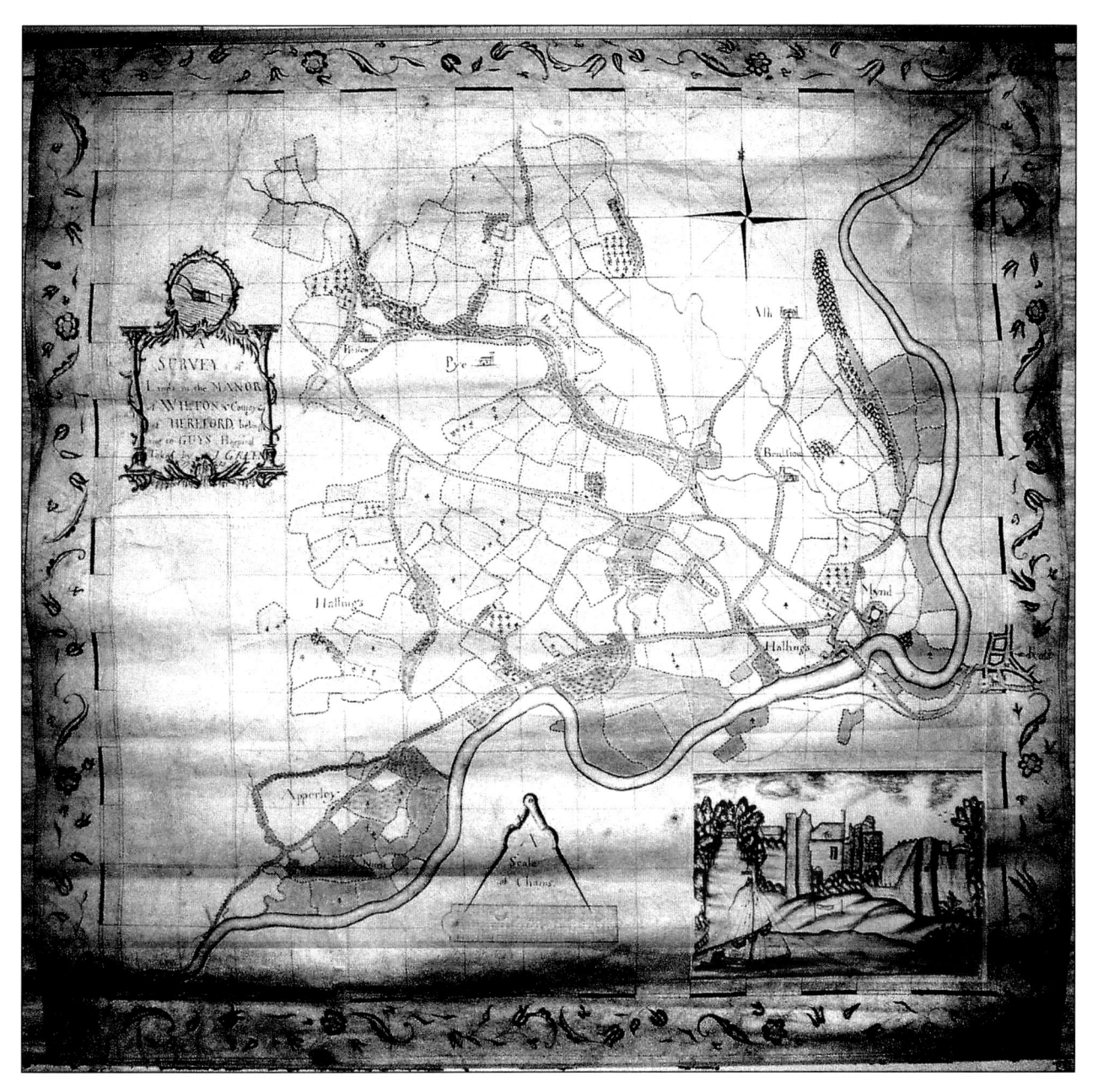

48. *Bridstow by John Green III 1755 (see also Plate 59).*

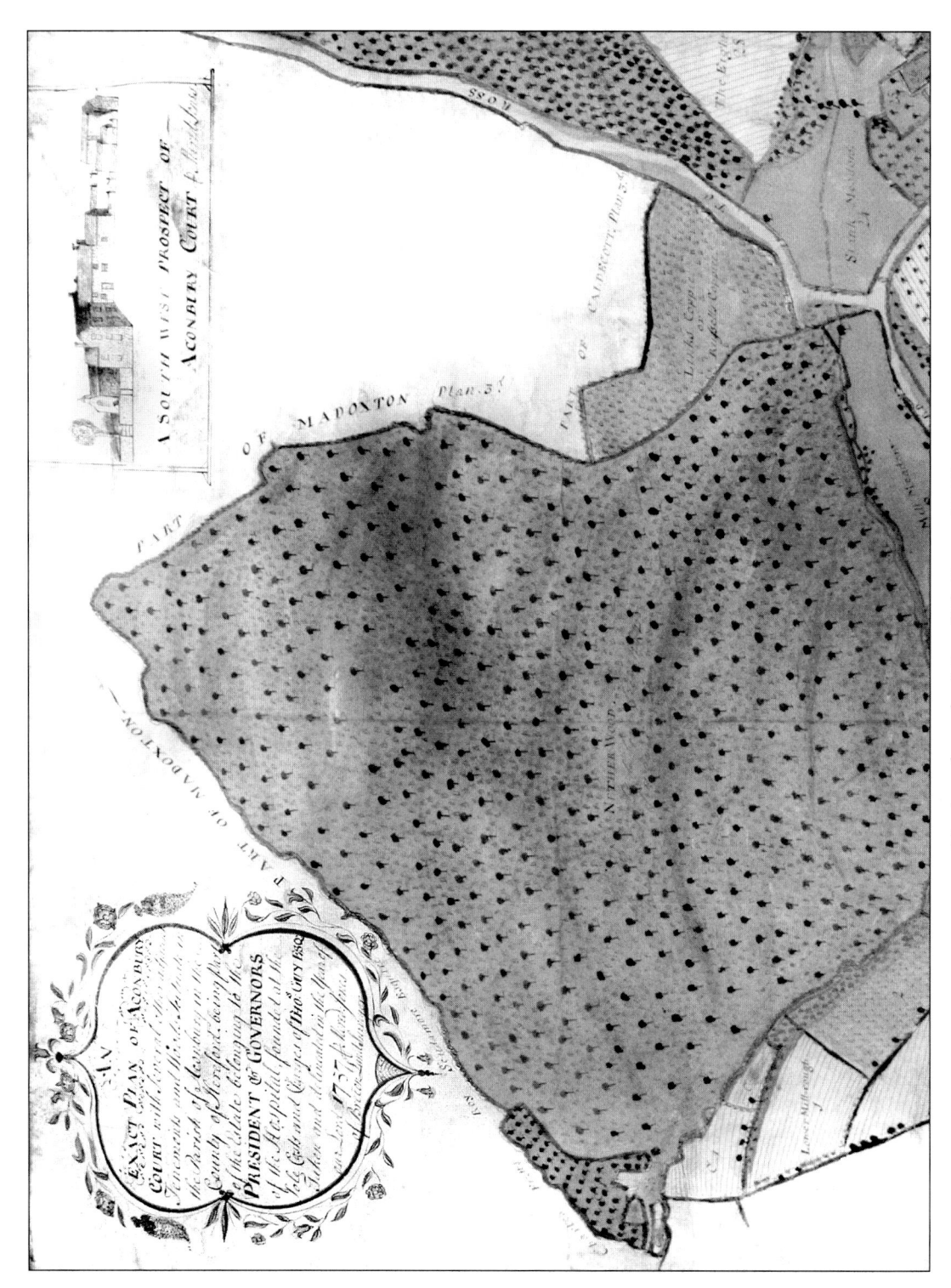

49. *Aconbury by Meredith Jones 1757.*

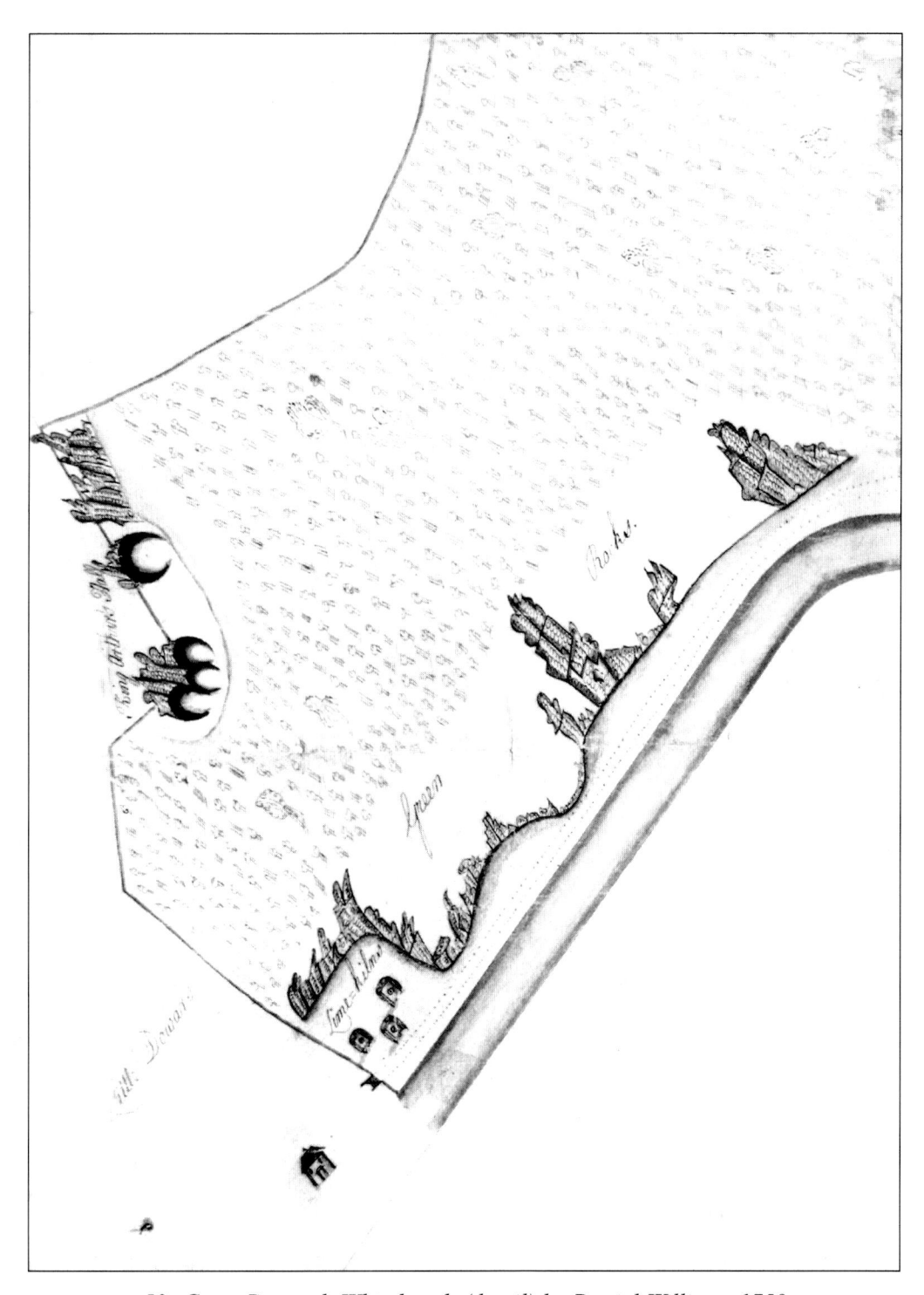

50. *Great Doward, Whitchurch (detail) by Daniel Williams 1758.*

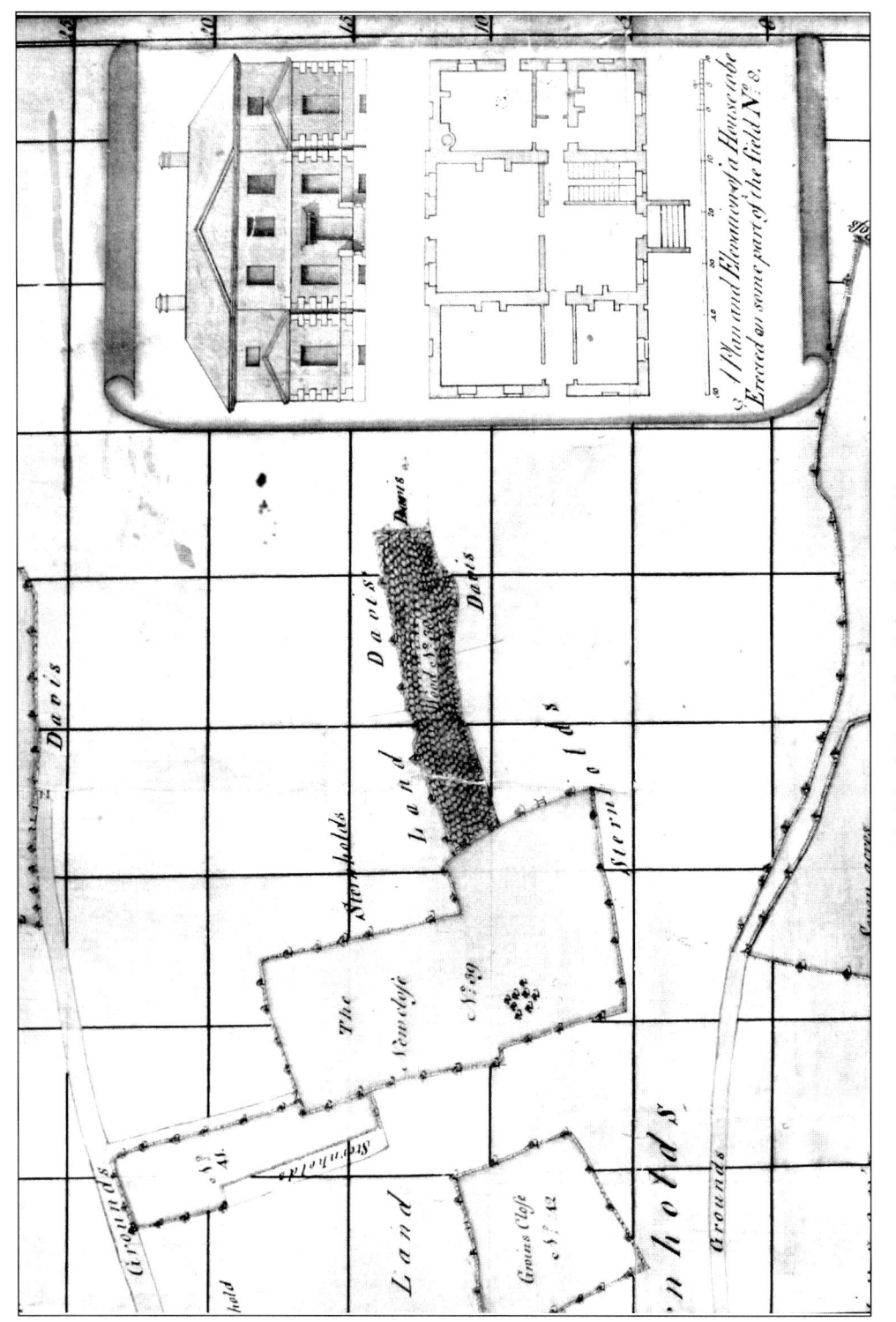

51. *Much Birch by John Bach 1768.*

1764
PRESTON ON WYE

'The DEMESNE lands Of the MANOR Of PRESTON...Being A FARM belonging to the DEAN and CHAPTER OF HEREFORD Drawn by I: Taylor'

226a. in centre and N of parish [SO 3841]. Shows buildings in block plan, fields with a few strips in Canon Common Meadow, orchards, woods, river Wye with islands [SO 374430], streams, 'drain', ponds. Marks church, Preston Court with gardens, parsonage, mill and mill pond. Two distinct tree plantings are shown in Court Orchard. Gives names of adjoining owners and field-names, including Nipsfield, Sandpit Lease, Pound Orchard, Upper and Lower Lakes (E of church).

Reference table with field-names and both statutory (226a.) and customary (339a.) acreage.

Title cartouche. Compass points extending over most of map. Wide ruled border.

SCALE: 88 yards to 1 inch [1:3168].

SIZE: 504 x 923. Ink on paper, linen-backed. The paper is browned and flaking, the reference table being especially badly affected.

SURVEYOR: I[saac] Taylor.

REFERENCE: DCA, 5740. Copied *c*.1830 with changes to Preston Court gardens and orchard; see 1784 CLEHONGER.

1765
THORNBURY

'A MAP of GREAT KYRBATCH LITTLE KYRBATCH and DUCKSLEY'

[197a., SO 622612]. Shows buildings in block plan, enclosed fields with field-names and acreage, woods. Woodland on N boundary partly cleared. Plot numbers added in a later hand, mostly identical with those on 1772 THORNBURY. Gives names of adjoining owners.

Title cartouche, rococo 'silverware' with foliage, a green man (?) and horse. Scale bar entitled 'A Scale of Gunter's Chain'. Compass rose. Rather crude draughtsmanship with heavy delineation of boundary hedges.

SCALE: 4 chains to 1 inch [1:3168].

SIZE: 367 x 587. Ink and grey wash on parchment.

SURVEYOR: Joseph Powel [*sic*].

REFERENCE: Shropshire RO, 1037/23/34.

1766
BYTON

No title. Endorsed 'Byton Mapp'

Three double-page sketch maps of lands (360a.) in the whole of the parish [SO 3764]. Show buildings in profile, enclosed fields and strips in open fields in simple outline, giving field-names and acreage, common land, roads. Marks the church, Common Pool [SO 369623], Staple Bar [Byton Hand, SO 369630]. Field-names include Byton Common, Park Wood, Holleys Hill, Burnt Orchard, Upper and Lower Chilley, Mountain Leasow [SO 367629]. Names adjoining owners.

No N point [E is at the top]. No scale. On the last pages are rough sketches with measurements and bearings of five fields at Shirley Lane.

SCALE: [*c*.1:6670].

SIZE: 410 x 310. The maps are drawn on facing pages of a gathering of 16 pages measuring 310 x 205. Ink and watercolour on paper.

SURVEYOR: Not named.

REFERENCE: Private.

1766
ALMELEY, LYONSHALL

'AN EXACT PLAN OF ALMELEY'S WOOTTON Together with several other Tenements herein particularly Named in the Table of Acres, lying in the Parish of Almeley, and Blackmoor Tenement, in the Parish of Lyonshall ... being the Estate of M[r]. Thomas Prichard. Taken and delineated in the Month of April, in the Year of our Lord 1766, By Meredith Jones'[1]

168a., consisting of Almeley Wootton (155a.) [SO 334525], Upper House at Almelely Wootton (1a.), The Batch (2a.), Wett-Moor (3a.), [Buttington, SO 339526], Hopleys Green house ¼a.], Cisley-Kimmin house (1a.) [SO 338522], all in Almeley, and Blackmoor (6a.) in Lyonshall [The Wood, SO 325537]. Shows houses on the estate in bird's-eye view, enclosed fields and a few strips in Buttington Field, orchards, woods, coppices, streams, roads with destination. Field-names in Almeley include Mill Close, Upper and Lower Berry Fields, The Hopyard, Merry Land, The Nursery by Meeting House Orchard, Quarry Field and Buttington Field common fields. Land-use is indicated by colouring. Gives names of adjoining owners.

Reference table gives plot numbers, field-names, customary (253a.) and statutory (168a.) acreage.

Within a frame is an 'EXPLANATION. The Meadows are distinguished by the deepest Green. Pasture by the lightest Green. And the Arable Land by the Yellow or Stone Colour. N.B. Sixteen Feet and an half Square make one Perch or Pole, Forty Perches one Rood, and four Roods one Acre. The Sta[tute Acre is the] same Proport[io]n to the Customary Acre [as 2] to 3 in all Arable Lands and Pasture; but Meadows are Computed by yᵉ days Math'.

Title cartouche with floral border. Ornate compass rose. Scale bar with dividers. Ruled border with spider's web patterns in top corners.

SCALE: 4 chains to 1 inch [1:3168].

SIZE: 730 x 530. Ink and watercolour on parchment.

SURVEYOR: Meredith Jones.

REFERENCE: NLW, Map 7701.

1. I am grateful to Lona Jones, assistant curator in the Department of Pictures and Maps of the National Library of Wales, for locating and providing photocopies of this and other maps for me.

1767
EYTON

'A Correct PLAN of an ESTATE at or near EYTON ... the property of Joh: Orton, Gent.'

81a, N and W of the church [SO 476615], dovetailing with 1782 EYTON. Shows buildings in block plan, fields, mostly enclosed but some strips in Castle Hill, hopyards, orchards, roads. Marks church and nearby pound. Gives names of adjoining owners.

Reference table drawn as a scroll, giving grid references to a grid superimposed on the map, field-names and acreage.

Cartouche and reference table. Compass rose. Scale bar.

SCALE: 4 chains to 1 inch [1:3168].

SIZE: 580 x 460. Ink and waterclour on parchment, backed with (?a contemporary) newspaper and linen. In poor condition.

SURVEYOR: Thomas Ward.

REFERENCE: HRO, F49/16.

1768
MUCH BIRCH

'A MAP of several Estates at Much Birch ... belonging to JN: WILLIAMS OF WORCESTER GENTᴺ BACH OF HEREF' (Plate 51)

322a. in N of parish. Shows houses in perspective or block plan, gardens, fields outlined in colour, orchards, woods, isolated trees, roads.

Reference table giving field-names, including Tump Field at Wormelow Tump, Tyth Barn near Ash Farm [SO 502308], acreage, land-use.

Title cartouche. Inset plan and elevation drawing of a small Classical house to be erected, but apparently not built, at SO 498316. Scroll with a note about meadow in Much Dewchurch. A grid is superimposed on the map. Scale bar with dividers.

SCALE: 2 chains to 1 inch [1:1584].

SIZE: 122 x 173. Ink and faint watercolour on paper, repaired and linen-backed.

SURVEYOR: [John] Bach, Hereford.

REFERENCE: HRO, J78.

1768
MUCH BIRCH

'Account of my Lands at the Farm Call.ᵈ at the Ash in Much Birch ... Map.ᵈ in 1768. This is not a very exact copy only intend.ᵈ to carry in Pockett to know the Lands ... Jam.ˢ Thomas Tentᵗ.'

Ash Farm [SO 502308]. 173a. in N of parish. Shows buildings in block plan, field boundaries, gates, roads. Gives field-names and acreage, including Tump Field and Tump Leasow at Wormelow. Marks Tithe Barn near Ash Farm and Wormelow Tump [SO 493305].

Crude draughtsmanship.

SCALE: [c.1:3520].

SIZE: 415 x 445. Ink and watercolour on paper.

SURVEYOR: Not named, but based on 1768 MUCH BIRCH by John Bach. Probably re-drawn by the owner in 1768, John Williams of Worcester.

REFERENCE: HRO, K68/17.

1768
MUCH BIRCH

'Acc.ᵗ of my Lands at the Hill farm in Much Birch ... Map.ᵈ in 1768. This is not a very Exact Copy only intend.ᵈ for the Pockell to know the Lands ... Samˡ Sandland Tentᵗ'

58a. adjoining the road from Hereford and Grove Lane [? near Hill Barn Field, SO 497315]. Similar to preceding map of 1768 MUCH BIRCH (HRO, K68/17).

SCALE: [*c.*1:3520].

SIZE: 244 x 302. Ink and watercoloour on paper.

SURVEYOR: Not named, but based on 1768 MUCH BIRCH by John Bach. Probably re-drawn by the owner in 1768, John Williams of Worcester.

REFERENCE: HRO, F68/17.

(1768)
SELLACK

'Map of Sellack Glebe'

118a. in N of parish. Shows buildings in block plan, enclosed lands and strips in open fields, ponds, river Wye, roads. Marks Sellack church [SO 565277] and [King's] Caple church [SO 559288], 'Tump' at King's Caple church. Field-names include Lough Pool Field; (the pool itself is marked but not named).

Map includes 10a. belonging to Mr Symonds and 2a. to Mr Woodhouse.

Reference table.

SCALE: [*c.*1: 6336].

SIZE: 357 x 210. Ink and watercolour on paper, in a volume, vellum and half-leather bound. Copied *c.*1830. Original map not traced. See also 1784 CLEHONGER.

SURVEYOR: Not named.

REFERENCE: DCA, 7004/1, p. 64.

1769
BROCKHAMPTON (by BROMYARD)

'A PLAN of the intended Improvements at BROCKHAMPTON Seat of BARTHOLOMEW RICH[D] BARNEBY ESQ[R] by Tho[s] Leggett 1769'

Brockhampton Park (?241a.) [SO 685551]. Shows Brockhampton House [built *c.*1765] and Warren Farm in block plan, walled kitchen garden, parkland with plantations and isolated trees, with shadows, laid out on a grid plan, a long pool [on the site of Lawn Pool].

'Picturesque' cartouche with the title between a pair of trees and incorporating a classical pavilion. Scale bar with dividers, set squares and protractor. Compass rose. Ruled border.

Reference table giving reference numbers, field-names, acreage (very faint).

SCALE: *c.*3.7 chains to 1 inch [*c.*1:2930].

SIZE: 672 x 910. Ink and grey wash on parchment, framed. Faint.

SURVEYOR: Thomas Leggett.

REFERENCE: National Trust, Lower Brockhampton. Reproduced, much reduced, in M. Hall, 'Brockhampton, Herefordshire', *Country Life*, 4 January 1990.

1769
LEOMINSTER

'BROAD MARSH FIELDS NEAR LEOMINSTER'

Note in later hand 'Cut out of Map of Alcaston Farm in Salop, sold to M.[r] Warren, measured in 1769 by R.P.P. Surveyor' and endorsed in the same hand 'Map of Land at the Broad N.[r] Leominster'.

Spittle House and 38a. [SO 494613] about 1 mile N of the town of Leominster. Buildings shown crudely in perspective, fields and woods outlined in colour. See also 1776 EYE.

Reference table gives acreage and field-names, including Bindam Field, Bindam Meadow.

Decorative border.

SCALE: [*c.*1:3168].

SIZE: 320 x 300 overall, being an L-shaped portion cut out of a large map. Ink and watercolour on parchment.

SURVEYOR: R[obert] P[arry] P[rice].

REFERENCE: HRO, AD4/128.

1769
TITLEY

'THE COLLEGE LANDS AT TITLEY...belonging to the Society of Winchester Made and Measur'd in the Year 1769 By RPP Surveyor'

154a. in centre of parish. Buildings shown in perspective include the church and in plan include Titley Hall. Lands belonging to Winchester College estate outlined in red and coloured [to indicate land-use]. Gives names of adjoining owners.

Cartouche. Scale bar. Compass rose drawn as a semi-circle with the east-west base-line named 'Equinoctal'. Decorative border, matching the decoration of the cartouche and scale bar.

Reference table gives land-use, plot numbers, field-names, acreage and annual value. Field-names include Norman Hill, Burchall [Burcher], Priory, Priory Pitts, Priory Wood and Priory Leesow, Brick Kilns Mead, Knaps, Shall Lands and Shell Field [Shawl].

SCALE: 4 chains to 1 inch [1:3168].

SIZE: 709 x 605. Ink and watercolour on parchment. Some later amendments in pencil including a note of lands exchanged with Edward Harley, 5th Earl of Oxford 1806. Formerly attached to rollers.

SURVEYOR: R[obert] P[arry] P[rice].

REFERENCE: Winchester, Winchester College Archives 21436.[1]

1. I am grateful to Miss Suzanne Foster, Deputy College Archivist for providing me with the details for this entry.

1770
AYMESTREY

[Reference book only; map not traced].

'REFERENCES to the TOWNSHIPS, of Conhope <or Covenhope> AND Aymstry: <In County of Hereford.> Survey'd by, A: Burns. 1770'.[1]

1,236a., giving plot numbers, names of tenants, field-names and acreage. Refers to 'the Maps of the Townships of Conhope and Aymestrey'.

SIZE: Volume, ½ leather-bound with stiff marbled covers, 9 fols.

REFERENCE: HRO, G39/Maps. Collection not catalogued and not available for research.

1. Words in < > inserted later in the same hand.

1770
LYONSHALL

'MAP of WHITTERN · NEXT END and BROOK Farms in the Parish of Lion's-Hall within the County of HEREFORD Belonging to RICHARD HOOPER Esq[r] Survey'd and Map'd by Jn.º Davies 1770'

210a. in NE of parish [SO 3357]. Shows buildings in block plan, fields with letter and number references and outlined in colour, mostly enclosed but a few strips in open fields, orchards, woods, 'watering pools', roads. Marks Sallow Bead [Sally Bed].

Reference table with field-names and acreage. Field-names include Hore Mear, Malt House Meadow, The Abbey Layes, Fish Pool Close, Sheriffs Moor, Spout Meadow, New Street Meadow, Lewis's Witchfield.

Decorative vase-like title cartouche. Compass rose. Scale bar with dividers. Ruled border.

SCALE: 4 chains to 1 inch [1:3168].

SIZE: 630 x 700. Ink and watercolour on parchment. Label on dorse, c.1900: 'From W.H.Cooke & Arkwright, Land Agents and Surveyors, 24, High Street, Mold To Boultibrooke estate, Map of Farms at Lyonshall belonging to R Hooper d/1770'.

SURVEYOR: John Davies.

REFERENCE: HRO, AB 97/23.

1770
YAZOR, MANSEL LACY, BISHOPSTONE, MANSELL GAMAGE, BYFORD, HUNT-INGTON, CLIFFORD, KING'S PYON, BURGHILL, DILWYN, STOKE LACY

'A Book of Survey Containing the Manors of Yazor, Mancellacy, Bishopstone, part of Mancell Gamage, part of Byford, with the contents and Yearly estimats of Uvedale Price Esq,[r] of Foxley … lands in each of them. Also. his Farms at Penland, Clifford, Wooton, Tillington, Dilwyn, Stoke: Lacey … Surveyed in the year of our Lord 1770.'

Estate atlas of 29 maps and 9 written surveys, with index and summaries, covering 5,321a., of which 4,579a. lay in Yazor, Mansel Lacy, Bishopstone and Mansell Gamage. The maps show buildings in block plan, fields or strips in open fields outlined in colour, orchards, woods, gates, streams, ponds, roads. Field tracks or footpaths are marked by pecked lines. The reference tables which follow each map give field-names, acreage and values. Land-use is indicated after the field-name but not consistently noted in a separate column. The short titles on the maps differ a little from the longer titles of the reference tables, in which the names of the tenants are given. Each map has a simple compass rose, not always accurately aligned, but no scale bar. The maps and surveys are numbered.

Uvedale Price (1747-1829), who promoted the 'picturesque' landscape, inherited the Foxley estate in 1761. This survey was carried out shortly after he came of age. The unnamed surveyor produced a somewhat old-fashioned, workaday volume, which served as a base from which Price enlarged the estate and extended its landscaping begun by his father. It contrasts markedly with the finely executed survey for remodelling the estate, which Price commissioned from Nathaniel Kent four years later: see 1774 YAZOR.

[i]   Title page.

[ii]   Index.

1
'Land in Hand'. 860a. Marks the capital messuage [Foxley, SO 414467] in the Foxley valley, flanked by Yazor Wood, Darkhill Wood, The Beaches [Wood]. Field-names include Whitwell Field and New Kitchen Garden.

2
'A Map of the Farm at Yersop' [SO 409476]. 428a. [Yarsop] N of Foxley, including orchards and Ladylift Pasture (68a.). Field-names include Pigeon House Close, Stoney Bridge Field, Whitepit Field, The Brickkilns.

3
'A Terrar [terrier] of the Common Fields of Yazor and by whom Occupied'. 187a. in eight open fields, with the strips differentiated by colour. Gives names of occupiers.

4
'Yazor Farm'. 306a. [Yazor Court, SO 404464] and lands on both sides of the road to Hereford. Marks the former church by Yazor Court. Field-names include Tennis Court Croft, Hop Garden, Park Meadow, Stank Pasture, Summer Pool Field, Sheep Pasture and Sheepcot, Reygrass Pasture and Reygrass Arable.

5
'Upperhampton Farm'. 265a. [Upperton Farm, SO 395469] and scattered enclosed lands. Marks Blacksmiths Shop at the roadside. Field-names include Summer Pool Field, Upper Clay Orchard, Serjants Field, 'The Sheep Slait called Darkhill' (67a.).[1]

6
'Moorhampton Farm', [SO 389468]. 126a. in small closes or open fields, mostly pasture.

7
No title. Three scattered tenements, totalling 35a. and Flour-de-Luce Homestead. Includes a small piece in Summer Pool Field.

8
'Concordance' (no map), a summary total of 2,149a. in Yazor.

9
'Mancell North Field'. 131a. in the open field, with strips differentiated by colour and tenants' names.

10
'Mancell West Field.' 116a. on N and S sides of road to Hereford. Includes 'Mancell West Field'. Shows strips in open field with colours to distinguish tenancies. Field-names include Grove Furlong on N side of road and Road Furlong and Gosefoot Hill on S side.

    [Partly reproduced in S. Daniels and C. Watkins, eds., *The Picturesque landscape. Visions of Georgian Herefordshire*, Nottingham 1994, plate IV].

11
'Mancell Common Mead' and 'Mancell South Field'. Respectively 23a. on the W side of a brook and 30a. as in [9] and [10].

12
'Lands at Mansellacey in the Occupation of Josiah Ridgeway'. 171a. in closes and open fields at Lower Common.

13
'Morgans' and 'Lands at Mancellacey in the Occupation of James Morgan'. 105a., in closes.

14
'Tho: Powell' and 'Lands at Mancellacey in the Occupation of Thomas Powell'. 46a., mostly pasture and meadow.

15
'Lands in the Occupation of John Goodes' and 'Lands at Mancellacey in the Occupation of John Powell'. Respectively 12a. meadow and 71a. enclosed lands, including a hopyard.

16
'Lands in the Occupation of Mrs Dene'. 103a., closes, including Blackwain meadow and pasture.

17
'The Mills' [SE of the church], and 'Lands in the Occupation of Sam, Davis … Timothy Mathews … Mr, Allen … John Foote', respectively 17a., 44a., 12a. and 7a.

18
'Lands in the Occupation of Samuel Bainham'. 66a.

19
Reference table (no map) of small holdings.

20
'The bounds of Buncehill Farm'. Bunshill Farm in detached part of Mansel Lacy, now in Bishopstone [SO 431425]. Shows house

21
'A Terrar of the Parsonage Lands at Mancellacey'. 79a., scattered in closes and strips in open fields.

'Concordance', (no map), a summary total of 1,255a in Mansel Lacy.

22
Reference table (no map) of 'The present Occupiers of each part of the Parsonage Lands'.

23
'Bishopstone Court Farm'. 355a. in Bishopstone. Marks 'The House &c within the Moat', dovecote, orchards, New Hopyard, Pin Garden Hop Ground, Horestone Croft [SO 412432] and Hore Stone Field.

24
'A Terrar of Bishopstone Common Fields'. Similar to (9-11). Lands in Horestone Field (48a., including furlong called Bridge Corner), Claypit Field (9a.), Hill Field (30a.), Puckwell Field (38a.).

25
'Bishen Fields'. Similar to (24). Lands in E of Bishopstone parish in Commonhill Field (23a.), Line End Field (56a., including Shop Furlong and Marsh Gate Furlong), West Field (43a., including Shop Furlong).

26
'Lands at Bishen in the Occupation of Timothy Mathews'. 82a. at Bishon [SO 423432] in E of Bishopstone.

27
'Lands at Bishen in the Occupation of John Foote'. 97a.

28
Reference table (no map) of eight small holdings.

29
'Concordance' (no map), a summary total of 761a. in Bishopstone.

30
'A Terrar of Sheton Fields'. Similar to (24). Lands in East Field (899a.) and Marsh Corner (16a.) at Shetton in Mansell Gamage.

31
'North Field'. 48a. at Shetton, as in (30).

32
'Lands at Sheton in the Occupation of Timothy Mathews'. 125a. at Shetton [SO 407450] in Mansell Gamage, mostly closes, including orchards and hopyards. Field-names include Crabtree Plot, Hazle Mead, Broad Oak Meadow, Emmanual Wody.

33
'Lands at Sheton in the Occupation of James Jones late Llewellins'. 45a. in scattered small plots including Emmanual Pasture.

34
Reference table (no map) of small holdings.

35
'Concordance' (no map), a summary total of 414a. at Shetton in Mansell Gamage.

36-38
'A Terrar and Yearly Estimate of lands in Byford' (no map). 127a. in closes and open fields in various tenancies. Includes 'The Boat House and Ferry' held by James Cook with land opposite 'On the other side of the Water'.

39
'Penland Farm in the Parish of Huntingdon'. Penllan Farm in Huntington [SO 252527]. 54a., closes and orchard. Marks houses in 'Huntingdon Town' and 'An Arable Croft North of the Town the Castle Green on the east'.

40
'Lands at Clifford' (no map). 21a. Field-names include arable by the 'side of Boat Dytch Road', Chantry Mead east of the village, Locksters Pool.

41
'Wooten Farm' [Wootton in King's Pyon, SO 422493] (no map). 98a. in closes and open fields. Field-names include Berd [Beard] Field, Sennery Croft. With the comment 'In Mrs Hulls Pasture Piece called Long Seech is a Parcell of Ground but how much is unknown. Note To have the Stone Examined while this Tenant lives' [Siege Meadow occurs in the tithe award].

42
'Tillington Farm in the Parish of Burgfield' [Burghill, SO 4745] (no map). 80a.

43
'Lands at Dilwyn' (no map). 97a., including hopyards and 'The Moat Orchard at the South East End of the Town' [SO 415544].

44
'An Admeasurement of a Farm in Stoke Lacey' (no map). 264a., including hop grounds.

SCALE: Not given. [Varies between c.1:3168 and c.1:6336].

SIZE: Maps, c.470 x 300, ink and watercolour on paper, calf-bound volume 540 x 380.

SURVEYOR: Not named.

REFERENCE: Private.

1. The word 'slait', for a sheep-pasture, however variously spelt, has not been found in the *Oxford English Dictionary* or among field-names in Herefordshire and Shropshire.

n.d. [c.1770][1]
LEDBURY

'The ARGUS An Estate situated in the Parishes of Ledbury and Donnington ... The Property of MICHAEL BIDDULPH Esq.r'

Argus Farm [SO 710353]. 88a., S of Ledbury town with small portion in Donnington. Shows buildings in block plan, fields (all enclosed), orchards, trees with shadows, roads, tracks across fields, gates.

Elaborate reference table and grid superimposed on the map, giving grid reference numbers, field numbers, field-names, statute and computed acreage, valuations, land-use, timber trees (oak, elm, ash).

Cartouches for the title and 'The Admeasurement & Scheme of Estimation' are delicately drawn. Grid with reference letters. Reference table. Compass points. Scale.bar Ruled and yellow watercolour border.

SCALE: 3 chains to 1 inch [1:2376].

SIZE: 650 x 635. Ink and watercolour on parchment.

SURVEYOR: Not named.

REFERENCE: HRO, G2/III/57.

1. The map dates from before Michael Biddulph's death in 1800 (Burke, *Landed gentry*, 1838). He probably succeeded his father Robert c.1760.

1771
BREINTON, HOLMER, LUGWARDINE

'Dignity Close and Lug Meadows An Estate belonging to the Treasure-ship of Hereford. 1771. Map 2.'

Three small areas, with plot numbers but no field-names or acreage, (1) adjoining the road from Breinton to Hereford, (2) adjoining S side of the road between Tupsley and Lugwardine and (3) the largest, adjoining the river Lugg and the road from Hereford to Lugg Mill [at SO 531419].

The small isolated D-shaped plot (2) adjoining the road between Tupsley and Lugwardine has not been identified. It may be either a piece of the meadow on the S side of the road adjoining the former course of the river Lugg, still marked by the meanders of the parish boundary, or a nearby plot in the parish of Hampton Bishop.

Compass points.

SCALE: [1:2362].

SIZE: 375 x 340. Ink and watercolour on paper, linen-backed.

SURVEYOR: Not named.

REFERENCE: DCA, D121.

1771
MONNINGTON-ON-WYE

No title. Atlas of the Monnington-on-Wye estates of J. Whitmore.

Four maps with reference tables. All have similar features, with buildings shown in perspective, arable and pasture fields distinguished by bold pink and emerald wash, hedges, woods and trees, river Wye with direction of flow, roads with destinations, some gates, more particularly where roads are gated.

Reference tables give plot numbers, field-names, acreage.

Rococo cartouches with patterns of foliage and cherubs. Grid lines superimposed. Compass roses. Scale bars with dividers. Coloured ruled borders. The penmanship is workmanlike but unsophisticated.

fol. 3
'To J: WHITMORE Esq The Following Maps of his ESTATE in the County of HEREFORD are dedicated By his most ob.t Serv.t J: BACH HEREFORD 1771'

Dedicatory note in a rococo cartouche of foliage incorporating two cherubs, one shooting an arrow at a dragon, the other apprently wearing pince-nez.

Grid. Yellow border.

fols. 4-5
[1] 'A MAP of the Manor of MONNINGTON 1771 BACH HEREFORD'

95a., mostly enclosed. Marks Monnington Court [SO 373434], church, islands [SO 375430] and weir [SO 375431] in river Wye, Monnington Walk, Brobury Scar.

133

Title cartouche. Grid. Three compass roses. Scale bar with dividers above and below. Pink border.

SCALE: 8 chains to 1 inch [1:6336].

fol. 6
Reference table.

fols. 7-8
[2] 'A Survey and MAP of Monnington Court ESTATE. J: BACH 1771'

334a.. mostly enclosed but includes a few strips in open fields. Shows Monnington Court and other buildings in perspective and bird's-eye view. Marks Mill [Mill Cottage].

Cartouche, mounted on a pedestal. Grid. Three compass roses. Scale bar with dividers. Emerald border.

SCALE: 4 chains to 1 inch [1:3168].

fol. 9
Reference table. Field-names include Boat Meadow [SO 373425], Beaten Flood Orchard, Millfield and Mill Meadow, Scar Bank, hop grounds.

fols. 10-11
[3] 'Wright's FARM in Monnington. J: Bach 1771'

289a. N of The Long Walk [Monnington Walk], mostly enclosed but some strips in open fields. Marks Wrights farm house near Monnington House.

Cartouche with five cherubs. Grid. Three compass roses. Scale bar with dividers. Pink border.

SCALE: 4 chains to 1 inch [1:3168].

fol. 12
Reference table. Field-names include Old Mill Meadow, Flax Ground, Fish Pool, Park Lawns.

fols. 13-14
'A Map of MATHEWS's farm in Monington. J. Bach. HEREFORD 1771'

154a. adjoining the turnpike road from Hereford to Brecon. Marks Mathews's farm house near Monnington House.

Cartouche, mounted on a pedestal. Two compass roses. Scale bar with dividers and protractor.

Yellow border.

SCALE: 4 chains to 1 inch [1:3168].

fol. 15
Reference table. Fields include some hop grounds.

SIZE: 550 x 440. Ink and watercolour on paper, 16 fols., leather-bound with gold tooling and raised bands.

SURVEYOR: J[ohn] Bach, Hereford.

REFERENCE: Hereford City Library, 912.4244, Woolhope Naturalists' Field Club collections. Given by Sir William Cornewall Bt., 1951.

1771
HOLME LACY, BOLSTONE

'A MAP of Homlacy and Bolston Estate of Charles Howard Junior Esqre and Frances Howard his Lady ... Containing as more particularly sett forth and mentioned in the Map and Reference making together 3862 $\overset{A\ \ R\ \ P}{0\ \ 11}$ English Statute Measure July 1771'[1]

Whole of the parishes of Holme Lacy and Bolstone [SO 8534] and the N part of the parishes of Little Dewchurch [Knap Green] and Ballingham [Hancocks Hill Farm] adjoining the river Wye. Buildings not marked, but shows the gardens and park at Holme Lacy House [SO 555350], outline boundaries of farms and fields, woods, river Wye with islands. Marks strips in Black Marsh.

Engraved title cartouche and compass rose stamped on the paper.

SCALE: 52 perches to 1 inch [1:10296].

SIZE: c.526 x 695. Ink and watercolour on paper. Repaired.

SURVEYORS: Richard Frizell and Charles Frizell, junior.

REFERENCE: BL, Add. MS. 36307, G24.

1. Charles Howard, (11th Duke of Norfolk 1786) married Frances Scudamore, heiress of the Holme Lacy estate in 1771. This is the earliest of the series of maps of the Holme Lacy estates in Herefordswhire and elsewhere made by Richard Frizell in 1771-75, which he brought together in an estate atlas in 1780 (see 1771-1775 HOLME LACY ESTATES, 1774-1775 HOLME LACY ESTATES and 1780 HOLME LACY ESTATES).

c.1771
YARKHILL

No title. Endorsed 'Monkhide. Map of two Copyholds called Badlands and Derry's The Upper Bach & Lower Bach'

A large area apparently N of the hamlet of Monkhide and on both sides of the main road between Hereford and Worcester

[SO 6144], lying in open fields including Upper and Lower Orles, Burleigh Field and Henacre Field. Other field-names include hopyards, Upper and Lower Batch, Batch Green, Burleigh Sling, Old Heath.

SCALE: [*c.*1:3168].

SIZE: 560 x 390. Ink on paper. Torn and crumpled.

SURVEYOR: Not named [?John Edwards].

REFERENCE: Gloucestershire RO, D936/E217, a bundle of estate papers including terriers by John Edwards 1771 and J. Stone 1790 and a rough sketch of Old House demesne [?Yarkhill].

[1771-2]
EASTNOR

'THE FARM AN Estate belonging to the Bishopric of Hereford situate in the Parish of [Eastnor] ... and exchanged by an Act of Parliament for an Estate situated in the Parish of Little Marcle ... belonging to CHARLES COCKS Esq.$^{r'1}$

[Eastnor Farm, SO 730371] and enclosed lands around Eastnor church and W of 'The Ridgeway Road'. Shows buildings in block plan, fields with plot numbers, orchards, trees, streams, pond, roads. Marks Poors Land and Glebe. Gives names of adjoining owners.

Reference book not traced.

Small rococo title cartouche. Compass points. Scale bar with dividers. Yellow border. Graticule. Crude colouring.

SCALE: 4 chains to 1 inch [1:3168].

SIZE: 737 x 580. Ink and watercolour on parchment.

SURVEYOR: Not named.

REFERENCE: Private.

1. The bishop of Hereford obtained a private Act of Parliament to enable him to make this exchange of lands in 1771 (11 Geo. III, c.9). Charles Cocks (1725-1806) of Eastmor was created a baronet 1772 and Lord Sommers of Evesham 1784.

1771-1775
HOLME LACY ESTATES

35 maps of the Scudamore family's estates in Holme Lacy, Amberley, Ballingham, Little Birch, Bolstone, Bridstow, King's Caple, Little Dewchurch, Hentland, Llangarron, Llanwarne, St. Weonards and Welsh Newton, made shortly after the marriage in 1771 of the Honourable Charles Howard, (11th Duke of Norfolk 1786) and Frances Scudamore.

The maps are all similar. They were drawn by Richard Frizell, mostly in August and December 1773, in ink and watercolour on small sheets of paper, many of which have prick points made by the draughtsman. Each has a blank engraved cartouche and compass rose stamped on the paper. The titles within the cartouche were added by hand in ink and contain the same wording, with minor variations. The standard formula of the title is given in full for the first map described below, but thereafter only the significant differences are noted. The compass roses are aligned to the top of the sheet, in a few cases incorrectly.

Each map is of a farm or small part of the estate and shows little more than the outline of the farm boundaries. These are marked in red, but most of the maps do not show the internal field boundaries or buildings. Roads are sometimes marked, painted in ochre. The names of adjoining owners or lands are noted, but only a few field-names.

The simple baroque cartouches incorporate flowers, which on most of the maps have been lightly painted in red. The maps are endorsed with reference numbers. A few are additionally endorsed 'X^d. [Examined] with new Survey'.

The maps have been listed here by date, followed by alphabetical order of parish. For each map details are given of variations in the titles, the area and features of the estate concerned, and the scale, size, date and reference of the map. All of them form part of a collection relating to the Duke of Norfolk's estates presented by Mr. D. H. Barry to the British Library in 1900. See also 1695 HOLME LACY ESTATES and [late 18th century] HOLME LACY ESTATES.

[1] LITTLE BIRCH
'A MAP of New Mill in the Parish of Little Birch and County of Hereford the Estate of the Hble Charles Howard and Frances Howard his Lady Containing 149: A R P 1: 17 Statute Measure laid down by a Scale of 52 perches in one Inch in Augt. 1773 by Rich^d. Frizell'

Marks outline boundary of New Mills Farm [SO 526301] in the E of the parish. Gives names of adjoining properties of Alston [Athelstans] Wood and Blewhinston [Blewhenstone].

SCALE: 52 perches to 1 inch [1:10296].

SIZE: 265 x 370.

REFERENCE: BL, Add. MS. 36307, G11.

[2] BOLSTONE
'A MAP of Bolston Court in the Parish of Bolston ... held by James Smith ... Containing 240. 0. 35... Augt. 1773'

Bolstone Court [SO 552328]. Includes 'Several Pieces of Col. Fitzroys Estate not known and which are Reserved out of this Number of Acres'. The compass rose is incorrectly aligned.

SCALE: 52 perches to 1 inch [1:10296].

SIZE: 272 x 377.

REFERENCE: BL, Add. MS. 36307, G15.

[3] BOLSTONE
'A MAP of Thump Farm in the Parish of Bolston...Containing 15a. 0r. 18p...Augt. 1773'

Tump Farm [SO 563323] in the E of the parish.

SCALE: 20 perches to 1 inch [1:3960].

SIZE: 268 x 374.

REFERENCE: BL, Add. MS. 36307, G14.

[4] HOLME LACY
'A MAP of Romsdon farm held by Isaac Chaddock in the Parish of Homlacy ... Con^g 129a: 2r: 6p ... Aug^t.'

Ramsden Farm in the W of the parish.

SCALE: 52 perches to 1 inch [1:10296].

SIZE: 271 x 372.

REFERENCE: BL, Add. MS. 36307, G47.

[5] LLANWARNE
'A MAP of Scudamores Hill in the Parish of Lanwarne ... Containing 235: 1: 0 ... Augt. 1773'

[Hills Farm, SO 489277]. A single large plot in the W of the parish adjoining the Hereford–Monmouth road and the parish boundary with Orcop.

SCALE: 52 perches to 1 inch [1:10296].

SIZE: 266 x 382.

REFERENCE: BL, Add. MS. 36307, G53.

[6] WELSH NEWTON
'A MAP of Welsh Newton Farm in the parish of Welsh Newton ... Containing 76a: 1r: 33p ... Novr. 1773'

Newton Farm [SO 499180] N and W of the village centre. Shows roads with destinations.

SCALE: 26 perches to 1 inch [1:5148].

SIZE: 270 x 375.

REFERENCE: BL, Add. MS. 36307, G61.

[7] AMBERLEY (MARDEN)
'A MAP of Amberley in the parish of Amberley ... Cont.g 213a: 0r: 6p ... December 1773'

Amberley [SO 5447].[1] 163a. enclosed lands on the W side of the road from Hereford and 40a. scattered in strips in open fields and 17a. in Lugg Meadow. The compass rose is incorrectly aligned.

SCALE: 20 perches to 1 inch [1:3960].

SIZE: 374 x 527.

REFERENCE: BL, Add. MS. 36307, G1. See also 1774 HOLME LACY ESTATES, AMBERLEY.

1. Amberley was a township and ancient chapelry in the parish of Marden. It was briefly a separate civil parish 1866-87.

[8] BALLINGHAM
'A MAP OF Bunways and Forge Meadows in the Parish of Ballingham ... 15a: 2r: 21p ... Copied from William Pyes Survey in December 1773'

SCALE: None [?1:5148].

SIZE: 275 x 375.

REFERENCE: BL, Add. MS. 36307, G10. See also 1775 BALLINGHAM, LITTLE DEWCHURCH.

[9] BALLINGHAM
'A MAP of Churches part of Carey Farm [SO 563314] in the Parish of Ballingham ... Containing 9: 1: 3 Copied from an old Map at Homlacy in December 1773'

One field of Thomas Church's farm forming part of Carey Farm.

SCALE: None [?1:5148].

REFERENCE: BL, Add. MS. 36307, G9. See also 1775 BALLINGHAM, LITTLE DEWCHURCH.

[10] BRIDSTOW
'A MAP of Two Pieces of Land Caled The Brooms in the Township and Parish of Bridstow...Containing 11: 2: 25 ... December 1773'

[The Broome, SO 569250]. Two plots of 6a. and 5½a., one adjoining the road to Ross.

SCALE: 26 perches to 1 inch [1:5148].

SIZE: 268 x 378.

REFERENCE: BL, Add. MS. 36307, G16.

[11] LITTLE DEWCHURCH, BALLINGHAM
'A MAP of Nap Green Farm and part of Billingham Meadows ... Containing all Together 160:1: 34 ... December 1773'
    Knapp Green Farm [SO 550317]. (130a.) in Little Dewchurch and 27a. in Ballingham Meadow.

SCALE: Knapp Farm, 52 perches to 1 inch [1:10296]; Ballingham Meadow 16 perches to 1 inch [1:3168].

SIZE: 265 x 370.

REFERENCE: BL, Add. MS. 36307, G8. See also 1775 BALLINGHAM, LITTLE DEWCHURCH.

[12, 13] HENTLAND
'A MAP of Altbough in the Parish of Hentland ... Containing 71a: 3r: 24p ... Copied from an old Map at Homlacy in December 1773'
    Altbough [SO 547300]. Scattered closes in the NE of the parish N of Hoarwithy church. Field-names include Bringwin, Black Pitts, Biblet Meadow and 'part of Coney Farm'.

SCALE: None [?1:10296].

SIZE: 380 x 535.

REFERENCE: BL, Add. MS. 36307, G20 and G21 duplicate, repaired.

[14, 15] HOLME LACY
'A MAP of part of Billingsby and part of Bolston in the Parish of Homlacy ... Containing 296a: 3r: 36p ... December 1773'
    Lands at Billingsley Farm [SO 535333] in the SW of the parish. Notes 'Ab^t. Eight acres of Charity Lands Intermixed Included but not Demised'.

SCALE: 52 perches to 1 inch [1:10296].

SIZE: 267 x 376.

REFERENCE: BL, Add. MS. 36307, G27 and G28 duplicate.

[16] HOLME LACY
'A MAP of Birton Farm &c held by M^r. John Price in the Parish of Homlacy ... Containing Together 157a: 0r: 28p ... Decemb^r. 1773'
    Two plots comprising Birton Farm [SO 570348] E of the church and Broad Meadow (103a.), adjoining the Deer Park. The compass rose is incorrectly aligned.

SCALE: 26 perches to 1 inch [1:5148].

SIZE: 270 x 375.

REFERENCE: BL, Add. MS. 36307, G29.

[17, 18] HOLME LACY
'A MAP of the Boat Piece Held By M^r. Gilbert Jones in the Parish of Homlacy ... Containing 56a: 3r: 29p ... Decemb^r. 1773'
    Two plots in Boat Piece adjoining the river Wye and the ferry crossing to Fownhope [SO 535357]. The compass rose is incorrectly aligned.

SCALE: 26 perches to 1 inch [1:5148].

SIZE: 270 x 374.

REFERENCE: BL, Add. MS. 36307, G30 and G 31 duplicate.

[19] HOLME LACY
'A MAP of part of Bochmarsh Farm and other Lands in the Parish of Homlacy ... now held by John Hutson ... Containing Together 305a: 2r: 15p ... Decemb^r. 1773'
    Bochmarsh [Bogmarsh, SO 537345] Farm (174a.) in the W of the parish and meadows by the river Wye (132a.) including 47a. in Boat Piece and 'Reynolds's House & Land Belonging to the Ferry'.

SCALE: 26 perches to 1 inch [1:5148].

SIZE: 382 x 537.

REFERENCE: BL, Add. MS. 36307, G 32.

[20, 21] HOLME LACY
'A MAP of Bower Farm in the Parish of Homlacy ... Containing Including half the River 271a: 3r: 14p ... December 1773'
    Bower Farm N of Holme Lacy village (260a.) and part of the river Wye (11a.).

SCALE: 26 perches to 1 inch [1:5148].

SIZE: 266 x 379.

REFERENCE: BL, Add. MS. 36307, G35, G36 duplicate.

[22] HOLME LACY
'A MAP of Dinedor Mill Farm and Fidos Meadow in the Parish of Homlacy ... Containing Together 94a: 3r: 17p ... December 1773'
    Dinedor Mill Farm [SO 542362] in the N of the parish; Fidos Meadow adjoining the Deer Park.

SCALE: 52 perches to 1 inch [1:10296].

SIZE: 271 x 380.

REFERENCE: BL, Add. MS. 36307, G39.

[23, 24] HOLME LACY
'A MAP of the Gannagh Farm &c in the Parish of Homlacy ... Containing Together 244a: 2r: 32p ... December 1773'
    [Gannah] Farm [SO 546335]. (137a.) in the S of the parish and Cunnigar to its N (72a.).

SCALE: 52 perches to 1 inch [1:10296].

SIZE: 270 x 374.

REFERENCE: BL, Add. MS. 36307, G41 and G42 duplicate.

[25] HOLME LACY
'A MAP of Holanton in the Parish of Homlacy ... Containing 244a: 3r: 19p ... December 1773'
    Hollington Farm in the SE of the parish [SO 562338].

SCALE: 52 perches to 1 inch [1:10296].

SIZE: 270 x 379.

REFERENCE: BL, Add. MS. 36307, G44.

[26] HOLME LACY
'A MAP of Peregrine Princes Farm of Bochmarsh ... Including part of the Cunniger Luggsmouth and Orles Meadows Containing in The Whole 246: 0: 9 ... Decb 1773'
    [Bogmarsh] Farm [SO 537345] 'Caled Henry Gainess farm' (157a.), Luggsmouth [SO 565370] and adjoining Orles Meadows on the right bank of the river Wye opposite the mouth of the river Lugg.

SCALE: 52 perches to 1 inch [1:10296].

SIZE: 270 x 375.

REFERENCE: BL, Add. MS. 36307, G33.

[27] HOLME LACY
'A MAP of William Reynolds's Farm in the Parish of Homlacy ... Containing 12a: 2r: 24p ... December 1773'
    One plot adjoining Boat Piece in the NE of the parish by the river Wye.

SCALE: [?1:5148].

SIZE: 270 x 380.

REFERENCE: BL, Add. MS. 36307, G46.

[28] TRETIRE [LLANGARRON]
'A MAP of Treyevan in the parish of Tretire ... Cont.g 49a: 3r: 10p Statute Measure Copied from an old Map at Homlacy in December 1773'
    [Tre-Evan is in Llangarron parish, SO 522219].

SCALE: [?1:5148].

SIZE: 380 x 538.

REFERENCE: BL, Add. MS. 36307, G57. See also 1775 TRETIRE, LLANGARRON, ST WEONARDS.

[29] KING'S CAPLE
'A MAP of part of King's Caple in the Parish of King's Caple ... held by John Roberts Esqr. Containing as by the Reference 123a: 1r: 1p ... May 1774'
    Lands in the W of the parish, showing fields outlined in red, with acreage, field-names and land-use, roads. Reference table gives plot numbers, field-names and acreage. Field-names include Redrails [SO 547385] on the left bank of the river Wye opposite Red Rail Farm in Ballingham, Bannetree Field, Castle Field [SO 562287], Eldernfield and Craize Oak, Windmill Hill [SO 554290], Fishpool.

SCALE: 26 perches to 1 inch [1:5148].

SIZE: 375 x 540.

REFERENCE: BL, Add. MS. 36307, G50.

[30, 31] BALLINGHAM
'A MAP of Ballingham Hall Farm in the Parish of Ballingham ... 434: 2: 31 ... 1775'
    Ballingham Hall [SO 576317].

SCALE: 26 perches to 1 inch [1:5148].

SIZE: 380 x 540.

REFERENCE: BL, Add. MS. 36307, G6 and G7 duplicate.

[32] KING'S CAPLE
'A MAP of part of King's Caple in the parish of King's Caple ... Containing by the Reference 147.1.4 ...' [undated, c.1774]
    Scattered lands. Reference table giving plot numbers, field-names, acreage. Ruled border.

SCALE: 26 perches to 1 inch [1:5148].

SIZE: c.475 x 634. Damaged and repaired.

REFERENCE: BL, Add. MS. 36307, G51.

[33] KING'S CAPLE
'A MAP of part of King's Caple held by Frances [sic] Woodhouse Esq[r] ... which Exc[l]usive of Littleloes Meadow which is not in this Map Contains 51a: 0r: 33p ...' [undated, c.1774]

Three small plots, one of which abuts the Hereford-Ross road, leased to Francis Woodhouse of Aramstone [d. 1791]. The compass rose is incorrectly aligned.

SCALE: 26 perches to 1 inch [1:5148].

SIZE: 270 x 375.

REFERENCE: BL, Add. MS. 36307, G52.

[34] LITTLE DEWCHURCH
'A MAP of Cary Farm in the Parish of Little Dew Church...held by the Widow Churchman ... Containing [blank] ...' [undated, c.1773]

Lands in the E of the parish on both sides of the Hereford-Ross road at Carey [SO 563314]. Gives field-names, land-use, acreage. Field-names include Quarryhead [SO 562309].

SCALE: 26 perches to 1 inch [1:5148].

SIZE: 270 x 377.

REFERENCE: BL, Add. MS. 36307, G17. See also 1775 BALLINGHAM, LITTLE DEWCHURCH.

[35] ST WEONARDS
'A MAP of Troughland Farm in the Parish of St Weonards ... Containing 123:0:27 ...' [undated, c.1773]

Trothland Farm [SO478224].

SCALE: 26 perches to 1 inch [1:5148].

SIZE: 271 x 374.

REFERENCE: BL, Add. MS. 36307, G55.

1772
MOCCAS, BREDWARDINE, DORSTONE, PETERCHURCH, CUSOP, CLIFFORD, CLODOCK [CRASWALL][1]

'A SURVEY of the MANORS of Moccas, Bredwardine, Grove, Radnor, Wilmaston, and Cusop, and a Farm at Crosswall; situate in the Parishes of Moccas, Bredwardine, Dorston, Peter-church, Cusop, Clifford, and Cluttuck, in the County of HEREFORD: belonging to Sir George Cornewall Baronet by John Lambe Davis. 1772'[2]

Moccas estate atlas of a fine character and convenient size. The six maps are spread through the volume, each being pasted to the dorse of a folio, folded to page size and followed by the relevant reference tables. The volume is paginated to page 31 and then foliated to the end. Most of the maps are dated 1772, but the surveys on fols. 103-106 were added later and are not mentioned in the introductory notes (p. xi). The accounts of chief rents and estate payments are dated 1773 (fols. 95, 101) and there are a few later undated notes about exchanges of lands.

The estate comprised some 4,684a. in Moccas, Bredwardine and adjacent parishes, with outlying lands in Cusop and Craswall. The maps mark buildings on the estate in block plan and other cottages and houses in conventional perspective drawings. Fields are outlined in green, with plot numbers, and waste lands such as Bredwardine Hill and Cusop Common are coloured with a pale green wash. Orchards, woods, trees with shadows, rivers and streams, pools, roads are marked. Names of adjoining landowners are given.

Each map has compass rose, scale bar and, except for [6] Craswall, a ruled border with decorative foliage at each corner.

The maps are all in ink and watercolour on parchment.

p. ix. Title page.

p. xi. Explanatory notes about abbreviations and contents.

p. xii. Map of Moccas, folded. See [1] below.

pp. 1-13. Reference tables, giving plot numbers, properties and field-names, tenants, land-use and acreage of tenanted land at Moccas.

p.14. Map of Peterchurch, folded. See [2] below.

pp. 15-17. Reference tables for lands at Peterchurch.

pp. 18-19. [Loose] Map of Dorstone, folded. See [3] below.

pp. 19-21. Reference tables for lands at Dorstone.

p. 22. Map of Bredwardine, folded. See [4] below.

pp. 23-31. Reference tables for Bredwardine.

fols. 33-67. Abstract of the previous reference tables, giving plot numbers, field-names and acreage arranged by names of tenants.

fol. 68. Abstract of the manors of Moccas, Bredwardine, Grove, Radnor and Wilmaston, folded, 335 x 670.

fols. 70-80. Abstract of freeholders in Moccas.

fol. 81. Map of Cusop, folded. See [5] below.

fols. 82-89. Survey of the manor of Cusop, giving plot numbers, field-names and acreage arranged by names of tenants.

fol. 90. Map of Clodock [Craswall], folded. See [6] below.

fol. 91. Survey of farm at Clodock [Craswall], giving plot numbers, field-names and acreage.

fol. 93. 'Waste Lands on the above Estates' (1,151a.), including Arthur's Stone Mountain in Bredwardine (124a.), Merbidge [Merbach] Point in Clifford (244a.) and Cusop Common (682a.).

fol. 95. 'Payments out of the Estate' 1773.

fols. 98-99. Index of tenants.

fol. 101. Chief rents 1773.

fols. 103-105. Particulars of Benfield Farm (207a.) and Upper Weston Farm (178a.) in Bredwardine purchased from Thomas Downs, giving plot numbers, field-names, land use, acreage.

[fol. 106]. Bottrell Farm (44a.) [N of Arthur's Stone, in Bredwardine] and Llanavan [Llanafon] Farm (29a.) in Dorstone.

[1] 'PARISH OF MOCCAS' (Plate 41)
Whole parish (1,079a.). Marks Moccas Court and church [SO 357435], strips in The Worth [SO 355421], The Lawn, Lawn Pool, The Little Park, Moccas Common. Field-names include Kites Hill Wood, Great Woodbury Hills, The Brickilns [SO 352428].

SCALE: 10 chains to 1 inch [1:7920].

SIZE: 395 x 434.

[2] 'Lands in PETERCHURCH'
Wilmaston [SO 341402] (117a.). Includes some strips in Peterchurch Field, Park Field, Peterchurch Meadow and Lower and Upper Wilmaston Fields. Other field-names suggest recent enclosure. Marks river Dore.

SCALE: 10 chains to 1 inch [1:7920].

SIZE: 272 x 370.

[3] 'Lands in Dorston'
743a. including Llanafon farm [SO 326415] and common land on Kyntyns Hill [Dorstone Hill] with boundary landmarks including two isolated yew trees. Marks Lodge Grounds [Cross Lodge], Bodcott [Lower Bodcott farm, SO 337431] with adjoining Kiln Piece. Inset, three detached plots near Snodhill and Vagar Hill.

SCALE: 10 chains to 1 inch [1:7920].

SIZE: 390 x 360.

[4] 'Lands in BREDWARDINE' (Plate 42)
1,158a., including common land on Bredwardine Hill and Arthur's Stone Mountain. Marks hopyards, orchards, river Wye and Bredwardine bridge, Old Court [SO 335448] and adjoining field called The Radnor (with the moat pencilled in a later hand), Old Castle [SO 312432], Cae Perthy, Bottrell and Cae Lloyd (one of two cottages opposite Arthur's Stone) [SO 318433], wells including 'Maiden Well, a bound stone' [SO 309449]. Field-names include Lawn Pool in Moccas, Mill Coppy [Coppice] [SO 343435], Castle Coppy, Castle Field, Castle Orchard and Castle Green S of the church, Kyntyns [Dorstone] Hill, Sundays Hill [SO 314450], Merbidge [Merbach], Clay Pitts, Mill Field [SO 335450], strips in Common Meadow [SO 325465], Horse Shoe Meadow (within the horseshoe bend of the river Wye, since cut through to become a pool named Horse Shoe Bend), Benfield Park, Lower and Upper Old Castle (on the river bank) [SO 325469], The Glisshire on N bank of the river [SO 323470].

SCALE: 10 chains to 1 inch [1:7920].

SIZE: 475 x 682.

[5] 'MANOR of CUSOP'
818a. Marks river Wye with islands [SO 230430] and The Gliss on N bank [SO 230429], Cusop Mill [SO 240414] and Cae Whipping Post [SO 238415], The Craggy [Craigau], and a quarry [?SO 266406]. Field-names include Mouse Castle Wood and Scudamore Common.

[The compass rose on the map is aligned incorrectly].

SCALE: 11 chains to 1 inch [1:8712].

SIZE: 340 x 640.

[6] 'A Survey of a Farm at Crosswall in the Parish of Cluttuck … belonging to Sir Geo: Cornewall Bar.[t] By John Lambe Davis 1772'

149a. [The Birches, SO 257381] in Craswall. Marks Brook Dulas, Parc y Meirch, Cae Quarry [SO 259384], a gate across the lane at the county boundary [SO 254382]. Field-names include Field of Battle [SO 265381].

Floral cartouche. Plain ruled border.

SCALE: 4 chains to 1 inch [1:3168].

SIZE: 417 x 510.

SURVEYOR: John Lambe Davis.

SIZE: Volume, 237 x 180, with folded maps, ink and watercolour on paper and parchment. xii + pp. 1-31 and fols. 32-111, of which the first 10 pp. and last 3 fols. are paper. Calf-bound, blind tooling and raised bands. Spine label 'HEREFORDSHIRE ESTATES OF SIR GEORGE CORNEWALL - DAVIS'.

REFERENCE: Hereford City Library 912.4244, Woolhope Naturalists' Field Club collections. Given by Sir William Cornewall, 1951.

1. Craswall, an ancient chapelry of Clodock, became an ecclesiastical parish in 1728 and a civil parish in 1866.
2. Velters Cornewall, M.P., who died in 1768, left the Moccas estate to his daughter Catherine. She married Sir George Amyand, bt., in 1771, when he took the surname Cornewall.

1772
STOKE PRIOR

'A Map of a Tenement and Lands at Stoke Prior belonging to The Right Honble Lady Fra.s Coningsby'

14a. [Hill Top, SO 530562] in E of parish. Shows buildings in perspective, scattered strips in open fields, orchards, roads, some crossing through open fields.

Reference table giving field-names, acreage, land-use.

Title cartouche. Compass rose. Scale bar. Rather crude draughtsmanship.

SCALE: 4 chains to 1 inch [1:3168].

SIZE: c.535 x 380. Red and black ink on paper, torn from a volume of surveys, p. 241.

SURVEYOR: J. Harris, junior.

REFERENCE: HRO, C69/5.

1772
THORNBURY

'A MAP of GREAT KYRBATCH LITTLE KYRBATCH AND DUCKSLEY…Survey'd by Joseph Powell 1772'

197a. in the N of the parish [SO 6261]. Shows houses in perspective, outbuildings in block plan, enclosed fields, hopyards, orchards, woods, an isolated hedgerow tree, gates. Gives names of adjoining owners.

'A TABLE of the CONTENTS' gives plot numbers, field-names, acreage. Field-names include Sallybed Mead, Coaney.

See also 1765 THORNBURY of the same estate. Since 1765 two small coppices or woods had been cleared and the woodland on the N boundary had been further reduced.

Title cartouche in rococo 'silverware' frame incorporating flowers and foliage. Scale bar. Compass rose. Ruled border.

SCALE: 4 chains to 1 inch [1:3168].

SIZE: 440 x 535. Ink and grey wash on parchment.

SURVEYOR: Joseph Powell.

REFERENCE: Shropshire RO, 1037/23/35.

1772-1774
COLLINGTON, BOCKLETON, EDWYN RALPH, LEOMINSTER, STOKE BLISS, THORNBURY, WACTON[1]

'Survey and Valuation of the Estate of Edmund Pytts Esq:r in the Counties of Hereford and Worcester'

Two volumes of a survey of the Kyre Park estate, Worcestershire [SO 6263], with lands in west Worcestershire and north-east Herefordshire. The farms and holdings are drawn separately on small maps followed by reference tables, statements of the landlord's and tenants' covenants and concluding with the surveyor's 'Remarks'. The leasehold tenancies are described first, followed by those on leases for lives, but within that broad division the arrangement of the individual holdings appears to be erratic and certainly not grouped by parish. At the end are general observations, recapitulations of contents and of rents, and an index. On the last page of the second volume is the note 'This Survey was made in the Years 1772, 1773 and 1774, by Nath Kent Of Fulham Middlesex'.

The maps, which differ in size by a few millimetres, are designed to fit the page by varying their scale. They show buildings in block plan, fields with plot numbers and boundaries, land-use, woods, streams, roads. Each has a simple scale bar, N point and ruled border. They are neatly drawn in ink and watercolour. The reference tables give plot numbers, field-names, which include hopyards, acreage, and the covenants are set out in detail.

Kent's 'Remarks' contain comments on the property at the time of his survey and recommendations for its improvement.

Herefordshire Maps

They include proposed exchanges of land to consolidate holdings, variations in boundaries for better husbandry, the effects of enclosure, drainage, clearance of scrub and fern, the enhancement of arable by liming and manuring, growing turnips, the management of alder groves for hop-poles and their protection against grazing, the eradication of moss, the removal of ant-hills and the revaluations that such improvements might bring about.

The estate straddled the county boundary and the parishes were formerly much intermingled. The ancient parishes of Bockleton and Stoke Bliss were divided between the two counties until the nineteenth century. The hamlet of Hampton Charles (Herefordshire) within Bockleton (Worcestershire) was created a separate civil parish in 1866 and remains in Herefordshire; the Herefordshire portion of Stoke Bliss was transferred to Worcestershire in 1897. Changes in the late nineteenth century were also made to the parish boundaries of Collington, Edwyn Ralph and Wacton. These are summarised in F.A. Youngs, *Guide to the local administrative units of England*, vol. II, (Royal Historical Society), London 1991. For convenience here, and because Kent's survey and recommendations should be read as a whole, both the Herefordshire and Worcestershire properties have been listed. Those in Worcestershire (in italics) are, however, only noted briefly.

Volume 1 Leasehold tenancies:

fol. 1
Ivington Woods, Leominster [SO 485550]. 27a. Camp Wood and Gorsty Hill Coppice. 6 chains to 1 inch [1:4752].

fol. 3
Park Hill Coppice, [Ivington in Leominster]. 14a. 6 chains to 1 inch [1:4752].

fol. 6
Edwins Wood, [Edwyn Ralph, SO 644587]. 146a. 6 chains to 1 inch [1:4752].

fol. 7
Stoke Coppice [Stoke Bliss].

fol. 8
Kyre Park and lands.

fol. 19
Upper Wick Farm [Rushwick], near Worcester.

fol. 24
Patchley Farm, near Worcester [?Peachley, Broadheath].

fol. 30
Ivington Mill, Leominster. 5½a. 6 chains to 1 inch [1:4752].

fol. 32
Berry Farm [Ivingtonbury in Leominster, SO 476569]. 328a. On good land but liable to flooding. Field-names include Berry Fields. 20 chains to 1 inch [1:15840].

fol. 40
Ivington Park Farm, Leominster [SO 479556]. 305a. Field-names include the Great and Little Camp. 20 chains to 1 inch [1:15840].

fol. 47
Fulham's Farm, Eastham.

fol. 56
Court House and other properties, Hanley [Child].

fol. 67
Townsend Farm, Edwin [Ralph, SO 646575]. 66a. Field-names include Broom Hill, Piece by the Mill. 20 chains to 1 inch [1:15840].

fol. 74
Wear Hop Yard [Edwyn Ralph *or Stoke Bliss*]. 2a. 6 chains to 1 inch [1:4752].

fol. 77
Pool Farm and other properties, Stoke Bliss.

fol. 113
Cheveridge Farm, Hanley [Child].

fol. 122
Pigeon House and Hill Pool Farms and other properties, Kyre.

fol 139
Frogend Farm, Stoke Bliss.

fol. 142
Butterley Farm, Edwin [Ralph, transferred to Wacton 1884, SO 613579]. 58a. Field- names include The Gritt. 20 chains to 1 inch [1:15840].

fol. 150
Bickley Farm, Knighton[-on-Teme].

fol. 154
Upper House Farm and other properties, Hanley [Child] and Stoke Bliss.

fol. 169
Collington Farm (in later pencil, Underhill), Collington, SO 650604]. 47a., including hopyards. 12 chains to 1 inch [1:9504].

fol. 176
The Batches, Collington. 7a. 6 chains to 1 inch [1:4752].

fol. 178
The Pound Farm, Collington [SO 644576]. 49a. Shows strips in open fields. Field-names include Onsulls Bridge, Pike Corner [Pie Corner SO 645613], Butts Field, Frogwell. 30 chains to 1 inch [1:23760].

fol. 186
Coombs Wood, Collington [SO 663603]. 15a. 6 chains to 1 inch [1:4752].

fol. 190
Cottages in Kyre and Stoke Bliss.

Volume 2

fol. 200
Small properties in Hanley, Kyre and Stoke Bliss.

fol. 210
Old Mills Land. 9a. meadow [adjoining Collington brook]. 6 chains to 1 inch [1:4752].

fol. 212
Ball Grove Tenement [?Collington]. 4a. in scattered plots. 20 chains to 1 inch [1:15840].

fol. 214
Small properties in Hanley and Kyre.

fol. 225
Collington Barn Farm, Collington. 51a. 12 chains to 1 inch [1:9504].

fol. 228
Garlands Farm, Edwin [Ralph]. 77a. Field-names include Black Venn, Sand Pitts Tillage, Sand Pitts, with banks that should be levelled and limed. 20 chains to 1 inch [1:15840].

fol. 235
Samuel Jones's Premises with Smith's Shop [smithy], [?Edwyn Ralph]. 1a. 6 chains to 1 inch [1:4752].

fol. 237
James Phillips's Tenement, Edwin [Ralph]. 7a. 6 chains to 1 inch [1:4752].

fol. 240
Bucknall [?Buckenhill] Mill, Edwin [Ralph, SO 654564]. 4½a. 6 chains to 1 inch [1:4752].

fol. 244
Collington Bank Farm, Collington. 19a. Field-names include Butts Field, Pike Corner Field [SO 645614]. 20 chains to 1 inch [1:15840].

fol. 250
John Bowcutts Farm, [Collington]. 18a. Includes hopyards. Field-names include Butts Field, Pike Corner Field [SO 645614]. 20 chains to 1 inch [1:15840].

fol. 254
Properties in Kyre and Stoke Bliss.

fol. 265
Brick House Farm, Edwin [Ralph, SO 647579]. 121a. Field-names include Sand Pitts, Sand Pitts Hopyard, Brickhouse Croft, Black Venn, Catchems end Field, Mondales, Townsend Croft, Pike. Comment that it 'has received great benefit from the late Inclosure and is now increasing in Value on that Account'. 20 chains to 1 inch [1:15840].

fol. 272
Netherwood, Thornbury [SO 634606]. 589a. in a compact holding. Shows hopyards, orchards, nursery, pool and streams. Field-names include Park Pasture, The Grove, Pikelane Cottage, Brick Barn Meadow and Hill, Bonfire Piece. Comment that the tenant had done a great deal of draining but that some land was troubled with moss and fern. 20 chains to 1 inch [1:15840].

fol. 281
Properties in Hanley, Kyre Common and Stoke Bliss.

Leases for lives:

fol. 291
Properties in Kyre and Stoke Bliss.

fol. 309
Upper Horton Farm, [Edwyn Ralph, transferred to Wacton 1884, SO 630585]. 144a. Field-names include Winbury Field. Comment that the holding 'is the worst State of any part of the Estate'. 20 chains to 1 inch [1:15840]

fol. 314
Woods Farm, Collington (in later pencil, Underhill) [SO 650604]. 126a. Field-names include Pike Corner Field, Deadmanshead. 30 chains to 1 inch [1:23760].

fol. 319
Properties in Stoke Bliss.

fol. 321
The Sallings [Sailings] Farm, Bockleton.

fol 324
Properties in Stoke Bliss.

fol. 328
Lower Horton, Edwin [Ralph, ?SO 635583]. 89a. Includes two nurseries, Sallow Bed, Quarry Croft. 20 chains to 1 inch [1:15840].

fol. 332
Properties in Hanley [Child].

fol. 335
Grafton Farm, Grafton (in a later hand, Bockleton).

SCALE: Various as noted above.

SIZE: Both volumes are 195 x 130 , with the maps varying slightly in size at about 170 x 105, calf-bound with blind tooling. The spine of Volume 1 is damaged but Volume 2 bears a spine label 'SURVEY OF M[R] PYTT'S ESTATE 2'. The two volumes are foliated successively 1-199 and 200-376. Ink and watercolour on paper.

SURVEYOR: Nathaniel Kent, Fulham, Middlesex.

REFERENCE: Worcestershire RO, (BA8683) 899:396.

1. Nathaniel Kent consistently used the spelling Edwin. Other sources vary but current Ordnance Survey maps favour Edvin Loach and Edwyn Ralph.

1774
ALMELEY, EARDISLEY, KINNERSLEY

'A PLAN of the Estate of ANDREW FOLEY ESQ.[R] at ALMELEY KINNERSLEY ERDERSLEY AND LYONS-HALL ... taken in 1774 by James King'[1]

Centre and S of Almeley [SO 3351] with adjoining parts in E of Kinnersley and 9a at Pennsylvania in S of Lyonshall with (inset) 'BOLLINGHAM FARM' in NE of Eardisley [SO 305529]. Shows buildings in block plan except for the perspective view of Almeley church, fields outlined in green and meadows with plot numbers in red, mostly enclosed but some strips in open fields, woods and trees with shadows from morning sunlight, gates, ponds, streams, roads with destinations, footpaths. Gives names of adjoining owners.

In N of Almeley field-names include Newport [Nieuport, SO 320521] and The Park, High Moors Wood with Welchmans Way and Poors Land Coppice immediately to its NE, Old Castle, Black Acre, Red Streak Orchard and Batch Common farther to NE; Pool Meadow S of Nieuport and Black House Meadow [SO 315519], Mill Orchard at Upcotts [SO 326508], Kinnersley Poors Land; Wingle Field [open field] NE of Almeley church, Wonton [Woonton] hamlet, which had its own open field system of Wall Field, Hill Field and West Field, Black Acre N of Hoplers [Hoppley's] Green, Upper Camp Wood [SO 363526] and Sparmage Common. In central and S part of Almeley and N of Kinnersley field-names include Quarry Close, Hop Yards, Almeley Parks; Croplow, New Church [Newchurch Farm] and Park, Church Yard (13a.) [SO 350507], Pidgeon House Field, Loggerson [Logaston] Common, Sallys Common, Mill Pound [SO 345519]. At Bollingham in Eardisley, Quiss Moor Coppice [Queestmoor], road to The Apostles. Marks motte in Court Orchard adjoining Almeley church [SO 333515], vicarage, Brick House [Bridge Farm] with Mill Orchard to its N.

Title cartouche with trailing flowers and leaves. Two compass points, both with an arrowhead for N and a cross for E [but the one for Bollingham incorrectly aligned]. Scale bar with floral decoration. Plain border. Neat and delicate penmanship.

No reference book or table.

Some later pencilled lines of roads and railway line have been added.

SCALE: 6 chains to 1 inch [1:4752].

SIZE: 983 x 1330; inset map of Bollingham 410 x 305. Ink and watercolour on parchment, 4 skins now separated, formerly attached to rollers. Grubby with some damp stains.

SURVEYOR: James King.

REFERENCE: HRO, G75/1.

1. The Newport estate, bought by Thomas 1st Lord Foley in 1712, passed by the marriage of Henrietta Maria, daughter of Andrew Foley, to Richard Francis Onslow. In 1863 Richard Foley Onslow sold it to the trustees of James Watt Gibson Watt and in the early twentieth century it was acquired by Herefordshire County Council. The Council subsequently transferred the map to the Herefordshire Record Office.

1774
BRIMFIELD

'A PLAN and Survey of Nunupton in the Parish of Brimfield ... belonging to M.[r] Tho.[s] Hill'

Nun Upton farm, (*c.*100a.) in E of parish [SO 543666]. Buildings shown in block plan with garden, fields, woods.

Reference table giving field-names, acreage, 'Explanation' of shading and colours.

Title cartouche in elaborate but crude frame. N point as a vertical ruled line also giving latitude. Scale bar.

SCALE: 3 chains to 1 inch [1:2376].

SIZE: 480 x 670. Ink and watercolour on paper. Ragged edges, stained by damp and colouring lost. The reference table is largely illegible. Repaired.

SURVEYOR: John Hope.[1]

REFERENCE: HRO, AH41/1.

1. The map is unusual, particularly the north point and 'Explanation'. It appears to be a late example of a country surveyor wishing to demonstrate the mystery of his craft.

(1774)
HEREFORD

'To Sir John Cotterell [and others named, including John Harris] Commissioners appointed under an Act of Parliament for Paving and Lighting the City of Hereford and for Inclosing the Commons & Waste Lands belonging thereto This Map of the said Commons & Waste Lands and the Allotments made by [the] said Commiss.rs is humbly dedicated by their most obed.t Servant John Bach the Surveyor'[1]

Parliamentary inclosure map. 154a., comprising old Widemarsh inclosure (76a.), new Widemarsh inclosure (64a.) and new Monk Moor inclosure (14a.). Shows new allotments, houses in perspective, streams, roads. Marks turnpike road to Leominster with turnpike house and bar [SO 511407], mills on Widemarsh brook [SO 508407] and at Monkmoor [SO 516404], the 'Horse Course' running inside the circular boundary of Widemarsh common and Tennis Court [SO 506407]. Gives names of adjoining owners or properties.

Tables of holdings and new allotments. N point. Scale bar.

SCALE: 4 chains to 1 inch [1:3168].

SIZE: 438 x 695. Ink and watercolour on tracing paper.

SURVEYOR: John Bach.

REFERENCE: HCL, 912.4244, Hereford plans.

1. Copy traced by Walter Pilley, 26 October 1876.

1774
SHOBDON, AYMESTREY, LINGEN

'The Rt: Hon: JOHN LORD Visct BATEMAN'S Property'.

A series of seven small maps of estates in Shobdon, Aymestrey and Lingen. All are drawn in a uniform style, showing buildings in block plan, fields outlined in colour with plot numbers, mostly enclosed and indicating land use, woods, trees, gates. Names of adjoining owners are given.

Each has a standardised design of cartouche and scale bar, differing only slightly in size. No compass points. Plain ruled borders. The penmanship is small and delicate. There is some fading of both ink and watercolour. The sheets of parchment are all a similar size, causing small differences in the scale of the maps, except in the one case where a much larger area was covered at a smaller scale. There are minor differences in the measurements of the borders of the maps. All were mounted on heavy brown paper, apparently when in the custody of Hereford City Library; only two remained mounted in 2002.

1. 'SHOBDON Court PARK'
Shows parkland with deer, formal gardens near Shobdon Court [SO 400626], pools. Marks Shobdon Arches but not the rebuilt parish church of 1752-6. Signed 'A. Burns 1774'.

SCALE: c.11½ chains to 1 inch [1:9108].

SIZE: 197 x 250.

2. 'UPHAMPTON'.
Area N of Shobdon Park and S of Shobdon Hill [SO 3963]. Adjoining lands include Downwood Common, The Moors.

SCALE: 9½ chains to 1 inch [1:7524].

SIZE: 189 x 240.

3. 'COVENHOPE'.
Adjoining area in parish of Aymestrey N of Shobdon [SO 4064]. Meer Hill Wood and Covenhope lands bordered by the river Lugg between Beach-bank Common [Beechenbank Wood] and Shobdon Hill.

SCALE: 11 chains to 1 inch [1:8712].

SIZE: 200 x 247, mounted.

4. 'LIDYCOT'.
Township of Ledicot [SO 414620]. Field-names include Charley Meadow, Qui Field, North Field.

SCALE: 11 chains to 1 inch [1:8712].

SIZE: 194 x 246.

5. 'THE BROOM'.
Lands in S of Shobdon parish [SO 401596]. The map is aligned with N on the left.

SCALE: 12 chains to 1 inch [1:9504].

SIZE: 193 x 246, mounted.

6. 'SHOBDON'.
Area S of village adjoining Near and Further Shobdon Fields [SO 4061].

SCALE: 12 chains to 1 inch [1:9504].

SIZE: 190 x 240.

7. 'MANOR OF LINGEN'.

Whole parish [SO 3667]. Shows church in bird's-eye view and other buildings in perspective or block plan, mostly enclosed fields but a few strips in open fields, stream [Lime brook], roads. Some evidence of recent enclosure. Field-names include Burtley Nowl [Birtley Knoll], which is shaded to indicate relief.

SCALE: 20 chains to 1 inch [1:15840].

SIZE: 224 x 250.

SIZE: Variable, as noted. Ink and watercolour on parchment.

SURVEYOR: A. Burns.

REFERENCE: HRO, G39/Maps. Collection not catalogued and not available for research in 2002.

1774

YAZOR, MANSEL LACY, BISHOPSTONE, MANSELL GAMAGE, BYFORD

'A Survey of Foxley and its Appendages in the County of Hereford. The estate of Uvedale Price Esq:r'

An estate atlas prepared for remodelling the Foxley estate [SO 4146] drawn up by Nathaniel Kent. The farms and holdings are drawn separately on small maps followed by reference tables, long and detailed statements of the landlord's and tenants' covenants and concluding with the surveyor's 'Hints to the Landlord'. The later part of the volume contains further general notes on covenants and improvements with surveys and advice on outlying properties that were not mapped.

The maps, which differ in size by a few millimetres, are designed to fit the page by varying their scale. They show buildings in block plan, fields with plot numbers, boundaries and land-use, woods, streams, roads. Each has a simple scale bar, N point and ruled border. Though small, they are detailed and neatly drawn in ink and coloured in watercolour. The maps are entitled with the names of the tenants, not the properties. As the reference tables give only plot numbers, acreage and the names of open fields, as Kent's compass points are not always accurate, and as the estate was extensively altered following this survey, it is difficult to identify many of the holdings. It is therefore helpful to compare this survey with the larger estate atlas drawn up four years earlier: see 1770 YAZOR.

Kent's comments concern good husbandry practices, including the repair of buildings and fences, the maintenance of gates and stiles, drainage, liming and manuring, limitations on the acreage sown with corn and breaking up pasture, the cleaning of meadows and removal of ant-hills, planting withies on 'Aquatic' land, demolition of unnecessary buildings, payment of rates and taxes. On p. 65 is a note 'The Premises are valued at £7 10s. 0d. but he [the tenant Richard Evans] would not give it and Ridgeway [the tenant of Upperton Farm in 1770] intimated that he was a Man which it would be imprudent for M:r Price to have any dispute with'.

fols. 2-13
Table of land in hand (1,425a.) and Josiah Ridgeway's Farm (328a.). At fol. 8 is a stub on which a missing map was originally tipped in.

fol. 14
'Map of J: Goods Farm'. 312a. 20 chains to 1 inch [1:15840]. Reproduced in S. Daniels & C.Watkins, eds., *The Picturesque landscape. Visions of Georgian Herefordshire*, Nottingham 1994, fig. 2.

fol. 20
'Map of M.r Allen's Allotment'. 13a. Marks Yazor old church. 4 chains to 1 inch [1:3168].

fol. 22
'Map of John Hargets Allotment'. 16a. 4 chains to 1 inch [1:3168].

fol. 25
'Map of Owen Davis's Allotment'. 7a. 4 chains to 1 inch [1:3168].

fol. 28
'Small Portions situate near the Clay Pits'. Several cottages and lanes. 6 chains to 1 inch [1:4752].

fol. 30
'Map of James Pritchards Allotment'. 5a. 4 chains to 1 inch [1:3168].

fol. 32
'Map of Tho:s William's Allotment'. 17a. 4 chains to 1 inch [1:3168].

fol. 35
'Map of W.m Welson's Premises'. 185a. 20 chains to 1 inch [1:15840].

fol. 42
'Map of Llewellyns Farm'. 41a. of scattered lands [at Shetton in Mansell Gamage, SO 407449]. 20 chains to 1 inch [1:15840].

fol. 45
'Map of George Gardners Farm'. [Mansell Court, SO 425454], 358a. Marks Mansel Lacy church and moated site. Wrongly aligned; N is at the bottom, not the top. Reproduced in Daniels & Watkins, *op. cit* at fol. 14 above, plate V.

fol. 52
'Map of John Pember's Premises'. 5a. 6 chains to 1 inch [1:4752].

fol. 54
'Map of Mary Mathews Premises'. 264a. S of Bishopstone Common [Bishon Farm, SO 423433]. 20 chains to 1 inch [1:15840].

fol. 60
Map of W:m Mosleys Premises'. 4½a. 6 chains to 1 inch [1:4752].

fol. 62
'Map of Susannah Parlour's, John Love's and Tho:s Watkin's Allotments'. 3a. of small plots at Byford Hill. 6 chains to 1 inch [1:4752].

fol. 65
'Map of Rich.d Evans's Premises'. 10a. 6 chains to 1 inch [1:4752].

fol. 67
No title. 5a. S of Byford Hill and Bishopstone Hill. 6 chains to 1 inch [1:4752].

fol. 70
'Map of Tho:s Davis's Allotment' and of 'Tho:s Prossers Allotment'. 1a. 6 chains to 1 inch [1:4752].

fol. 72
'Map of Sam.l Davis's Premises'. 37a. 6 chains to 1 inch [1:4752].

fol. 75
'Map of Tho:s Preec's Premises'. 20a. 4 chains to 1 inch [1:3168].

fol. 78
'Map of James Pritchards Premises'. Lands N of Lower Common. Marks Mansel Lacy church. 20 chains to 1 inch [1:15840].

fol. 83
'Map of Benjamin Baynhams Premises'. 137a. in Mansel Lacy open fields. 20 chains to 1 inch [1:15840].

fol. 88
'Map of W:m Lloyds, John Pembers, & James Bainhams Allotments'. 5a. 6 chains to 1 inch [1:4752].

fol. 91
'Map of Thomas Meredith's Allotment'. 1a. 6 chains to 1 inch [1:4752].

fol. 93
'Map of John Birche's & W:m Higgins's Allotment'. 3a. opposite Mansel Lacy church. 6 chains to 1 inch [1:4752].

fol. 95
'Map of Bainhams Farm, divided between T: Powell and [blank]'. 70a. and 54a. 20 chains to 1 inch [1:15840].

fols. 99-134
Notes on covenants and improvements, with surveys and contents of properties of outlying estates (no maps).

fols. 135-140
Index of names.

fol. 141
'This Estate was Surveyed Modelled and Let by me upon the Agreements contained in this Book in the Year 1774. Nath. Kent Fulham 20th Sep.r 1774'.

SCALES: 4 chains [1:3168], 6 chains [1:4752] and 20 chains [1:15840] to an inch.

SIZE: Volume, 195 x 120, leather-bound, with raised bands and spine label 'FOXLEY'. The maps vary slightly in size at about 102 x 177. Ink and watercolour on paper.

SURVEYOR: Nathaniel Kent, Fulham, Middlesex.

REFERENCE: Private.

1774
WESTHIDE

'Plan of lands at Westhide to be Exchanged between the College of Gloucester and the Revd. Mr. Gwillym'
15a. S of Kiming [Kymin] Brook in the N of the parish [SO 585446]. Shows strips in an open field or meadow, crossed by a bridle road. Marks 'Team Road'. Gives names of adjoining owners.
Compass rose. Scale bar. Ruled border.

SCALE: 2 chains to 1 inch [1:1584].

SIZE: 305 x 390. Ink on paper.

SURVEYOR: John Harris, 3 November 1774.

REFERENCE: Gloucestershire RO, D936/E217.

1774-1775
HOLME LACY ESTATES

Four parchment maps of Scudamore family estates in Amberley, Ballingham, Holmer and Tretire by Richard Frizell. Three of these are based upon the paper drawings noticed above; see 1771-1775 HOLME LACY ESTATES. The exception is the map of Holmer. The maps have Frizell's standard engraved cartouche and compass rose stamped on them. Their titles follow the wording of the map of Amberley and only significant variations have been noted below.

[1] March 1774
AMBERLEY (MARDEN)[1]
'A MAP of Amberly in the parish of Amberly … the Estate of the Honble Charles Howard & Frances Howard his Lady containing 204. 2. 9. Statute Measure Laid down by a Scale of 26 perches in one Inch in March 1774 by Rich[d]. Frizell'

163a. enclosed lands and 40a. in strips in the common meadows [SO 5447]. Field-names include Church Orchard, Worgan's [Wergins's] Meadow.

Reference table arranged by names of tenants, giving plot numbers field-names, acreage.

Similar to Frizell's drawing dated December 1773 [BL, Add. MS. 36307, G1] except for a different layout of the strips in the meadows.

Engraved cartouche and compass rose. Scale bar. Ruled border.

SCALE: 26 perches to 1 inch [1:5148].

SIZE: 512 x 610. Ink and watercolour on parchment.

SURVEYOR: Richard Frizell.

REFERENCE: BL, Add. MS. 36307, G2.

1. Amberley was a township and ancient chapelry of Marden. It was briefly a separate civil parish 1866-87.

[2] 1775
BALLINGHAM, LITTLE DEWCHURCH
'A MAP of part of the Estates of the Honourable Charles Howard and Frances his Lady in the Parish of Ballingham &c … Containing 772a: 2r: 24p …'

Ballingham Hall Farm [SO 5731], Carey Farm [SO 5731], Although Farm [adjoining Kilforge in Little Dewchurch; probably Altbough Farm, then in Hentland parish, now Hoarwithy]. Shows field boundaries in outline, with plot numbers.

Reference table arranged by names of tenants gives plot numbers, field-names, acreage. Field-names include Bereval meadow, Claypitt Hill Tops, Quaryfields [SO 573315], Quary Head [SO 562309], Forge Meadows, Capler Wood, Blackpits [about SO 553327], Nap [Knapp] Green in Little Dewchurch.

Engraved cartouche and compass rose. Ruled border.

SCALE: 26 perches to 1 inch [1:5148].

SIZE: 785 x 915. Ink and watercolour on parchment.

SURVEYOR: Richard Frizell.

REFERENCE: BL, Add. MS. 36307, G5.

[3] 1775
HOLMER
'A MAP of the Bircott Farm in the parish of Homer and Liberty of the City of Hereford … Containing 129a. 0r. 22p …'

Burcott Farm [SO 523422] straddling but mostly N of the Roman road in the SE of the parish. Shows fields, mostly enclosed, with boundaries in outline, roads.

Reference table gives plot numbers, field-names, acreage.
Engraved cartouche and compass rose.

SCALE: 26 perches to 1 inch [1:5148].

SIZE: 550 x 563. Ink and a little watercolour on parchment.
Damaged with parts missing, grubby. Repaired.

SURVEYOR: Richard Frizell.

REFERENCE: BL, Add. MS. 36307, G48.

[4] 1775
TRETIRE, LLANGARRON, ST. WEONARDS
'A MAP of part of the Estate of Charles Howard and Frances Howard in the Counties of Hereford and Monmouth Containing separate as in the Reference making together 399.0.10 …'

White House Farm (178a.) in Tretire [SO 518245], Tryevan [Tre-Evan] (51a.) and Tredunnock [SO 522210] (129a.) Farms in Llangarron, Trepenkennet [Trippenkennett Farm] (41a.) in St. Weonards [SO 501225], showing field boundaries in outline. Marks Llangunnock Mill near Trippennkennett.

Reference table gives plot numbers, field-names, acreage. Field-names at White House include Pitt Close, Parsonage Meadow, Hopeyard.

SCALE: 26 perches to 1 inch [1:5148].

SIZE: 470 x 610. Ink and watercolour on parchment.

SURVEYOR: Richard Frizell.

REFERENCE: BL, Add. MS. 36307, G62.

1775
BOSBURY, CODDINGTON

'Map of the Inclos'd part of the TOWNEND ESTATE belonging to RICHARD BRYDGES ESQ:R in the Parishes of Coddington and Bosbury'

139a., scattered lands in E of Bosbury and N of Coddington parishes [Lower Townend Farm, SO 702442] and Old Country. Shows buildings in perspective, fields with reference nunbers and outlined in colour, woods, Gives names of some adjoining owners.

Reference table on a separate sheet of torn parchment, giving some field-names, land-use, acreage.

Title cartouche. Compass rose. Scale bar. Simple decorated border.

SCALE: 4 chains to 1 inch [1:3168].

SIZE: 535 x 710. Ink and watercolour on parchment. About one-third of the left-hand side of the parchment which contained the reference table has been torn off.

SURVEYOR: W. Hull, Kempsey, Worcestershire.[1]

REFERENCE: HRO, D96/89 and 90.

1 The surveyor's style is rather similar to that of the Dougharty family of Worcester in the 1730s to mid 1750s.

1775
ELTON

'A Map of the Manor and Estates...belonging to Edward Salwey Esq.r Survey'd by R. Sankey Oct: 1775'

Central and W part of the parish [SO 4570] and a group of 17 detached fields to the E. Shows buildings in perspective, farm buildings, appparently with some accuracy, enclosed fields with plot numbers and distinguishing arable and pasture, woods, hedges, streams, ponds, roads and gates across roads. Marks church [SO 458710], [Elton Hall] with a large barn, [Marlbrook Hall]. Gives names of adjoining owners, including Elton Common.

Title (faintly written) in a cartouche within two trees. Scale bar with dividers. Small simple compass rose.

SCALE: 4 chains to about 0.9 inches [c.1:2850].

SIZE: 740 x 1715. Ink and grey wash on parchment.

SURVEYOR: R. Sankey.

REFERENCE: Shropshire RO, 1141/123.

1775
LITTLE HEREFORD

'A MAP of the MANOR of LITTLE HEREFORD … belonging to Dansey Dansey Esq …'

Whole of the manor N of river Teme [SO 5568]. Shows buildings in block plan, fields with field-names, orchards, woods, isolated trees with shadows, streams, ponds, roads, bridges. Marks church [SO 554680], mill and The Bylet [island E of the church], footbridge opposite church, road bridge with three piers. Field-names include The Park, Park Field, Bleathwood Common, Leddidge Side [on bank of Ledwych Brook].

Floral cartouche. Compass rose. Scale bar within frame. Decorated border.

Reference book, not traced.

SCALE: 17 perches to 1 inch [1:3366].

SIZE: 1245 x 1680. Ink and grey wash on parchment.

SURVEYOR: Joseph Powell.

REFERENCE: HRO, C94/129.

1775
WACTON

'A MAP of the Upper House Farm: in the Parish of Wackton the property of R.C. Hopton Esq.r measured & planned by John Aird 1775'

[Upper Wacton Farm, SO 620569]. 18a. in scattered closes and strips in open fields. Buildings shown in block plan, fields outlined in colour, orchard and isolated trees with shadows, gates.

Reference table with plot numbers, land use, acreage and field-names, including Hopyards.

Neat floral cartouche. Compass rose. Scale bar. Yellow border.

SCALE: 5 chains to one inch. [1:3960].

SIZE: 320 x 545. Ink and watercolour on parchment.

SURVEYOR: John Aird.

REFERENCE: HRO, R93.

n.d.[c.1775]
MOCCAS

'[A] SU[RVEY of the] Manor[s of] Mocc[as and Bredwardine in the Parishes of] Moccas, B[redwardine, Dorstone and]

Peterchurch [in the County of] HE[REFORD Belonging to] Sir Georg[e Cornewall]'[1]

Whole of the parishes of Moccas [SO 3543] and Bredwardine [SO 3344], NE part of Peterchurch and E of Dorstone. Shows buildings on the estate in block plan and other buildings in conventional perspective, fields, mostly enclosed but some strips in open fields, outlined in grey with reference numbers and field-names, distinguishing arable, woods, river Wye, ponds, roads. Relief indicated by hachuring. Marks Moccas church [SO 357433] and Court with formal gardens, The Walk, The Park, The Warren, The Lawn, Lawn Pool, Lilla Pool and Moccas Common; in Bredwardine marks the church in perspective view [SO 335445], Old Court, Bredwardine bridge and Turner's Boat; in Dorstone marks Arthur's Stone, [SO 318431] Arthur's Stone Mountain [Merbach Hill], much of it open land but some enclosed fields, and Kystyns Hill [Dorstone Hill]; in Peterchurch marks Golden Valley, Wilmaston [SO 341402] and Nag's Head inn [not named]. Shows the Horseshoe Bend in the river Wye [SO 334460] before it was cut through. Small areas in Dorstone are drawn in four insets at the top of the map.

Title cartouche, mostly missing, with flowers and ribbons. Large compass points. Coat of arms and crests of Cornewall quartered with Amyand. Scale bar. Ruled border.

SCALE: 4 chains to 1 inch [1:3168].

SIZE: 1610 x 2330. Ink and watercolour on parchment. Badly torn with large parts flaked away. Faint; damaged by dirt and damp. Repaired by the National Library of Wales 1962-3.

SURVEYOR: Not known. [Possibly John Lambe Davis; see note 1].

REFERENCE: HRO, C62.

1. Undated. The date and the name of the surveyor were probably written on one of the missing parts of the title cartouche. Sir George Amyand, 2nd baronet [1748-1819] married in 1771 Catherine Cornewall [1752-1835], only daughter and heir of Velters Cornewall of Moccas, and took her surname. Features on the map suggest that it was made within a short time of John Lambe Davis's surveys of 1772 and certainly before Lancelot 'Capability' Brown's recommendations of 1778 were put into effect and the agreement for the enclosure of Moccas Common in 1787.

1776
BREINTON

'A MAP of upper Breinton Farm ...'

114a. in N of parish [SO 464403]. Shows buildings in block plan, enclosed fields, gates, roads, tracks or footpaths. Fields coloured with a pale wash and stronger outline of boundaries. Field-names include Fieldfares Pasture. Gives names of adjoining owners.

Reference table giving field-names, values and acreage 'with mounds' [boundaries] (114a.) and without mounds (100a.). The table includes strips in open fields at King's Acre and adjoining the road from Hereford to Brecon, which are not shown on the map.

Title in small octagonal frame. Compass points. Ruled border.

SCALE: [c.1:3725].

SIZE: 322 x 434. Ink and watercolour on parchment.

SURVEYOR: Written in another contemporary hand, 'Survey'd in May 1776 by Tho.s Smith. Surveyor'.

REFERENCE: DCA, D122.

1776
EYE

'MAP and Survey of an Estate in the Parish of Eye...belonging to Adam Ward of Bircher Esq.r taken July 1776' Endorsed 'Broad Meadows'

37a. in S of parish [SO 4960]. Shows buildings in block plan, enclosed fields, orchard, roads, gates. Marks Portley Moor Lane, Leominster's Brook, Spitefull [Spittal] Ridge. Field-names include Old Bandham. See also 1769 LEOMINSTER.

Reference table with field-names, acreage.

Simple compass rose. Scale bar. Ruled border.

SCALE: 2 chains to 1 inch [1:1584].

SIZE: 375 x 463. Ink and watercolour on paper, torn and ragged.

SURVEYOR: Not named.

REFERENCE: HRO, AD4/129.

1776
MARDEN

'A Survey of the Glebe Lands and Tithes of Marden ...' with inset 'A Plan of the Homestead and Meads and a Terrier of the Field Lands 1776'.

Survey listing the glebe lands and tithes of the parish of Marden [SO 514472]. Inset plan shows house in block plan and adjoining 10a. in The Street, apparently at SO 515470. Fields outlined in colour. Marks gates.

Compass rose. Ruled border.

SCALE: c.500 yards = 4½ inches [1:4000].

SIZE: 160 x 108 [of inset plan]. Ink and watercolour on parchment.

SURVEYOR: Not named.

REFERENCE: DCA, 4753.

1776
ROWLESTONE

'A SURVEY of a Tenement in the Parish of ROWLSTON … Survyed and Mapped By Edward Thomas Anno 1776'

53a. in SE of parish [SO 3726]. Shows buildings in block plan, enclosed fields, a copse, roads and lanes with destinations, streams with direction of flow. Gives names of adjoining owners. Marks house and barn [at Edward's Barn, SO 381265], Grist Mill [SO 375265], 'Old Watercourse for Watering the Land',[1] Gridall Brook [Cwm Brook]. Relief indicated by hachures.

Reference table gives plot numbers, field-names, land-use, acreage. Field-names include Parson's Hill, Cae Pistill, Pound Meadow described as a 'Watering Meadow'.

Title cartouche, rococo with foliage, flowers and a heron. Compass rose. Scale bar of Gunter's chains. Ruled border.

SCALE: 4 chains to 1 inch [1:3168].

SIZE: 380 x 270. Ink and watercolour on paper.

SURVEYOR: Edward Thomas.

REFERENCE: NLW, Vol. 28, 'A Book of Maps of the Estate of Thomas Edwards Esq. In the Several Parishes therein mentioned within the Counties of Glamorgan, Monmouth and Hereford Surveyed and Mapped by E. Thomas Anno 1776', p. 71 [map] and p. 72 [reference table].[2]

1. This appears to be rare documentary evidence for a scheme for drowning water meadows. Rowland Vaughan (1559-1628) of Newcourt, Bacton, reputedly invented the practice of drowning meadows in England. His original scheme in the Dore valley between Peterchurch and Bacton dated from c.1588. There is a reference to a contemporary scheme at Ewyas Harold, the parish adjoining Rowlestone. *Rowland Vaugham, his booke*, published 1610. Republished and prefaced by Ellen Beatrice Wood, London 1897.
2. I am grateful to Lona Jones, assistant curator in the Department of Pictures and Maps of the National Library of Wales for her help in compiling this catalogue entry.

1777
BREDENBURY

'A PLAN of Lands at the RED HILL in the Parish of Briddenbury'.

Small area in centre of parish at Red Hill [SO 618556]. Shows buildings crudely in perspective, fields, hopyard, orchard, trees, gates, roads. Gives field-names.

Simple rectangular title frame. Scale bar. Ruled border. Exceptionally small, but containing all the principal features of larger estate maps.

SCALE: 4 chains to 1 inch [1:3168].

SIZE: 120 x 160. Ink on parchment.

SURVEYOR: J. Harris of Newton, 26 March 1777.

REFERENCE: HRO, F99/A/32.

1777
CLIFFORD

'A MAP OF UPPER MIDDLEWOOD. An Estate in the Parish of Clifford…belonging to M.r William Higgins Gent. Mapped and Measured in the Year 1777'.

116a. in E of parish [SO 298447]. Shows buildings in perspective, enclosed fields, orchards, woods, river Wye, streams, roads. Marks The Island [meadow on the E bank of the river Wye, probably at SO 245461].

Reference table gives acreage and field-names, including Quarrel Close.

Title cartouche and reference table in ornate frames, the same patterns repeated in the border. Compass rose. Scale bar.

SCALE: 3 chains to 1 inch [1:2376].

SIZE: 735 x 815. Ink on parchment.

SURVEYOR: Not named.

REFERENCE: HRO, A66/5.

1777
EYTON

'A Plan and Content of two Meadows in the Parish of Eyton belonging to Adam Ward of Bircher Esq.r taken Jan.ry 1777'.

6a. in SE of parish on W side of the turnpike road from Leominster to Ludlow at The Broad, immediately N of Leominster [SO 490605]. Fields outlined in colour, gate.

Reference table giving acreage.

Ruled border.

Endorsed 'Scandells Meadows' and, in a later hand in pencil, 'Leo.ter Broad'.

SCALE: [*c.*1:3168].

SIZE: 190 x 200. Ink and watercolour on paper.

SURVEYOR: Not named.

REFERENCE: HRO, F76/III/17.

(1777)
MORETON-ON-LUGG

[No title. Moreton Court estate].

904a. covering whole parish [SO 5045]. Shows buildings in block plan, but the church crudely in perspective, fields with field-names, acreage and boundaries in outline, mostly enclosed but some strips in open fields, hopyards, river Lugg, streams, roads. Marks New House [Brockhouse Farm], Field-names include Hylas Hill [SO 495467], Blacksmiths Croft, Deadmantree Field. Notes 'mears plowed up' on Hylas Hill and small areas of 'Computed Acres of Crows' in many fields.

Reference table containing summary contents of farms.

Bears the note 'Taken from a plan of the Moreton Property as it was in the year 1777. This Plan was delivered to me by M[r]. Keysall's surveyor, M[r]. Joseph Powell, when I was employed by the late Rev[d]. M[r]. Francis Woodcock, and the Revd M[r]. Sam[l]. Picart, to endeavour to ascertain the land belonging to their respective prebends. Henry Price'.[1]

Compass rose.

SCALE: [*c.*1:3727].

SIZE: 1000 x 670. Ink on paper, linen-backed.

SURVEYOR: [?Joseph Powell]. Copied by Henry Price *c.*1805.

REFERENCE: HRO, G23/3.

1. Henry Price of Hereford was active between 1800 and 1825. Revd Francis Woodcock held the prebend of Moreton Magna 1783-1818 and Revd Samuel Picart the prebend of Moreton Parva 1805-35. Price must therefore have copied the map between 1805 and 1818, probably soon after 1805. He does not actually state that Joseph Powell made the original map of 1777, only that it had been delivered to him by Joseph Powell. There were two surveyors of that name in Herefordshire, active respectively 1765-1801 and 1819. The note probably refers to the elder Joseph. Powell, whose relationship with his namesake, if any, has not been traced.

1778
ASHPERTON

'A MAP of HANSNETT HILL Farm in the Parish of Ashperton ... belonging to Richard Cope Hopton Esq[r].

13a. in NE of parish [SO 656428]. Shows buildings conventionally in bird's-eye view, arable and pasture fields, orchards. Gives names of adjoining owners.

Reference table gives acreage, land-use, field-names.
Title in rectangle. Compass rose. Scale bar.

SCALE: 2 chains to 1 inch [1:1584].

SIZE: 324 x 392. Ink on paper, linen-backed.

SURVEYOR: John Harris, Newton [near Leominster], September 1778.

REFERENCE: HRO, E3/456.

1778
HOPE UNDER DINMORE (now BIRLEY WITH UPPER HILL)

'A MAP of the YOKE ESTATE in the Parish of HOPE under Dinmore ... belonging to Charles Berrington Esq[r]. Surveyed in April 1778. by John Harris of Newton'

Yoke Farm (117a.) at Upper Hill [SO 469529]. Shows buildings in bird's-eye view, enclosed fields, distinguishing land-use, with plot numbers and field-names, land-use, hopyards, orchards, coppices, woods, gates, roads with destinations. Marks hill of Yoke Wood by hachuring. Gives names of adjoining owners.

Note, 'NB where the Fences belong to other Farms, they are marked on the outside, with broken lines; thus - - - - - - -'.

Reference table gives plot numbers and field-names grouped according to land-use, acreage.

Rococo cartouche incorporating a horned cow and a seated bagpiper. Compass rose. Scale bar. Ruled border. The reference table is drawn as a scroll. A finely drawn map.

SCALE: 4 chains to 1 inch [1:3168].

SIZE: 500 x 665. Ink and grey wash on paper, linen-backed.

SURVEYOR: John Harris, Newton [near Leominster]. In the margin, 'J. Harris Delin.'.

REFERENCE: Worcestershire RO, (BA869) r705:24/1369.

1778
LEINTHALL EARLS

'A MAP of Leinthall Earls Manor ... the Property of Rd. P. Knight Esqr. Map'd by James Sherriff 1778'

Most of the parish between Gatley Park and N of Leinthall Hill [Common]. Shows buildings, including the house of Gatley Park, in block plan, pink for dwellings or black, enclosed lands and some strips in open fields with plot letters and numbers, distinguishing arable and pasture, coppices and woods, roads. Gives names of adjoining owners, including William Greenly.

Cartouche with 'picturesque' landscape view. Compass points. Ruled border.

SCALE: [*c.*1:3960].

SIZE: *c.*685 x 1160, framed. Ink and pale watercolour on parchment.

SURVEYOR: James Sherriff.

REFERENCE: Private

[1778] LEINTHALL EARLS

'A MAP of an Estate situate at Leinthall Earls … the property of W.M GREENLY ESQ.R'

265a. in centre of parish [SO 450680] S of Gatley Park and N of Leinthall Hill [Common]. Shows buildings in block plan, enclosed lands and some strips in open fields, with field-names and distinguishing between arable and pasture, woods, roads. Gives names of adjoining owners including Richard Payne Knight (1750-1824). Marks church in perspective. Field-names include School Land, Pound Field, Portway, Poors Land, Oaker Wood, Parkfield, Wallstye.

Summary table of lands.

Cartouche with 'picturesque' landscape view incorporating two peasants and a goat, trees and castle. Compass points. Ruled border.

SCALE: [*c.*1:3335].

SIZE: 597 x 950. Ink and grey wash on parchment.

SURVEYOR: Not named. [James Sherriff].

REFERENCE: HRO, F76/B/239.

1779 [BODENHAM]

'Survey and Valuation of Estates the Property of the Heirs at Law of the Late Thos. More Esqr. [in Shropshire] … also of an Estate at Bowley in the County of Hereford'

An atlas of the estate, 1,347a., of which only four of the Shropshire maps remain in the volume. The whereabouts of the others, including that of Bowley in Bodenham [SO 541523], has not been traced.

REFERENCE: Shropshire RO, D3651/B/5/5/1/9].

1779 RIVER WYE

[1]

'A PLAN of the RIVER WYE, from the City of HEREFORD to Tinterne Wear a little below Brockware, together with several propos'd New Cuts for the better improvement of the NAVIGATION Survey'd in 1779 p[er] Rob. Whitworth'

Shows the course of the river Wye for 58 miles 1 furlong and a fall of 147 feet 9 inches, with its side streams, rocks, weirs, ferries, bridges and 'rovings' [? 'rowings', private crossings or landing places]. Relief indicated by hachuring. Marks towns and villages, churches shown in perspective, principal country seats, mills and roads. Other landmarks include warehouses at Hereford, coal wharf, lime works above Huntsham and iron furnace at Lydbrook.

Tables (5) of landowners affected by the proposed new cuts, giving parishes, plot numbers, names and lengths of the cuts through their lands.

SCALE: [2 inches to 1 mile. 1:31680].

SIZE: 475 x 1375. Ink and grey wash on parchment. The bottom border has been trimmed, thereby probably losing the scale bar.

SURVEYOR: Robert Whitworth.

REFERENCE: HRO, Q/RWn/1.

[2]

'A PLAN of the R[IVER WYE from the] City of HEREFORD to Tinterne W[eare a little] below Brockware, together [with] several propos'd New Cuts for the better improvement of the NAVIGATION Survey'd in 1779 by Robt. Whitworth'

As above, but rather roughly drawn and lacking the reference tables. A ruled border incorporating the scale bar.

SCALE: 2 inches to 1 mile [1:31680].

SIZE: *c.*460 x 1390. Ink and grey wash on paper. Repaired, fragments missing.

SURVEYOR: Robert Whitworth.

REFERENCE: HRO, Q/RWn/2.

[3]

'No 1 A PROFILE and SCALE of LEVELS along the BARGE CHANNEL of the RIVER WYE, from the City of HEREFORD to WILTON BRIDGE near ROSS. Survey'd in July 1779; by Robt. Whitworth'

Detailed survey of the profile of the river bed, water levels, the high flood watermark of 1 November 1771, proposed new cuts and dams. Measures the total length of the improvements as 30 miles, 7 furlongs, 1.60 chains.

SCALES: [1] Length, 1320 feet to 1 inch or 4 inches to 1 mile [1:15840]; [2] Falls and depths, 10 feet to 1 inch [1:120].

SIZE: 470 x 2870. Ink and watercolour on paper, frayed at top and bottom.

REFERENCE: HRO, Q/RWn/3.

1779
WHITCHURCH

'A PLAN and PROFILE of that most dangerous part of the RIVER WYE at NEW WEIR; also of a propos'd CANAL to avoid that dangerous place'

Small area on right bank of river Wye at New Weir [SO 561157], near Symonds Yat. Shows buildings in block plan, including forge and slitting mill. Marks salmon weir, existing and proposed locks. With two elevations or 'profiles' of the proposed canal along the left bank of the river below the salmon weir. Marks 1771 flood level.

Scale and key to reference letters.

SCALE: 1000 feet = 4$^{15}/_{16}$ inches [1:2437].

SIZE: 310 x 455. Ink and watercolour on paper.

SURVEYOR: R. Whitworth.

REFERENCE: HRO, T90/1.

1780
CROFT

'A MAP of GORWELL and WHITTINGTON'S ESTATE in the Township of NEWTON in the Parish of CROFT … belonging to Mr Jos.^{h} Harris'.

26a. [Gorwells, SO 5053] and 39a. [Whittington's] mostly in the open fields of Newton, a detached part of the parish of Croft between Leominster and Hope under Dinmore. The surviving part of the map covers only Gorwells in the E of the township. Shows buildings in block plan, strips (many of reversed S-shape) in open arable fields, orchards, woods, roads and tracks. Gives names of adjoining owners. Marks river Lugg. Field-names include Shire Lane end, a small pasture in the centre of the township.

Reference table, with field-names, acreage, land-use.

No scale. No compass points: W is at top of the sheet.

Endorsed 'Corporation Hereford'. [In a later hand] 'Croft. Map of Gorwell & Whittington's Estates. 1780'.

SCALE: [c.1:5280].

SIZE: c.480 x c.460. The area in the W of the township is missing. Ink on paper, cloth-backed, damaged and repaired.

SURVEYOR: Not named [John Harris].

REFERENCE: HRO, R45/1.

1780
DOWNTON ESTATES

No title. A fine atlas of the estates of Richard Payne Knight consisting of 5,616a. in Downton, Leinthall Starkes and Elton, including 131a. in Clungunford, Shropshire and (not mapped) 1,203a. of Mocktree, Aston and Elton Commons; other estates in Leintwardine, Hope-under-Dinmore, Stretford, Aymestrey, Kingsland, Dilwyn, Leominster, Eardisley in Herefordshire, Stow, Neen Sollers, Aston Botterel, Llanvair Waterdine, Cainham, Bitterley, in Shropshire, Hartlebury in Worcestershire, Heyop and Llangunllo in Radnorshire, Church Stoke in Montgomery and Brompton, Shropshire.

All the maps contain similar features, showing buildings in block plan, fields with plot numbers and distinguishing arable and pasture, orchards, rough ground, woods, hedgerow trees with shadows, rivers, pools, roads, tracks, bridges. Adjoining owners are named.

Reference tables, within a ruled border, are either on the following page or pages or immediately below the smaller maps. They contain columns for plot numbers, field-names, land-use, acreage, and observations. There are only a few observations, in the surveyor's hand or added later in pencil, .

Many of the maps have cartouches in the 'picturesque' style of landscape drawings with the title on a building or stonework in the foreground and a castle standing by a river or lake and church spire in the background. Other cartouches have rococo brooch-like frames. The titles of the maps are given below as they appear in the cartouches and in a few instances these differ slightly from the titles written in upper case in the accompanying reference tables. A few have the surveyor's name and date. Even fewer have a scale bar or compass rose. Most are aligned with north or north-east at the top. All are enclosed within a ruled border.

Richard Payne Knight (1751-1824), antiquarian, promoter of the 'picturesque' movement, MP for Leominster 1780-84, inherited the Downton estate in 1772. He designed and by 1778 completed the building of Downton Castle. At the same time he began landscaping the park and between about 1780 and 1785 erected the bridges and other features in the gorge of the river Teme. This survey was commissioned in the context of Knight's improvement of his whole estate, leading up to the Leintwardine and Downton inclosure act 1799.

1. Frontispiece. 'A SOUTH WEST VIEW OF DOWNTON CASTLE IN THE COUNTY OF HEREFORD'.
Undated. View of Downton Castle, [built 1772-78], seen from Castle Bridge over the river Teme, with figures and farm animals in the park. Ruled border. 240 x 382. Ink and grey wash on paper laid on parchment, tipped in.

2. A. 'DOWNTON CASTLE DEMESNE' (Plate 52)
873a. surrounding Downton Castle and to S of river Teme, including W part of Bringewood, shown as a sheep walk, and Hunstay Hill. Shows formal garden N of the Castle, parkland and plantations. Field-names include Limekiln Hays.
Cartouche, a 'picturesque' scene with the title on a stone tablet. Reference table includes observations dated 1783 and some later pencil additions. Scale bar of 8 chains to 1 inch [1:6336], decorated with draughtsman's tools of dividers, parallel rule, set square and penknife. 408 x 530. Parchment.

3. B. 'MONSTAY FARM'
120a. [SO 468732] and 92a. rough grazing in Brindgwood [Bringewood] Chase. Field-names include Limekiln Piece.
Cartouche with title in a rococo frame. 194 x 25. Parchment.

4. C. 'Maryknowl Farm'
348a. [Mary Knoll, SO 483736]. Field-names include Vinalls, Brickiln Piece.
Cartouche, 'picturesque' scene with title on a stone classical garden folly and dated 'I.S. 1780'. 412 x 255. Parchment.

5. D. 'DEEP WOOD FARM'
295a. [Deepwood Farm] Field-names include Well Meadow, Yew Tree Meadow, Mill Coppice, Compting House Meadow [SO 453748].
Cartouche with title on a stone slab propped against a garden folly. 410 x 250. Parchment.

6. E. 'BRINDGWOOD TENEMENTS' (Plate 53)
36a., showing Bringewood Forges [SO 455750], Compting House Yard and Tin Mill House, river Teme and mill leat E of Downton Castle. 83 x 253. Parchment.

7. F. 'SHOP HOUSE'
7a. [SO 449754] at crossroads NE of Downton Castle. 69 x 74. Parchment.

8. G. 'LOUSE FARM'
75a. N of river Teme and Bringewood Forge. Marks farm house [SO 450755]. Field-names include Battle Field Leasow [SO 450759]. 93 x 251. Parchment.

9. H. 'BROMFIELDS WOOD GATE'
8a. lying between Brakes Farm, Lodge Farm and Louse Farm. 68 x 74. Parchment.

10. I. 'LODGE FARM'
95a. [Old Lodge, SO 450760]. Field-names include Brick Kiln Piece, Burrow Close. 160 x 250. Parchment.

11. K. 'HULL PIKE FARM'
67a. [Hillpike Farm, SO 438762] and a cottage at Mocktree Common. Field-names include Quarry Piece. 133 x 250. Parchment.

12. L. 'BRAKES FARM'
253a. [SO 444754] adjoining Mocktree Commmon. Marks farm house [SO 447752]. Field- names include 'Barn at Hempbutt', Maunds Bank, Stonebrook Field, Quarry Field.
Cartouche, floral with a ribbon bearing the inscription 'To R.d P. Knight Esq.r DOWNTON CASTLE'. 252 x 457. Parchment.

13. M. 'GRAVEL FARM'
80a. [SO 441751] adjoining Mocktree Common. Marks Blacksmiths House on new enclosure from Mocktree Common [SO 440749]. 168 x 251. Parchment.

14. N. 'COP HALL FARM'
204a. [Cophall Farm] adjoining Mocktree Common. Field-names include Cockshut Bank, Limekiln Piece, The Tumps. 189 x 249. Parchment.

15. O. 'THE POOLS FARM'
156a. [SO 436738]. Field-names include Kiln Piece, The Parks, Great and Little Chimney, Vaukway, Spout Piece.
Cartouche, 'picturesque' with title on a plinth including 'James Sherriff Delin.t 1780'. Notes in the Observations column of reference table. 407 x 250. Parchment.

16. P. 'LOWER HALL FARM'
237a. scattered in three blocks of land, some by the river Teme. Marks farm house at Downton-on-the-Rock.
  Cartouche, rococo with a flower. 408 × 250. Parchment.

17. Q. 'UPPER HALL FARM'
182a. in two blocks of land adjoining Lower Hall Farm. Field-names include The Postholes.
  Cartouche with a scroll set against a tree, including 'J. Sherriff Delin. 1780'. 405 × 247. Parchment.

18. R. 'Downton Farm'
181a. S of Downton-on-the-Rock [SO 428734]. Field-names include Tottridge, Pedlars Rest, Dunn Field.
  Cartouche, 'picturesque' with a stone slab leaning against a tree. 407 × 297. Parchment.

19. S. 'Leinthall Starks Farm'
192a. and 52a. on Leinthall Starkes Common in E of parish of Leinthall Starkes [SO 4369]. Field-names include Brickiln Field, Fish Pool Ground.
  Similar cartouche to no. 16 above. 405 × 245. Parchment.

20. T. 'Petchfield Farm'
292a. in centre of parish of Elton. Field-names include Hopyard, Fish-pool Piece, Old Limekiln Piece, Ditch Close [SO 453704], Limekiln Piece, Meer Field.
  Cartouche, 'picturesque' view of a ruined castle. 404 × 246. Parchment.

21. U. 'PIGEON HOUSE FARM'
190a., scattered lands in Elton adjoining Petchfield Farm. Field-names include Tinkerbridge.
  Cartouche with rococo frame. 405 × 244. Parchment.

22. V. 'EVENHAY FARM'
26a. on fringe of Elton Common in E of Elton parish [SO 463700]. 178 × 243. Parchment.

23. W. 'LONG LEINTHALL FARM'
109a. [in Leinthall Starkes, SO 435697]. Field-names include Mortimers Stocking. 410 × 243. Parchment.

24. X. 'MONNINGTON'S FARM'
31a. [on S side of village of Leinthall Starkes, SO 435696]. 210 × 245. Parchment.

25. Y. 'BRAND HILL FARM in the PARISH of CLUNGUNFORD and COUNTY of SALOP'
131a. in E of parish adjoining Aldons Mind Common. Field-names include Limekiln Ground, May Hill. 208 × 247. Parchment.

[This concludes the first part of the volume comprising the surveys of the Downton Castle estate. A summary table lists the farms and their total area of 5,616a. 0r. 09p. This area includes Mocktree Common, in Downton 774a. 1r. 12p., Aston Common in Pipe Aston 298a. 1r. 14p. and Elton Common in Elton 219a. 3r. 22p., which were not mapped].

26. 'A MAP of Leintwardine Township in the County of Hereford' (Plate 54)
Whole parish. Shows the village [SO 4074], enclosed fields mostly to N and E of the village, strips in open fields to S and W including Kinton and Langley common fields, river Teme, roads. Marks Brandon Camp [SO 400724]. Field-names include Quarry Piece, Brick Closes, The Vineyard.
  Reference tables (7 pp.) arranged by farms, Seedley Farm (197a.), Brandon Farm (127a.), Rowleys Farm (69a.), Bedfords Farm (61a.), Marlow Farm (44a.), Court House Farm (98a.), Kinton Farm (154a.), Rees Farm (30a.) and a number of small tenants.
  Cartouche, 'picturesque', of a column and ruined temple with the title on a stone slab leaning against it, on a separate stone slab the inscription 'Ja.^{s} Sherriff. Delin.^{t}', a castle in the distance.
  Compass rose, with E at top. 422 × 720. Parchment.

27. 'The Manor of GATTERTOP in the COUNTY of Hereford'
387a. in the parish of Hope under Dinmore [SO 481539]. Place-names include Ivington Camp [SO 483547], Camp Field, Ramshill. Map aligned with SE at top.
  Cartouche with rococo frame. 405 × 295. Parchment.

28. 'Stretford Estate'
265a. in Stretford and Monkland. Shows church in bird's-eye view, some strips in open fields, Stretford Brook. Field-names include Hopyards, Sandpitt Fields, Tavern Croft. Gives destination of roads.
  Cartouche, 'picturesque' view of a ruined castle with the title written on a round tower. 404 × 245. Parchment.

29. 'Shirley Farm in the Parish of Aimstrey [Aymestrey] …'
136a. in W of parish [SO 384653]. Marks river Lugg and Lime Brook. Field-names include Dearvoles [Deerfold] Bridge. Gives destination of roads. Map aligned with S at top.
  Cartouche with title written within a tree. 180 × 243. Parchment.

30. 'A MAP of LOW-TOWN FARM in the PARISH of KINGSLAND …'
282a. [Lawton], SO 444594] in S of parish. Shows open and enclosed fields. Field-names include Mill Field. Gives destination of roads.

Cartouche with rococo and floral frame. Compass rose. 407 x 320. Parchment.

31. 'VENDMOOR FARM in the Parish of Dilwyn'
143a. in E of parish [SO 432549]. Field-names include Hundred Field [W of Venmore Bridge SO 439550], Canal Orchard W of Vendmoor farm house, Stony Bridge Meadow. 163 x 245. Parchment.

32. 'LURKINHOPE FARM in the Parish of STOW and COUNTY OF SALOP'
158a. [Lurkenhope Farm] in W of parish of Stowe and N of Knighton, Powys. 207 x 205. Parchment.

33. A. 'Wood Farm in the Parish of Neen Sollers and County of Salop'
159a. in two blocks of land in Neen Sollars. Field-names include Romans Leasow.
Cartouche, 'picturesque' with trees. 406 x 245. Parchment.

34. B. 'New House Farm in the Parish of Neen Sollers …'
114a.
Cartouche, 'picturesque' with a stone-built folly amid trees. Inscribed 'Ja.s Sherriff 1780'. 406 x 244. Parchment.

35. 'CHALCOTT FARM in the Parish of ASTON BOTTERELL and COUNTY OF SALOP'
241a. Field-names include Cinder Hill Piece, Brickiln Leasow.
Cartouche with title on a stone with foliage, a castle in the distance. Inscribed 'Jas Sherriff 1780'. 405 x 243. Parchment.

36. 'LOWER MITTON AND TIN MILL FARM in the Parish of HARTLEBURY and COUNTY of WORCESTER'
90a. adjoining Hartlebury Common. Field-names include Tin Mill, Flood Gate Meadow, Jenny- hole with forges, yards etc, Lower Mitton Forges, gardens, Byletts, Far Mill Leasow. 190 x 242. Parchment.

37. 'Quismore Estate'
241a. [Queest Moor in Eardisley, SO 307522]. Marks moat [SO 310517]. Field-names include Pound Coppice and Closes.
Cartouche, 'picturesque' with a cottage by a large boulder. 405 x 243. Parchment.

38. 'BLACK HALL ESTATE in the Parish of LLANVAIR WATERDINE and COUNTY of SALOP'
178a. N of village. Field-names include Kiln Close. 228 x 241. Parchment.

39. 'Weston. Pen Y Bryn and LITTLE DITCHES Situate in the COUNTIES of Salop & Montgom.y'
392a. in three blocks of land at Weston Madoc (306a.) and Pen-y-Bryn (63a., including Town Field) in a detached part of Church Stoke, Powys and Little Ditches (23a.), named after Offa's Dyke, in Brompton and Rhiston, Shropshire.
Cartouche, 'picturesque' drawing of a gravestone with foliage. Compass rose. 405 x 246. Parchment.

40. 'Upper Hall Estate in the Parishes of Hayup and Llanguntley in the County of Radnor'
389a. comprising Upper Hall Farm in Heyop with lands, woods (63a.) and Sheepwalk (101a.) in Heyop and Llangunllo.
Cartouche, 'picturesque' view of ruined castle. Compass rose. 403 x 241. Parchment.

41. 'LAND lying near LEOMINSTER'
56a. of scattered lands at The Marsh. 405 x 243. Parchment, but not matching with the parchment of the rest of the volume.

42. 'A MAP of the Clee Hill Waste in the Parishes of CAINHAM and BITTERLEY in the County of Salop with the Cottages and Enclosures thereon underneath the whole of which the Mines and Minerals belong to the R.t Honb.le E.d L.d Clive and R.d P. Knight Esq.r in equal undivided Moieties'.
290a. in Snitton Township of Bitterley and 693a. at Bennetttsend and Caynham. Shows one farm (63a.) [Nine Springs Farm], Angel Inn [SO 576759] with lands (24a.), new inclosures (75a. mostly unspecified) and an extensive scatter of cottages on Clee Hill. Marks Road Gate, Knowlbury Gate, Horse Ditch, Cop Stone [at about SO 597737], turnpike road from Cleobury to Ludlow, Bensers Brook.
Cartouche, 'picturesque' with a large monument in a landscape scene. Scale bar on the base of the monument. Scale: 16 chains to 1 inch [1:12672]. 403 x 242. Ink and watercolour on paper.

SCALE: 8 chains to 1 inch [1:6336], unless otherwise noted above.

SIZE: Maps vary in size as noted above. Ink and grey wash on parchment, with one map ink and watercolour on paper, in a volume 452 x 285, bound in red morocco with raised bands, head bands, gilt lettering 'DOWNTON CASTLE' on the spine.

SURVEYOR: James Sherriff [of Birmingham].

REFERENCE: HRO, BL 35.

1780
HOLME LACY ESTATES

'Maps OF THE ESTATE of the Right Honourable the Earl AND Countess OF Surrey IN THE COUNTIES OF HEREFORD * GLOCESTER * BUCKINGHAM AND Monmouth surveyed by Richard Frizell 1780'

A fine atlas of the estates (16,654a.) of the Scudamore family of Holme Lacy, based upon the surveys carried out by Richard Frizell 1771-75 (see 1771-1775 HOLME LACY ESTATES and 1774-1775 HOLME LACY ESTATES). There is an index of place-names at the beginning and a summary table of the estatess at the end. The reference table and map of each property on the estate appear on a double-page spread, the reference table on the left hand page opposite the map on the right. The neatly drawn reference tables give field-names and names of tenants, land-use and acreage. Below many of the tables Frizell added brief remarks about the state of farms and their potential for improvement.

The maps have prick holes made in preparing the drawings and the boundaries of the properties have spaced symbols [? to indicate places where bearings were taken]. All show buildings with some appearance of pictorial accuracy. Most are drawn in profile with a few important buildings in bird's-eye view, for example Holme Lacy House and churches. The buildings are coloured pink with turquoise roofs. Fields are coloured in pale green for pasture and meadow or pinkish-brown for arable. Also shown are woods and orchards with shaded trees, rivers, roads, sometimes with directional pointing hands. The individual maps and reference tables do not have titles and only some maps have scale bars, drawn at scales of 25 or 26 perches to one inch [1:4950 or 1:5148] and 50 or 52 perches to one inch [1:9900 or 1:10296]. Very few have a compass or north point.

Title page (as above), with the arms of Charles Howard, Earl of Surrey [11th Duke of Norfolk 1786].

fols. 1-4
Index of places, with page [folio] numbers.

fols. 5v-6
Holme Lacy Demesne. 383a. Shows Holme Lacy House accurately in bird's-eye view, with deer park (412a.), woodland and coppice (70a.), orchards, timber yard, five fish ponds and bowling green. N point indicated by a naked man standing on a barrel marked 'Cider' with a raised flagon in his right hand and an arrow at arm's length in his left.

fols. 7v-8
Pound Farm, Holme Lacy. 155a.

fols. 9v-10
Bower Farm, Holme Lacy. 380a., enclosed lands held in small holdingss except for the principal farm (91a.) and meadows (77a.). Field-names include Boat Piece.

fols. 11v-12
Mill Farm (part) and Bower Farm, Holme Lacy. 425a., of which Mill Farm comprised 100a. Field-names include Sallow Garden. Comment that Mill Farm was in good order but without a house, 'the one that is on it is gone to ruins'.

fols. 13v-14
Cunnyger Farm, Holme Lacy [SO 5334]. 545a. Includes cottages, a small wood and orchard. Among few field-names is Cannon-dale. Comment that 'Part of Cunnygar Farm is very convenient for corn grain for the Demesne. It was formerly part of the Park'.

fols. 15v-16
Hollanton's [Hollington] Farm (part), Holme Lacy. 295a. between the Deer Park and river Wye, adjoining Handcocks Mill [c. SO 565328]. Includes hopyard (2a.), common (1a.), 'fallen' woods (30a.), orchard (2a.).

fols. 17v-18
Hollanton's Farm (remaining part, 325a.) and Burton's Farm (65a.), with part of the river Wye (9a.) and the previous entry brought forward, a total 709a. 'very good land'. Marks Holme Lacy church in bird's-eye view and 'Strand', the island at SO 572351.

fols.19v-20
Ganna [Gannah] Farm, Holme Lacy adjoining the Deer Park (87a.) with part of Hollantons now added to Ganna, Bowson [Bolstone] Wood (268a.) and Bolston Farm, Bolstone (116a.). 523a. Comment that it was 'fit for dairy and Tillage' but the house was 'greatly gone to Wreck, & shou'd be let to an Occupier who would reside'.

fols. 21v-22
Bellingham [Ballingham] Hall, Ballingham. 529a. Marks Ballingham church in perspective. Much of the land was arable and orchard but includes common [SO 575325]. Field-names include Claypit Orchard, Mill Meadow [SO 565306], Wear Meadow [SO 573309], Yearly's Boat [SO 584314] on the right bank of the river Wye and Capler Wood on the left bank. Comment that the farm is 'beautifully situated on the river Wye, and almost surrounded by it' with a good house. Scale: 26 perches to 1 inch [1:5148].

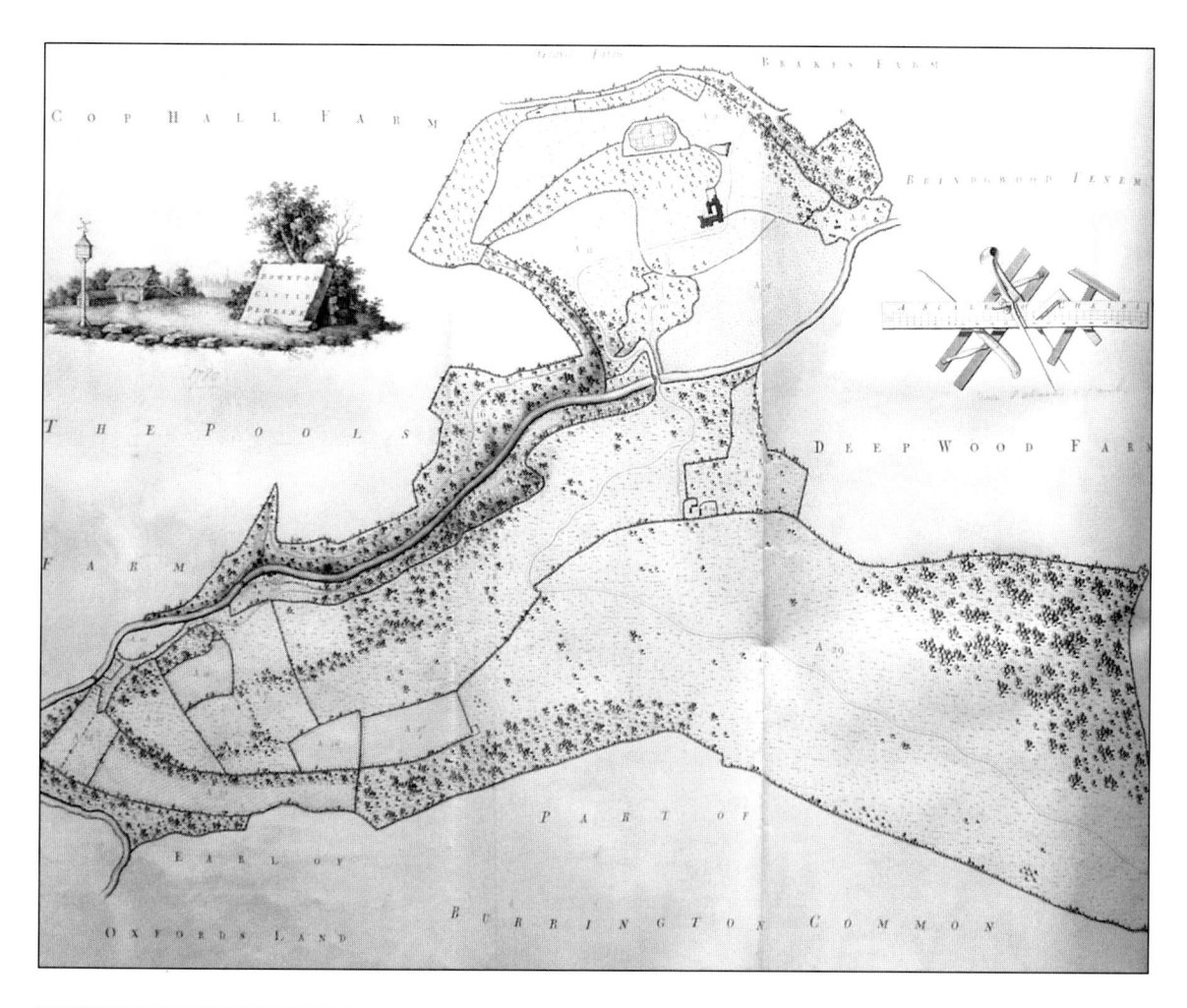

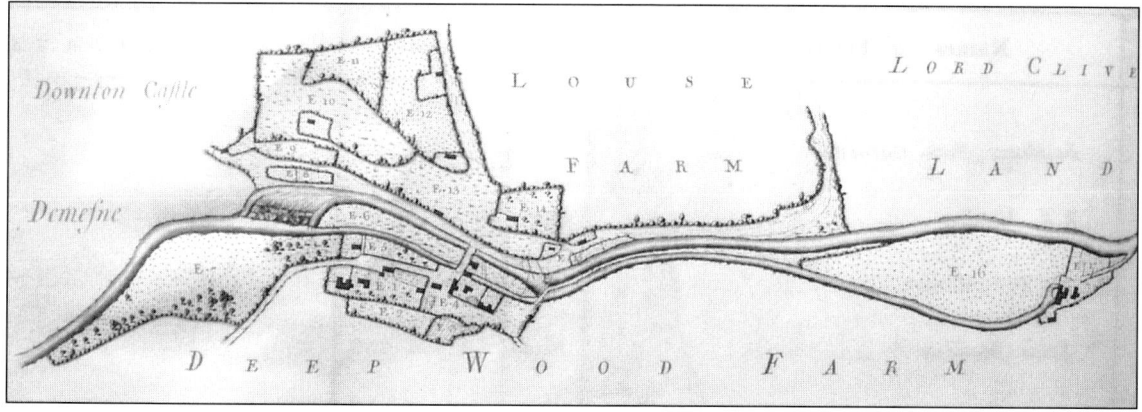

Downton estate atlas by James Sherriff 1780.
52. *Downton Castle demesne (top).*   53. *Bringewood Furnace (bottom).*

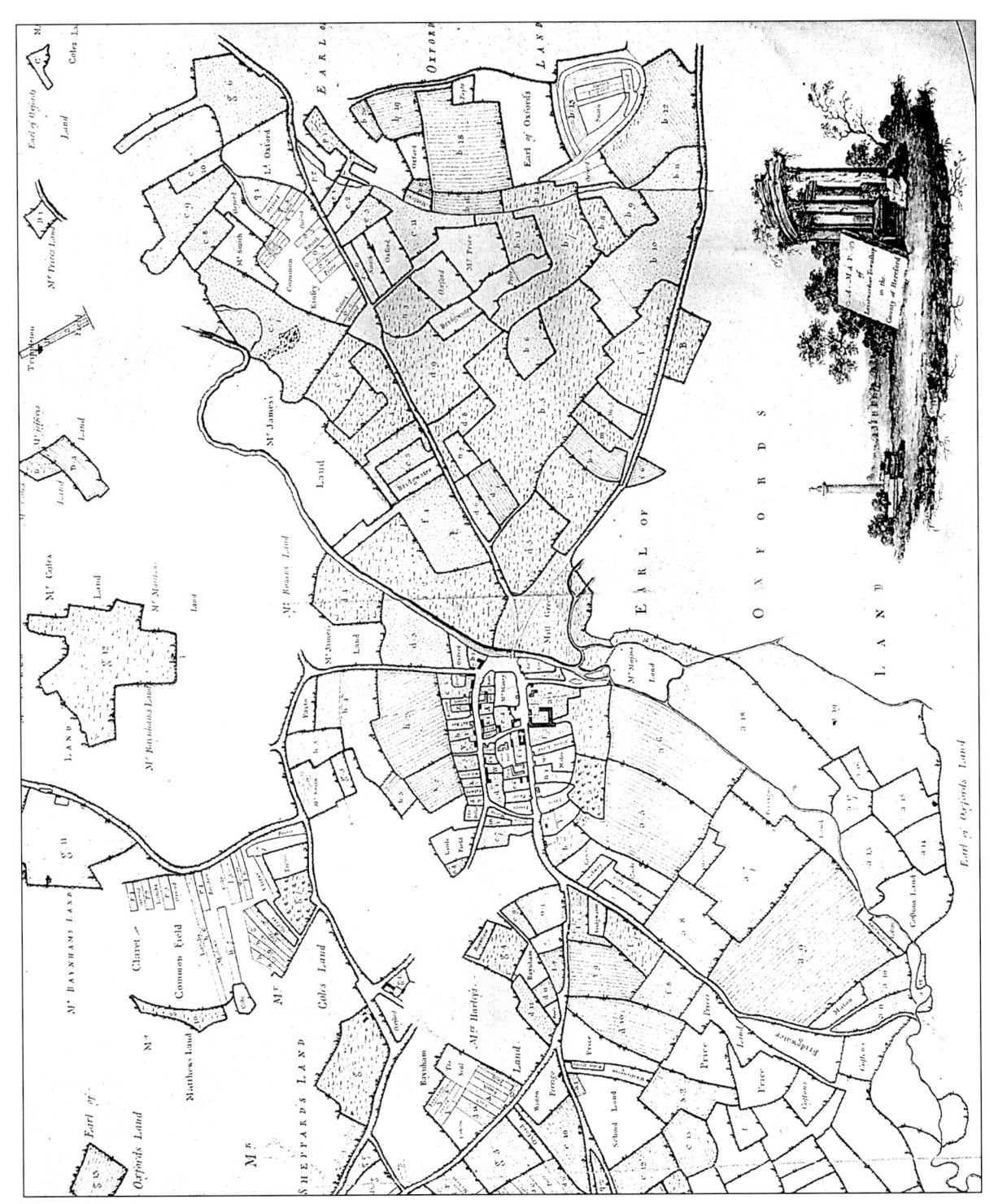

54. Leintwardine in the Downton estate atlas by James Sherriff 1780.

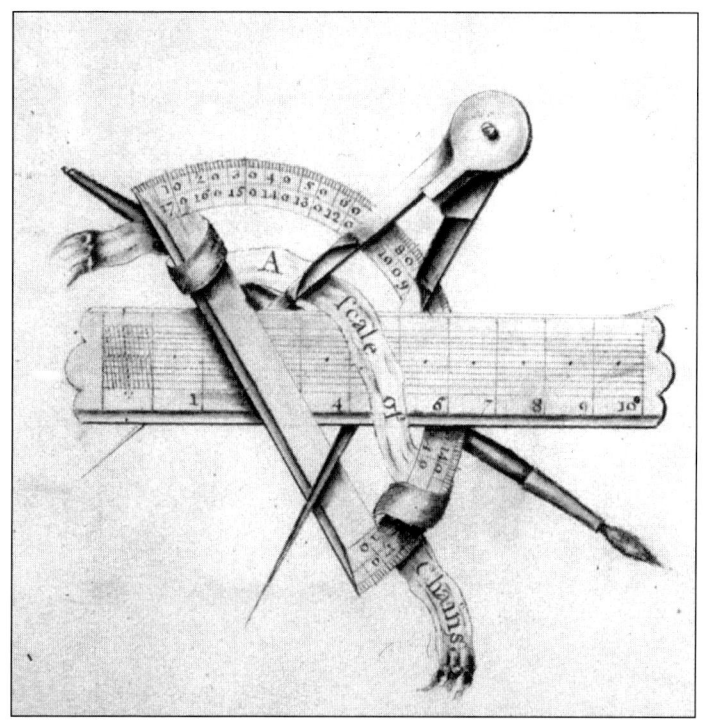

Almshall, Pipe and Lyde by John Harris 1784.

55. *Top left : Cartouche.*
56. *Top right : Drawing instruments.*
57. *Bottom left : Reference table.*
58. *Above : North and east points.*

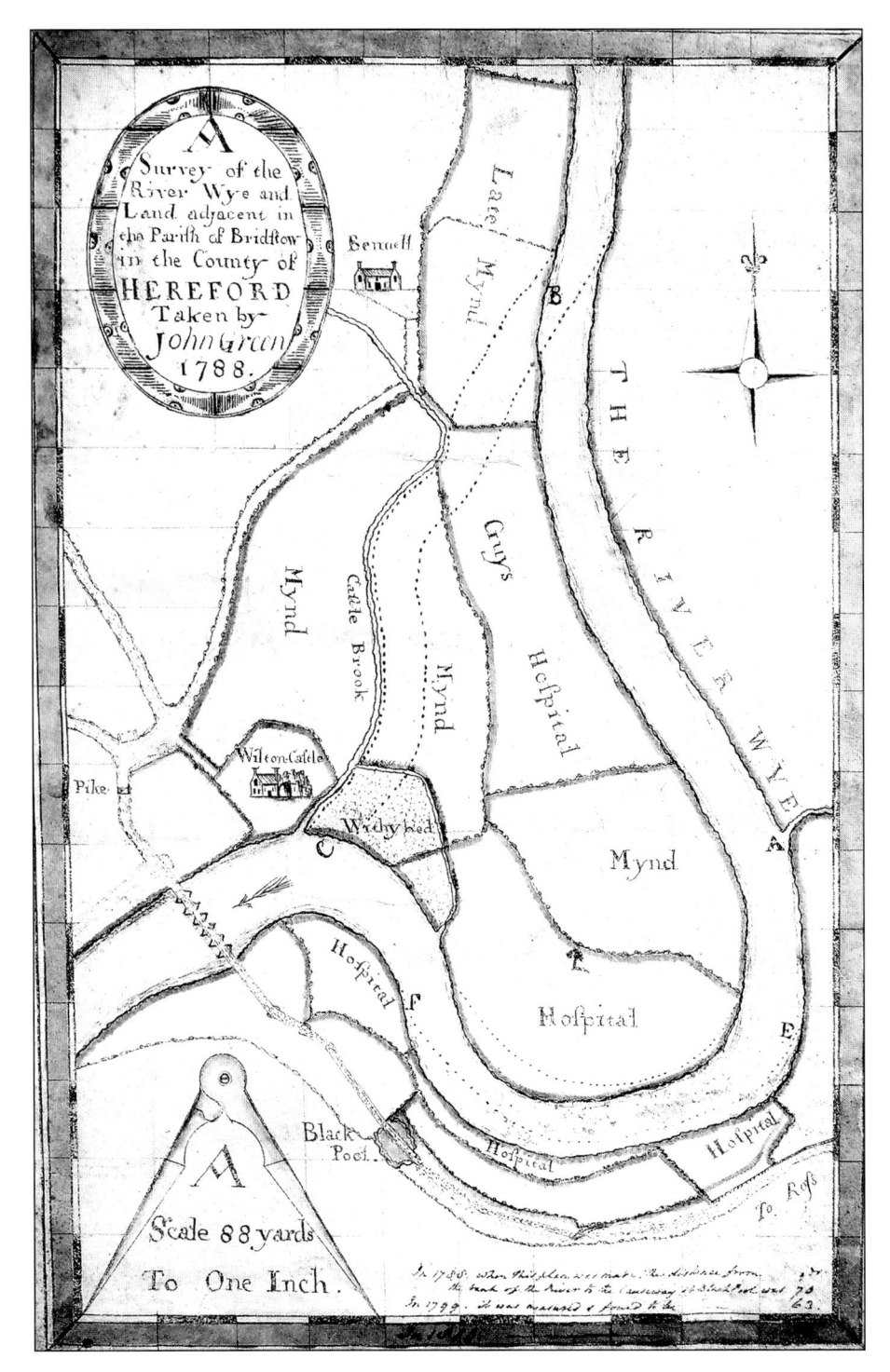

59. *Bridstow by John Green IV 1788 (see also Plate 48).*

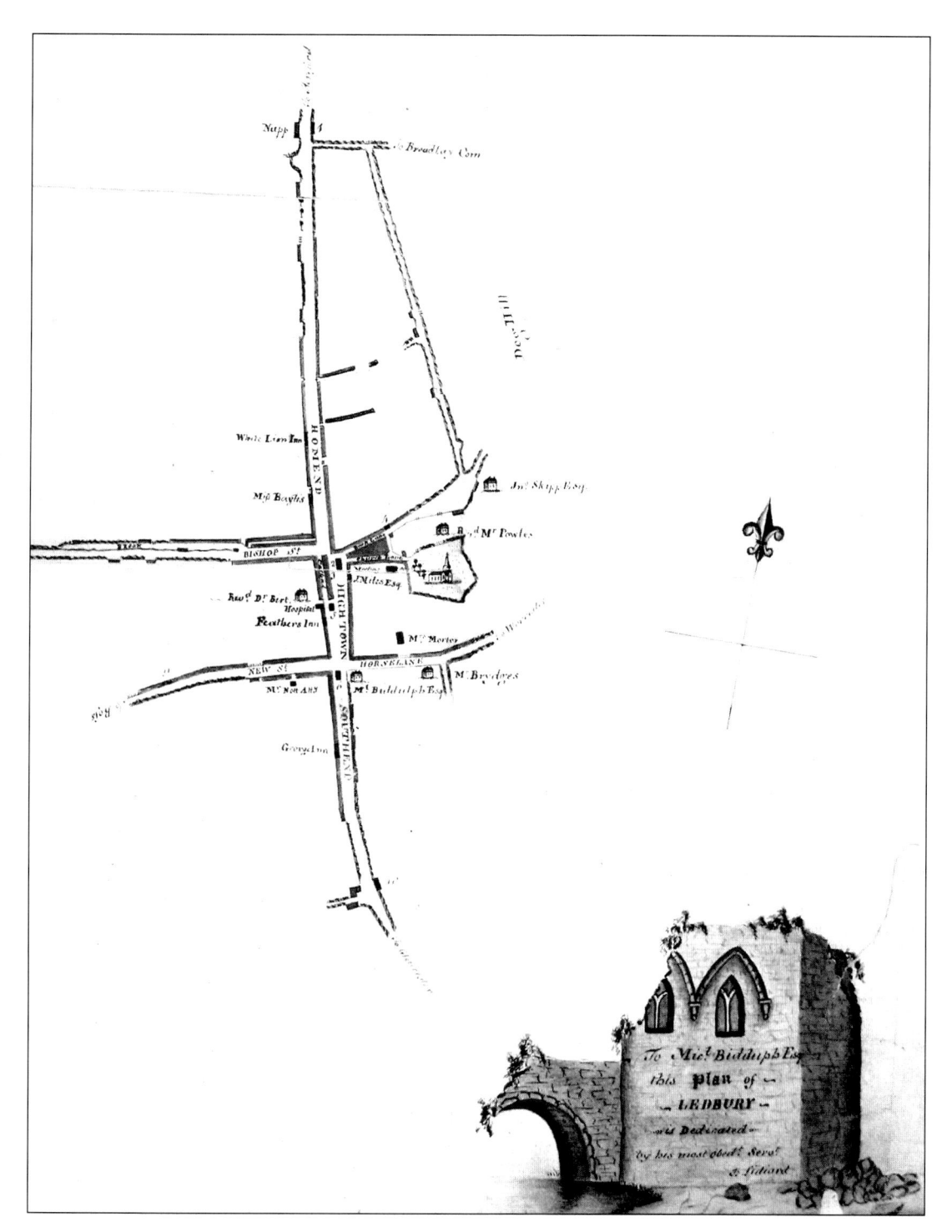

60. *Ledbury by John Lidiard 1788.*

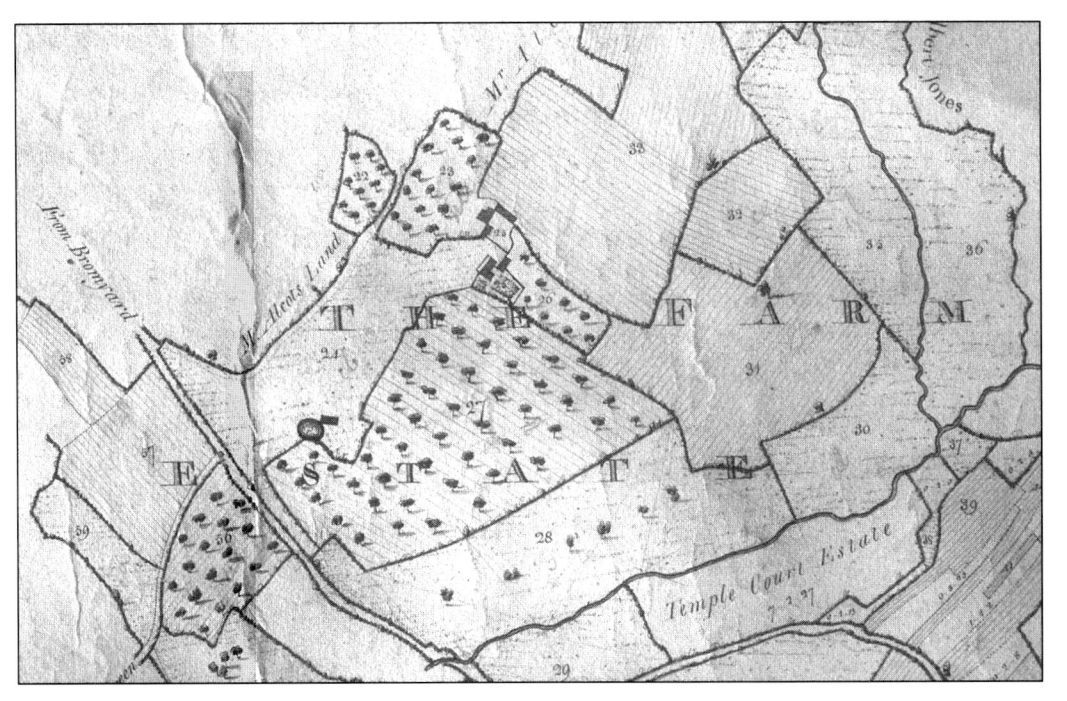

61. *Bosbury by Joseph Powell, an accomplished surveyor and draughtsman, 1791.*

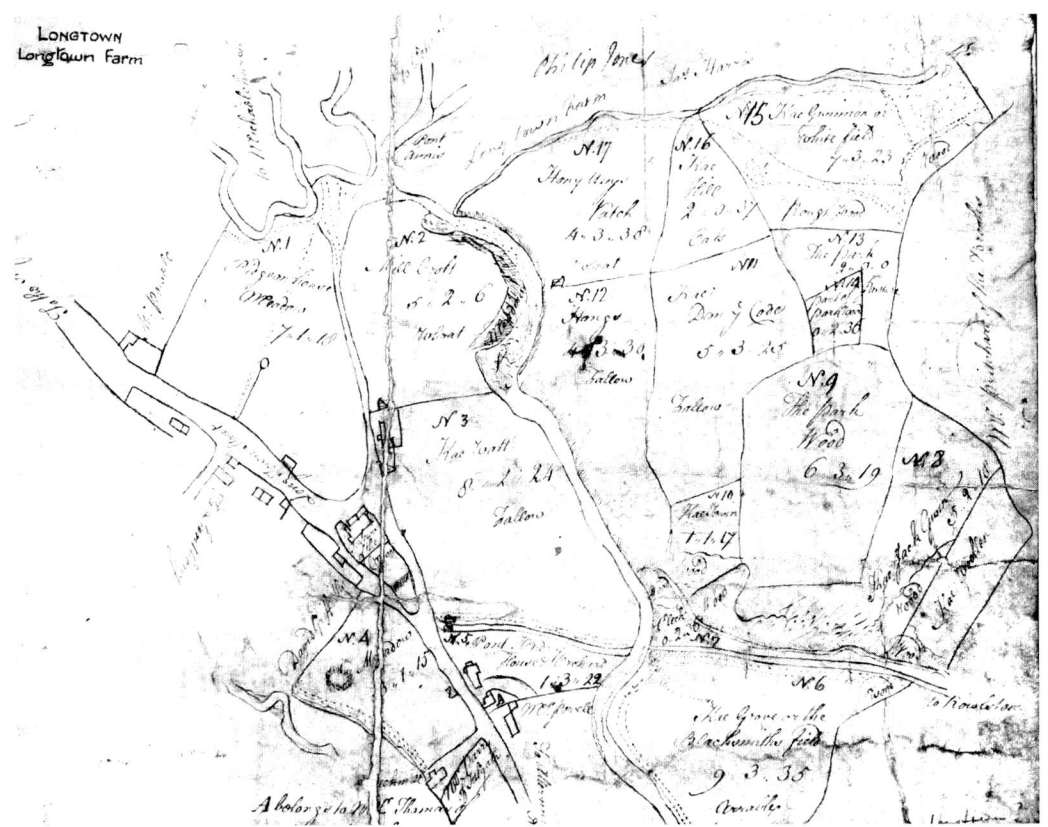

62. *Longtown by an anonymous land-measurer, late 18th century.*

63. Stretton Sugwas
by
James Cranston
1794.

Comparison with Meredith Jones's map of 1757 (see Plate 36) shows that the piecemeal inclosure of the open fields between the Hereford-Brecon and Roman roads was completed in the second half of the eighteenth century.

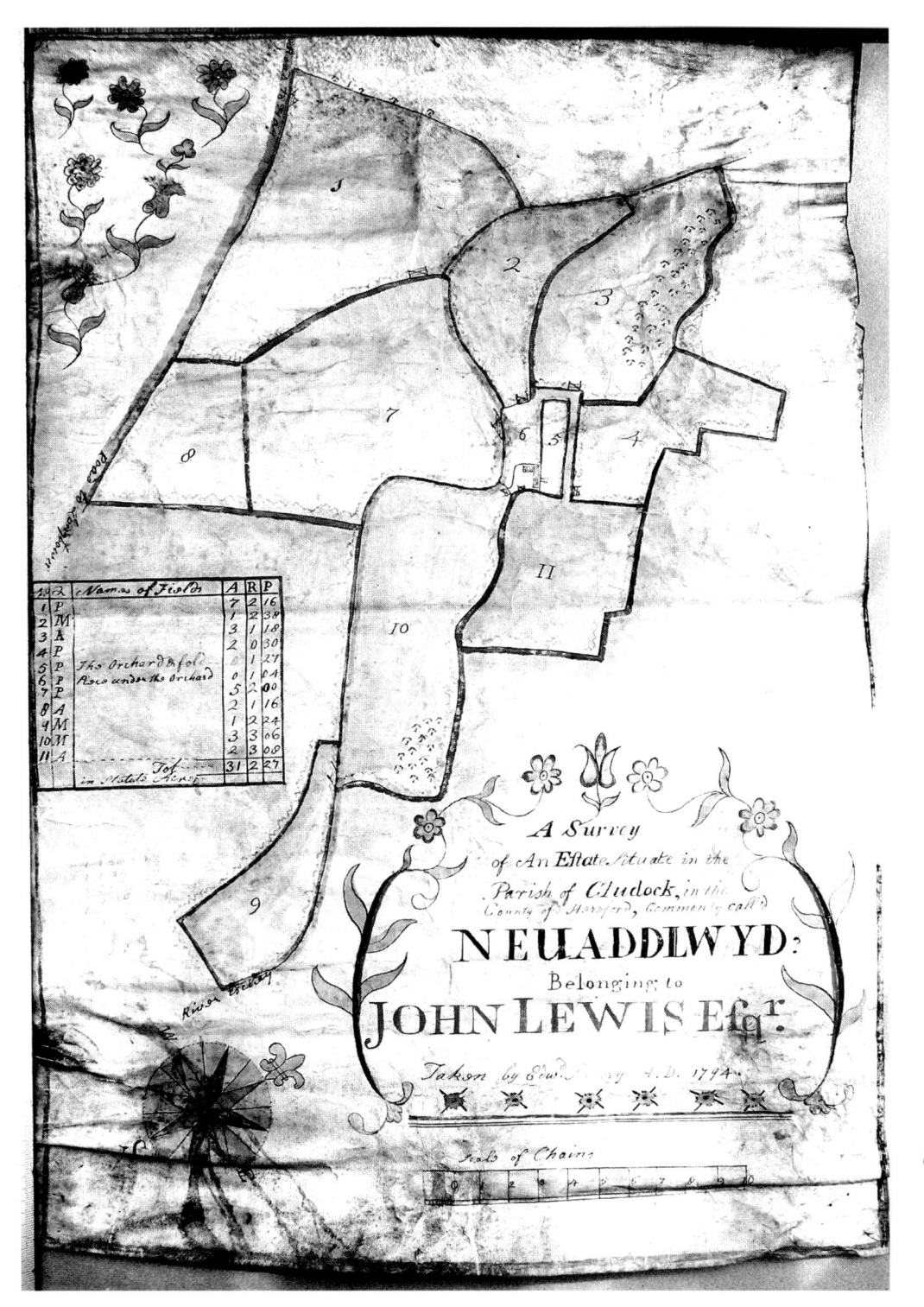

64. Neuadd Lwd, Longtown by Edward Penry 1794.

fols. 23v-24
Carey Farm (40a.), Barnfield (13a.), Bellingham Hall (part, 100a.), Hadbough alias Altbough Farm (72a.) in Ballingham, Little Dewchurch and Hentland [Hoarwithy]. Field-names include Blackspits, Forge Meadow, Quarry Head. Comment that Carey Farm is 'pleasantly situated commanding a fine prospect ... the soil is fit for feeding dairy and tillage, and produces large crops of wheat, Oats & potatoes ... it may be greatly improved at small expense'.

fols. 25v-26
Tryeven [Tre-Evan], [Llangarron] (50a.), Whitehouse Farm, Tretire (178a.), Trepenkennet [Trippenkennett], St Weonards (41a.). 270a. Tre-Evan consisted of scattered lands. Field-names include Sawpit at Tre-Evan, a hopyard at White House and Mill Meadow at Trippenkennet. Comment that White House Farm 'wants manuring and is improvable' and that both it and Trippenkennet are convenient to market. Scale: 26 perches to 1 inch [1:5148].

fols. 27v-28
Pothethar [Prothither] lands, Little Dewchurch. 123a. Marks mill. Scale 25 perches to 1 inch [1:4950].

fols. 29v-30
Hill Farm (336a.) including Scudamores Hill Coppice, (100a.), Blewhenstone House (53a.), Llanwarne, New Mills Farm and Alstons [Athelstans] Wood (215a.), Little Birch. 753a., including 'valuable woods'. Scale: 52 perches to 1 inch [1:10296].

fols. 31v-32
Scattered arable and pasture, King's Caple. Field-names include Red Rails Laqueries, Windmill Hill [SO 554290], Caple Tump.

fols. 33v-34
Scattered lands and strips in Allen and Upper Allen Fields, Brockhampton (by Ross).

fols. 35v-36
Park Farm (139a.), Chantry Farm (108a.), Spent Farm (20a.), Pockles [?Phocle] Land (40a.), Lyndor Farm (80a.), Leights Farm (94a.), Lower House Farm (149a.), Old Gore Farm (127a.) and woodland, Foy. 797a., some in open fields. Marks river Wye. Scale: 50 perches to 1 inch [1:9900].

fols 37v-38
Towns-end Farm, Brampton Abbots. 127a. Closes and orchard (15a.), the remainder in open fields. Scale: 26 perches to 1 inch [1:5148].

fols. 39v-40
Prill Farm (108a.), Deans Place (127a), Much Marcle [Yatton]. 242a. Comment that 'The land is in great order', a very rich farm with a good house, orchard and two hopyards.

fols. 41v-42
Manor of Yatton including Chapel Farm (192a.), Baynhams Farm (97a.). Field-names include commons, Westnors End, Old-bury Hall, Prill Farm, Cross roads [Crossway], Walshes [Welsh] Court, Yatton Wood (103a.), Cawborough [Coldborough] Park Coppice (111a.) and other woodland.

fols. 43v-44
Baytons Farm (60a.), Felhampton (171a.), Upton Bishop and lands (12a.) in Bridstow. 296a. Felhampton 'includes a large young orchard'.

fols. 45v-46
Callow Farm (123a.), Walford, Walsh Newton Farm (83a.), Welsh Newton, Hopestile Farm (28a.) [parish not identified]. 231a. Field-names include Ryegrass Field. Comment that Callow Farm is 'two miles from Ross by the ferry'. [The ferry near Callow Farm led to Goodrich. Ross was a straightforward overland journey by road].

fols. 47v-48
Abbey Dore (part). 976a. Includes Lower House Farm (246a.). Contains two good houses and large, choice meadows. Place-names include Quarres [Quarrels] Green. Scale: 52 perches to 1 inch [1:10296].

fols. 49v-50
Cock-yard Farm (138a.), Jury Farm (242a.), Abbey Dore. 541 a. Scale: 50 perches to 1 inch [1:9900].

fols. 51v-52
Dore Wood, Abbey Dore. 139a. including twenty-two years growth (92a.), young coppice fallen in 1770 (30a.), Below Grove fallen in 1770 (11a.).

fols. 53v-54
Bircott [Burcott] Farm, Holmer. 127a., much intermixed in open fields. 'An old house, and good Orchard and offices, which want repairs'.

fols. 55v-56
Vern Farm, Bodenham. 341a. held by two tenants. Field-names include hopyards, Hawthorn-hill, Whithardy Common Meadow, Portlane-head. With title of a cherub holding up a picture frame inscribed 'VERN FARM'

fols. 57v-58
Vauld Farm, Marden (83a.), Troughland. [Trothland], St Weonards (136a.). Total 223a. Some scattered lands, hopyards and orchards. N point in a drawing of a sheaf of corn with a scythe, hay rake and billhook.

fols. 59v-60
Maund Court, Bodenham (186a.). Lands partly enclosed and partly in open fields. Field- names include The Shuthooks, Upper Stamebridge, Quarrel Hill, Cockshoot Meadow, Connygars.

fols. 61v-62
Lower Priddleton (109a.), Upper Priddleton (250a.), Humber. 361a. Marks the road to Leominster from Bromyard. Field-names include Sheanbridge [Steens Bridge] Meadows. Advises exchanges of lands to consolidate both farms, describing Lower Priddleton as 'a fine place for a Gentleman's residence'.

fols. 63v-64
Bodenham. 312a., including Derrydales [Dudales] Hope and woods (140a.), Stone- house Farm. Field-names include hop-yards, Pidgeon House Meadow, Mill Meadow, Quarrel House, Marsh Meadow.

fols. 65v-66
Marden [Amberley]. 105a., including some common meadow Comment that it 'has a good and profitable Orchard…and would be a pleasant place for a Gentleman's residence'. Scale: 26 perches to 1 inch [1:5148].

fols. 67v-68
Swinstons [Swanstone] Court, Dilwyn. 214a. Comment that it is good for dairy and very improvable. Scale: 26 perches to 1 inch [1:5148].

fols. 69-70
Lands in scattered strips in Lake and Chester Common Meadows (39a.), with a good farm house 'fit for rearing young cattle and dairy'. Scale 25 perches to 1 inch [1:4950].
[The parish has not been identified. On tithe maps of c.1840 the field-name Lake Meadow is found in Marden and Chester or Chestern Way in Weobley and Dilwyn respectively].

fols. 71v-72
Lands in scattered strips in Ricklow and Perry Common Fields (66a.). [The parish has not been identified, but it was probably near Hampton Court as one of the adjoining owners was Lady Coningsby. On tithe maps of c.1840 the field-name Ricklow could not be traced; Perry Stub Field occurs in Bodenham].

fols. 73v-76
2,697a. in open fields of Broadfield and West Field in Northleach with Eastington, Gloucestershire. Scale: 50 perches to 1 inch [1:9900].

fols. 77v-78
Fredunnock [Tredunnock] (94a.) adjoining the river Usk and the parish of 'Lanhisan', Monmouthshire [Gwent] [Llantrisant, not Llanishen near Trelleck]. N point with a sheaf of corn and farming tools]. Scale: 25 perches to 1 inch [1:4950].

fols. 79v-82
Stoke Hammond (1,551 a.), Buckinghamshire. Scale: 26 perches to 1 inch [1:5148].

fol. 86
Total summary of the estates, 16,654a.

SIZE: Ink and watercolour on paper, 527 x 365. Leather-bound with gold tooling and a red title-label on the front cover. The front cover is detached and the spine missing. 537 x 373.

SURVEYOR: Richard Frizell.

REFERENCE: Private.

1780
MATHON, COLWALL, CRADLEY

'A MAP of the estate of Mr Edw.d Holder and Ann his Wife, in right of her, in the parishes of Mathon, Colwall and Cradley in the Counties of Worcester and Hereford'[1]

252a. [Park Farm, SO 761449] in E of Mathon and Colwall and Stiffords Homestall and land adjoining Cradley Brook. Shows houses in block plan, enclosed fields outlined in colour, hopyards, orchards, woods, roads, footpaths, gates. Gives names of adjoining owners.

Reference table gives acreage, land-use, field-names, including The Parks, Court Furlong, Long Marl Pits, Limestone Acre, Fernhill Bank and hopyard, Upper Peters Park, Cockshut.

Floral title cartouche in pen and ink. Compass rose. Scale bar.

SCALE: 3 chains to 1 inch [1:2376].

SIZE: 748 x 1195. Ink and watercolour on paper, linen-backed.

SURVEYOR: J. Aird.

REFERENCE: HRO, B9/6.

1 The rural part of Mathon was transferred from Worcestershire to Herefordshire in 1897, but the easternmost part of the parish comprising the built-up area of West Malvern was retained in Worcestershire. The county boundary now passes through Park Farm.

n.d. [c.1780]
BROMYARD (NORTON)

'NORTON COCKLAYE & SANDY-CROSS Farms belonging to Tho:s COLLEY ESQ:R In the Parish of Bromyard …'

179a. [Lower Norton Farm, SO 685570] in NE of parish and in S of Tedstone Delamere near Hill Cross. Shows buildings in block plan, gardens at Norton, fields, hopyards, orchards, woods, streams, ponds. Marks Blacksmith's House and shop at Sandy Cross. Gives names of adjoining owners.

Reference table gives field-names and acreage.

Pen and ink title cartouche. Compass points and scale in rather florid penmanship. Plain border.

SCALE: 22 inches to 1 mile [1:2880].

SIZE: 530 x 740. Brown ink on parchment.

SURVEYOR: Not named.

REFERENCE: HRO, M46/48.

[c.1780]
MADLEY

'CUBLINGTON CASTLE FARM. In the parish of Madley…'

174a. W of the village centre [SO 406384]. Shows buildings in block plan, enclosed fields, ponds, trees, gates, roads, tracks. Marks motte and pond, formerly part of a moat. Gives names of adjoining owners and open fields. Field names include Castle Meadow, Longbridge Field, Barkway Street [the road from Madley village to Brampton] and 'Stony Street A Roman Road'.

Reference table with plot numbers, field-names with notes about husbandry, gross and net statutory and computed acreage.

Title cartouche on a stone building with pediment, flanked by trees and a cornucopia. Compass points extended across the map. Scale bar. Ruled border. Decorative initial letters of field-names.

SCALE: 8 chains to 1 inch [1:6336].

SIZE: 310 x 390. Ink on paper, linen-backed.

SURVEYOR: J. Taylor, Ross.[1]

REFERENCE: DCA, 4768.

1. The surveyor has clearly written his initial 'J' as occasionally did Isaac Taylor, for instance on his printed map of Hereford city 1757. No other relevant J. Taylor has been traced and although the DCA catalogue dates the map c.1800 it may safely be attributed to Isaac Taylor, who died in 1788.

1781
FOWNHOPE

'Plan of Exchanges of Glebes between the Rector and Vicar of Fownhope, those marked V for those marked R the Numbers referring to a general Map deposited in the Archives of the Cathedral Church of Hereford. N:B: The Parcels uncoloured are private Property Freehold map'd and measur'd with a View to future Exchanges'[1].

8a. of scattered plots [SO 5734], some enclosed and some strips in open fields, orchard. There are no distinctive landmarks to identify their location.

No compass point. No scale bar. Ruled border.

Attached to a deed of exchange between (1) Revd. James Birt, vicar of Fownhope, (2) James Kinnersley of the Lynch, Herefordshire and (3) the Dean and Chapter of Hereford, 29 November 1781.

SCALE: [c.1:3168].

SIZE: 328 x 402. Ink and watercolour on parchment.

SURVEYOR: Not named.

REFERENCE: HRO, AT69/9.

1. The 'general map' has not been traced in the Dean and Chapter's archives.

1781
LEINTWARDINE

'A MAP of the GRANGE ESTATE in the Parish of LEINTWARDINE … the Property of Theophilus Richard Salwey Esqr. Survey'd April 1781. John Harris of Newton Delin.'

The Grange estate, being the site of Wigmore Abbey [SO 410713]. Shows buildings in bird's-eye view, farm buildings, formal gardens, pools, dovecote, enclosed fields with plot numbers, pasture, meadow, hopyards, orchards, gates, roads with destination. Gives names of adjoining owners. Marks Green Lane and the turnpike road to Knighton. Field-names include Beatons, a former open field NW of the Grange divided into closes, Vines Hopyard and Vines [SO 406711], The Breeth, Red Pool Meadow [SO 412712], Pidgeon Yard and Pidgeon House Orchard [SO 411712], Dunstall, Park Field [SO 395720] and Leasow, Princes Leasow.

Title cartouche, rococo, incoporating tree trunks, foliage and a classical pediment. Compass rose. Scale bar with dividers, protractor and two drawing pens. Ruled border.

SCALE: 4 chains to 1 inch.

SIZE: 636 x 720. Ink and grey wash on parchment. The top left-hand corner has been torn off wthout loss to the working map.

SURVEYOR: John Harris, Wickton.

REFERENCE: Shropshire RO, 1141/126.

1782
EYTON

'A MAP of an ESTATE at EYTON … belonging to SAM. WARING Esq.'

Land E and SW of the church [SO 474616] dovetailing with EYTON 1767. Shows fields with reference numbers to a missing reference table. Marks roads.

Elaborate cartouche. Compass rose. Scale bar. All neatly drawn with a fine pen.

SCALE: 6 chains to 1 inch [1:4752].

SIZE: 360 x 445. Ink on paper, linen-backed. Fragmentary.

SURVEYOR: J. Powell, Bridgnorth.

REFERENCE: HRO, F49/18.

1783
BREINTON

'A PLAN of the FREEHOLD and COPYHOLD LANDS and also of LEASEHOLD LANDS Held under the DEAN & CHAPTER and CUSTOS and VICARS of the CATHEDRAL CHURCH OF HEREFORD late in the Possession of Richard Aubrey Esq.r'

Central part of the parish between King's Acre road and the river Wye [SO 4740]. Shows buildings in block plan, fields, mostly enclosed, 'hop garden', orchards, woods, river Wye, pools, roads.[1] Marks church with 'Camp' immediately to S, Lower Breinton farm and Upper Breinton farm [Upper Hill Farm]. Shading indicates the steep wooded slope N of the river and the Breinton Camp. Field-names include Brick Close, Sandpit Field, Green Lane Copse and The Foul Slough on S side of King's Acre road. [SO 480410].

Title cartouche of a classical pediment flanked by urns. North point . Scale bar. Faint ruled border.

SCALE: 3 chains to 1 inch [1:2376].

SIZE: 938 x 725. Ink and watercolour on parchment.

SURVEYOR: John Haywood, London.

REFERENCE: HRO, R23.

1. Note the London surveyor's use of the Kentish 'hop garden' instead of the Herefordshire hopyard or hop-ground.

(1783)
DONNINGTON

'A MAP of DONNINGTON FARM…the property of Richard Hill Esre. Measured and Maped by Thomas Buckle in April one thousand seven hundred & eighty three' (copy)

107a. in S and W of parish [Home Farm, SO 708339]. Shows buildings in block plan, strips in open fields and enclosed orchards outlined in colour, roads. Marks river Leadon. Gives names of adjoining owners. Contains copying error of road to 'Sedbury' [Ledbury].

Reference table gives field-names and acerage.

Compass points. Scale bar. Some later annotations in pencil.

Endorsed 'Plan on indenture of 2 August 1826 The Rev.d E. Freeman & others to Richd. Webb & Trustee'.

SCALE: 3 chains to 1 inch [1:2376].

SIZE: 645 x 800. Ink and watercolour on oiled paper (copy), with photocopy.

SURVEYOR: Thomas Buckle.

REFERENCE: HRO, R60/1.

1783
PEMBRIDGE

'Plan for division of Northwood'

Parliamentary inclosure map sealed with the award under an Act of Parliament 1780 for the inclosure of Northwood Common in Noke township [SO 358595]. 150a. in W of parish between Strangward [Strangworth] Forge and Noke. Names the sites of the Forge, Lord Oxford's weir in the river Arrow, Noke mill.

No compass point. No scale bar.

The surveying expenses (attached sheet) amounted to £15 10s 4d.

SCALE: [c.1:7040].

SIZE: 290 x 480. Ink and watercolour on parchment.

COMMISSIONERS: William Preece of Aymestry, Richard Turberville of Pembridge, Roger Edwards of Lugwardine.

SURVEYOR: John Harris, landmeasurer.

REFERENCE: HRO, Q/RI/39.

1783
ULLINGSWICK, PENCOMBE

'Map of the General Leasehold & Copyhold Estates in the Manor of Ullingswick … Held under the Dean and Chapter of Gloucester'[1]

Whole of the manor (913a.) [SO 5949]. Shows buildings in block plan and the church in perspective, arable open fields with plot letters and numbers, orchards, rough grazing, woods, gates, streams, ponds, roads with destinations. Gives names of adjoining owners and parishes.

Title cartouche of ribbon and flowers. Compass rose. Scale bar with pen, dividers and protractor. Doubled ruled border incorporating upper and lower case grid references a-l, A-L.

SCALE: 4 chains to 1 inch [1:3168].

SIZE: 1360 x 1380. Ink and watercolour on parchment.

REFERENCE BOOK: 'Book of Reference to the Deans Mapp. Survey and Valuation of the Several Leasehold and Copyhold Estates in the Manor of Ullingswick…held under the Dean and Chapter of Gloucester'. Arranged by tenancies with grid letter references and field plot numbers, field-names, land-use, acreage, value per acre, annual value, with some later annotation in pencil. Field-names include hop-yards, Broxash Field, Bebury Field, Linnetts Bridge. The estate included Sydnal [Sidnall] Farm (73a.) in Pencombe [SO 595515]. The title page finishes with a flourish in the shape of a swan. 380 x 250, quarter-leather and marbled paper-bound, 32 written folios. Unsigned [J. Harris].

SURVEYOR: J. Harris.

REFERENCE: Gloucestershire RO, D1740/P3 (map) and D1740/E4 (reference book).[2]

1. The map is examined in H.L. Gray, *English field systems*, 1959 and J.E. Grundy, 'Ullingswick: a study of open fields and settlement patterns'

in D. Whitehead and J. Eisel, *A Herefordshire miscellany*, Hereford 2000, pp. 286-300. I am indebted to Joan Grundy for drawing the map to my attention.
2. The Dean and Chapter of Gloucester's estate papers also contain two undated rough sketch-maps of Ullingswick, drawn on contemporary thin paper, and an undated valuation, possibly a summary of the 'Book of Reference' 1783 but with different plot numbers.

1783
WHITNEY-ON-WYE, CLIFFORD

No title. An atlas of maps in S of Whitney and N of Clifford, with a terrier. Most of the maps cover two facing pages and are numbered, each with a separate title. The maps show buildings in profile or block plan, enclosed lands with a few traces of strips, hedges, woods, streams, roads. Fields are numbered with reference to the terrier. Changes in the course of the river Wye between Clifford and Whitney, most notably in 1730, and the alignment of most of the maps with S at the top, may have contributed to the critical remark on p. 2.

p. 1
'The CARRIER Farm in the PARISH of WHITNEY … Survey'd by Edward PENRY. 1783'. [Carriers House] and 27a. [SO 273482] in Whitney. Shows buildings in block plan, hedges, wood, roads. Inset reference table with field-names and acreage. Field-names include The Sholly, The Rules. Oval cartouche, rectangular reference table and border with crude decoration. Compass rose, and the word 'SOUTH' at the foot of the map. [Despite this the map appears to be aligned with W at the top].

p. 2
[Title page]. Comment that 'This Map or Plan of the Estate at Whitney & Clifford Herefordshire is very Bad & Not at all Correct 1784'

pp. 3-6
Terrier of Whitney Court Farm (640a.), Boat Side Farm (52a.), Apperley's Farm (110a.), Mill Farm (8a.), Sheep Cott (353a.), Castleton (626a.), Whitney Wood (86a.), Common Wood (13a.), The Lordship upon the Hill (14a.), Clifford Castle (17a.), with other small plots making a total of 1,919a.

pp.7-8
1. 'WHITNEY COURT' [SO 266476]. Shows fields within Whitney Park 'now divided into 7 Pieces' and at Perry Hill.

pp. 9-10
2. 'WHITNEY COURT'. Field-names include Stow Meadow, Stow Hop Yard, Rainbow Meadow.

pp. 11-12
3. 'WHITNEY COURT'. Shows Old Court in block plan with a formal garden and house at New Court in profile. Field-names include meadows, orchards and hopyards, Boat Field. Marks road to Hereford and road 'to the Bridge' [Lyde bridge, SO 269473].

p. 13
4. 'WHITNEY COURT'. Field-names include 'Platts or Old House part of Boat Farm', Kiln Ground.

pp. 14-15
5. 'BOAT SIDE FARM', Whitney [SO 269473]. Marks the river Wye, 'Whitney new Stone Bridge over the Wye built in the Year 1783' [destroyed by flood in 1795] with toll-house cottage and adjacent land 'given by Act of Parliament', two roads to Brilley through Whitney Wood and past Freemans Wood and Common Wood, road 'to the Bridge' [Lyde bridge] and the church [SO 267475]. Field-names include glebe, Clifford Common [meadow at SO 265473], Timber Yard.

p. 16-17
6. 'APPERLEY'S FARM', Whitney. Field-names include Park Stile, Quarry Close, Upper and Lower Mill Plot.

p.18
[7]. 'MILL FARM PART OF APPERLEY'S'. Includes Tinkers Plot and strips in Rules Field.

pp. 19-20
8. 'SHEEP COTT FARM'. Lands of Sheepcote farm, Clifford and to its W, formerly in Whitney. Marks road [to Whitney bridge] and common meadow of Clifford and Whitney along the river Wye. Field-names include Rainbow Meadow, Great and Little Senary.

pp. 20-21
9. 'SHEEP COTT'. Marks Sheepcote farm house in profile, river Wye and Whitney bridge, with lands on both sides of the 'New Road to the Hay'. Field-names include Black Pool, Boat fall Ditch.

pp. 22-23
10. 'CASTLETON' Old Castleton Farm, Clifford [SO 285456] and lands, mostly riverside meadow. Marks buildings in profile. Field-names include The Gliss, Day House Meadow, The Tump [motte and bailey at SO 283457], The Nocks.

pp. 24-25
11. 'CASTLETON'. Field-names include Sheep Walk, Cwm y Hull.

pp.26-27
12. 'CASTLETON'. Field-names include Sheep Walk, The Lawns, Anthony's Park.

pp. 27-28
'THE LORDSHIP UPON THE HILL'. [Probably in Clifford W of Castleton]. Marks small scattered enclosures with five cottages and two other buildings in profile, two springs with streams. In a later hand one boundary is marked 'Brow of Hill'.

p. 29
'CLIFFORD'S CASTLE'. Marks two cottages and The Castle [SO 243456] in profile.

p. 30
'CASTLETON SHEEP WALK'. A cottage and small plot only.

p. 31
'Whitney Wood Survey'd by Edward Penry 1783'. Shows Whitney Wood and the roads to Brilley, Whitney bridge and [Lyde] bridge. Marks cottages on the edge of the wood and the division of the wood into the plots of freeholders and tenants. Compass rose. Framed border.

SCALE: [*c.*1:3620].

SIZE: 380 x 250. Ink and watercolour on paper, soft leather-bound, with flap.

SURVEYOR: Edward Penry.

REFERENCE: University of Wales Bangor, Archives Department, Whitney and Clifford estate papers 424.

1784

CLEHONGER

'A MAP of the VALLETTS, SHARK-HOUSE and late PHILPOTTS'S Estates in the Parish of Clehonger and Allensmore ... held by Lease under S.t Ethelbert's Hospital in Hereford'[1]

The Valletts [SO 462388], 53a. enclosed lands in N of parish, Shark-House (22a.) and late Philpott's estates (11a.) mostly scattered strips in open fields near Cage Brook and Gorsty Common in W of parish. Shows buildings in block plan, distinguishes arable, pasture, meadow, orchard, coppice. Marks river Wye, roads.

Title cartouche with surround of leaves and flowers. Separate reference tables for each property, drawn as scrolls. Compass N point only, drawn as a (flag) standard. Scale bar with ruler, dividers, protractor, pen and paintbrush. Ruled border.

SCALE: 4 chains to 1 inch [1:3168].

SIZE: 600 x 760. Ink and grey wash on parchment.

SURVEYOR: John Harris, Wickton.

REFERENCE: DCA, 4745/25.

1. A copy of this map at a scale of 8 chains to 1 inch [1:6336] features in a survey book of the Dean and Chapter of Hereford's property compiled c.1820 (DCA, 7004/1). In this copy the reference tables are put together and the name of the original surveyor is not recorded. It is followed by a map of lands exchanged at Shark-house 48a. and Gorsty Common 16a. (pp. 51-3). The volume is of paper, watermarked 1828, vellum and half-leather bound. Several numbered pages are missing.

   The volume continued in use until 1838. Its fly-sheet bears the inscription 'Rich.d Underwood Castle St [Hereford] Janry 24 1842'. It contains the following maps:

pp. 21-2	Lulham Court and the Toddage, Madley, 1789. For catalogue entry, see 1789 MADLEY.
pp. 29-30	Canon Bridge, [Madley], n.d.
pp. 37-38	Gorwall prebend, Woolhope, 1819.
pp. 39-40	Manor of Blackmarston, Hereford, 1835.
pp. 41-2	Mr Hodge's lands, Preston on Wye, 1824.
pp. 43-4	Manor of Preston on Wye, 1764. For catalogue entry, see 1764 PRESTON ON WYE.
pp. 47-48	Castle Crabb, Cock-Shot and Coed Evan-y-Bodda, Disserth, co. Radnor, 1779.
pp. 51-3	The Valletts, Shark-House and late Philpott's's, Clehonger, 1784. For details, see this entry, 1784 CLEHONGER.
pp. 55-6	Manor of Blackmarston, Hereford, 1784. For catalogue entry, see 1784 HEREFORD ST MARTIN'S.
pp. 59-60	Almshall, Pipe and Holmer, 1784. For catalogue entry, see 1784 PIPE AND LYDE, HOLMER.
p. 61	[No title]. Canon Pyon, n.d.
p. 64	Glebe, Sellack, 1768. For catalogue entry, see 1768 SELLACK.
p. 67	Upper Hall, Ledbury, n.d.
p. 68	Derndale, Canon Pyon, n.d.
pp.70-1	Derndale, Canon Pyon, n.d. [unfinished].
[unpaginated]	Breinton.
	Canon Pyon.
	Woolhope and Brockhampton.
	Preston on Wye.
	Fownhope.

(1784)
HEREFORD, ST MARTIN'S & ST JOHN'S

'MAP of the Manor of Winaston situate at BLACKMARSTON near the CITY of HEREFORD being an undivided Estate, part held by Lease (under St. Ethelbert's Hospital and the College of Hereford) and the Residue a Freehold the PROPERTY OF James Donethorne Esq.[1]'

144a. mostly in open fields W of Blackmarston Farm [in Hunderton and Newton Farm area, SO 4938]. Shows buildings in block plan, strips in open fields, roads. Marks river Wye and outlying piece of meadow in the Wergins adjoining the river Lugg. Field-names include Gallows Tump on open land [at junction of Belmont Road and Hunderton Road, SO 501388], Great and Little Gallows Fields, Newtons Common Field, Vineyard Croft [SO 505387], Upper House [SO 505389].

Reference table with field-names and acreage.
Compass point. Scale bar.

SCALE: 8 chains to 1 inch [1:6336].

SIZE: 353 x 435. Ink and watercolour on paper, in a volume, vellum and 1/2-leather bound. Copied c.1820. Original map not traced. See also 1784 CLEHONGER.

SURVEYOR: Not named.

REFERENCE: DCA, 7004/1, pp. 55-6.

1. For an almost identical but larger map, possibly drawn later, see [c.1795] HEREFORD, ST MARTIN'S & ST JOHN'S.

1784
HEREFORD, ST OWEN'S

'A MAP Of Three Pieces of Meadow Land Scituate Without Saint Owens Gate in the Parish of Saint Owen Hereford The Property of Rich:d Jones, Gent'

29a. immediately outside St Owen's Gate [on N side of St Owen's Street between the Gate and Ledbury Road, SO 516396]. Marks barn near St Owen's Gate in bird's-eye view. Shows field boundaries, gates, roads and footpaths. Gives names of adjoining owners, including Hospital Gardens [of St Williams's Hospital and/or St Giles Hospital]. Marks Town Ditch, Scut Mill Lane [now Ledbury Road]. Field-names include [St Giles] Chapel Piece. See also 1787 HEREFORD, ST OWEN'S.

Reference table gives field-names, acreage.
Compass points. Scale bar.

SCALE: 4 chains to 1 inch [1:3168].

SIZE: 213 x 265. Ink on parchment.

SURVEYOR: William Jones.

REFERENCE: HRO, AE11/1.

1784
PIPE AND LYDE, HOLMER

'A MAP OF ALMSHALL Estate situate in the PARISH'S of Pipe and Holmer ... Held by Lease under S.t ETHELBERT'S Hospital in Hereford' (Plates 55-58)

52a. mostly in the open fields of Fernhill Field and Holmer Field adjoining the Hereford-Leominster road. Shows Almshall House in block plan [SO 506436], gates, ponds, roads and husbandry road. Distinguishes arable and meadow.

Reference table gives plot numbers, land-use, field-names, acreage.

Floral title cartouche. Reference table drawn as a paper with curled corners laid on the map. N and E points indicated by the head and tail of a half-coiled snake. Scale bar with dividers and protractor,. Ruled border. Good draughtsmanship.

SCALE: 4 chains to 1 inch [1:3168].

SIZE: 365 x 550. Ink and grey wash on parchment.

SURVEYOR: John Harris, Wickton.

REFERENCE: DCA, 4696/27. Copied c.1830; see 1784 CLEHONGER. A schedule entitled 'Contents of Almshall Estate in the Parishes of Pipe & Holmer' by John Harris, giving field-names, acreage, land-use, adjoining landowners, relates to its lease to James Hereford Esq.,1784: DCA 4696/28.

1784
STAUNTON-ON-WYE

'A Map of the Division of Staunton's Common and the Meddles's situate in the Parish of Staunton upon Wye ...'

Parliamentary inclosure map attached to the award under the Act of Parliament 1784 for the inclosure of the N part of the parish [SO 3546] including Old Lake, New Lake, Staunton Medleys and land adjoining Norton Medleys. Shows inclosure allotment boundaries in outline, roads.

Compass rose. Scale bar.

SCALE: 4 chains to 1 inch [1:3168].

SIZE: 660 x 590. Ink on parchment, very faint.

COMMISSIONERS: Richard Price of Kington, James Kinnersley of Eardisland, William Griffiths of Hereford.

SURVEYOR: John Harris of Wickton, land surveyor.

REFERENCE: HRO, Q/RI/50.

[c.1784]
BREINTON

'The COURT. AN Estate situated at LOWER BREINTON ... belonging to the Treasurer of the Cathedral Church of Hereford'.

Areas at (1) Lower Breinton [SO 473397] in S of parish adjoining the river Wye, (2) on N side of Green Lane [at SO 470408], (3) a small close adjoining the Roman Road [in the parish of Stretton Sugwas or Holmer] and (4) strips in an open field at King's Acre. Shows buildings in block plan, enclosed and open fields with plot numbers, orchards, woods, roads. Gives names of some fields and adjoining owners. Marks lands held by Mr Aubrey, either as owner or tenant.

A roughly drawn map, perhaps made in connection with the dispute between the Treasurer and Richard Aubrey. See also 1784-5 BREINTON.

Floral title cartouche. Compass points. Scale bar (lacking figures). Ruled border.

SCALE: [c.1:8740].

SIZE: 328 x 425. Ink and watercolour on parchment.

SURVEYOR: Not named.

REFERENCE: DCA, D123.

1784-5
BREINTON

'A MAP of LANDS allotted in this Award as belonging to the Treasurer of the Treasureship of the Cathedral Church of Hereford, and formerly leased to Richard Aubrey Esq.r of Cle<h>onger in the County of Hereford, and his predecessors'

134a. in Lower Breinton [SO 473397] and Green Lane area with, additionally, strips in open fields measuring 9 statutory acres or 14 'computed' acres. Shows buildings in block plan, enclosed lands and strips in open fields, outlined in colour, woods, ponds, roads. Marks church, 'Camp' immediately SW of church and ford across river Wye SW of church, 'Court house' [SO 470396]. Gives names of adjoining owners. Field-names include Hongret Common Field 'now' Crabtree Field, Middle and Upper Bringhill, Sandpit Field, Beckmarsh Common Field.

Attached and sealed with the award of 30 December 1784 to determine ownership of intermixed lands all in the occupation of Richard Aubrey as owner of some and tenant of others belonging to Dr William Parker, as treasurer of Hereford cathedral. Recites difficulty of distinguishing their lands because fences were long destroyed and that settlement had been reached by arbitration after examining records. The award set out husbandry ways for access and required existing banks to be thrown down, new ditches opened and stone landmarks erected. It imposed responsibility for their maintenance on each of the two parties and ruled that the costs of the survey were to be divided equally.

Folded loose inside the award, a sealed list of the fences to be maintained by the respective parties, 18 October 1785.

Compass points. Scale bar.

SCALE: 3 chains to 1 inch [1:2376].

SIZE: 960 x 565. Ink and watercolour on parchment.

SURVEYOR: John Haverfield, Kew, surveyor and Thomas Hardwick of Brentford, Middlesex, surveyor.

REFERENCE: DCA, D649.

1785
HOLME LACY

'A MAP of the POUND ESTATE situate in the Parish of HOM-LACY ... The property of the R[t]. Hon'ble the Earl and Countess of Surry'

Pound Farm [SO 555353] (62a.) in the centre of the parish. Shows buildings in block plan, fields with plot numbers and land-use, trees, gates, roads including the turnpike road from Hereford, ponds, 'the Fish Pools'. Gives names of adjoining owners.

Reference table gives plot numbers, field-names, land-use, acreage. Field-names include Lower and Upper Radway.

Title cartouche in a rococo frame. Scale bar with dividers. Compass points extending over the map. Fine penmanship. Ruled border. The reference table drawn as a curling piece of paper lying on top of the map.

SCALE: 4 chains to 1 inch [1:3168].

SIZE: 547 x 649. Ink and grey wash on parchment. Endorsed 'Plan No. 2'.

SURVEYOR: 'John Harris of Wickton Del[t]. July 15th. 1785'.

REFERENCE: BL, Add. MS. 36307, G45.

1785
LEA, WESTON UNDER PENYARD

'SURVEY OF THE ESTATES OF MAYNARD COLCHESTER, ESQ:[R]'

Estate atlas of the estates of the Colchester family of Westbury-on-Severn, Gloucestershhire, mostly in Westbury-on-Severn. Each map is preceded by its reference table. The maps mark houses belonging to the estate in block plan and other buildings in bird's-eye view. Churches are drawn in perspective. Shows fields with plot numbers, coloured in pale brown or green to indicate land-use, woods, isolated trees, roads with destinations, gates, streams, ponds. Gives names of adjoining owners.

Compass rose with E cross. Scale bar. Ruled border in pale brown watercolour. Fine and meticulously drawn small maps. Some later comments in pencil.

The reference tables give plot numbers, field-names, acreage, land-use, value, timber trees, surveyor's comments, of which there are few.

pp. 120-122
The Lea. Reference table. Includes 265 timber trees.

pp. 123-124
[Missing].

p. 125
'THE LEA'. 140a. of enclosed fields in the village and centre of the parish. Marks church, village cross [SO 661218], Red Lion House, Poor Lands. Field-names include Pound Close [SO 658219], Rock Meadow [SO 659218], Tumps Knoll and The Tump [SO 662219].

pp. 128-130
The Lea. Reference table. Includes 361 timber trees.

p.133
'THE LEA <& Weston>' [added c.1837]. Newhouse Farm [SO 649212] and 127a. of enclosed fields, with a few traces of former strips, lying to the W of the village. Includes two fields [10a.] in Weston under Penyard. Marks Wharton Farm. Field-names include Sandy Way, Burn'd Oak, Goose Tump immediately SW of Newhouse Farm, Far Jews Land.

p. 135
The Lea. 'Part lies in the Parish of Weston' [added c.1837]. Reference table. Includes 15 timber trees.

p. 137
'THE LEA <& Weston>' [added c.1837]. 22a. in W of Lea parish. Marks parish boundary, Ponselt [Pontshill] Marsh [SO 640220].

p. 139
The <Lea> struck through and replaced c.1837 by 'Aston Ingham & Newland & Lea'.

p. 141

'The LEA'. 14a. of scattered lands adjoining the roads N and S of Lea Line. Marks the hamlet of Lea Line [SO 668213].

SCALE: 8 chains to 1 inch [1:6336].

SIZE: Volume 220 x 275, calf-bound with blind tooling, raised bands and two clasps, the front cover detached. About half filled, with 176 numbered pages. Maps 200 x 245, ink and watercolour on paper. A few tears and a little staining.

SURVEYOR: John Merrett, Gloucester.

REFERENCE: Gloucestershire RO, D2123.

1785
MUCH MARCLE

Endorsed 'Plan of the Rye Meadow Lands'
 Photocopy of outline plan of scattered plots [28¼a] [SO 654341]. Gives adjoining field-names.
 No compass points. No scale bar.

SCALE: [c.1:2534].

SIZE: 315 x 195. Ink on paper [original map].

SURVEYOR: Not named.

REFERENCE: HRO, AS8/4.

1785
WELSH NEWTON

No map.[1] Reference book only.
 'A BOOK of REFERENCE To the Plan of the Parish of Welsh Newton ... Survey'd For the Coheiresses of the Compton Estate and Daughters of the Late JOHN BERKELEY Esqr.
 Whole parish [SO 500180]. Gives plot numbers, field-names, land-use, acreage. Properties include Callow [Cally] Hills, Penbridge [Pembridge] Castle, Treworgan, Tromhaid [Tremahaid], Newton, Mill Farm, Gwaun-hirion [Gwenherrion], St Wolstan.

SIZE: 205 x 135. Paper in marbled covers, 56 pp., of which 26 pp. are blank.

SURVEYOR: Mathew Williams.

REFERENCE: HRO, O68/I/106.

1. The map is not among the Berkeley family muniments at Berkeley Castle, Glos.

[c.1785]
PRESTON-ON-WYE

PLAN of PRESTON-COURT and other intermix'd Land situate in the Parish of PRESTON upon WYE ...'
 Whole parish [SO 3841]. Shows buildings, including church, in block plan, open and enclosed fields, orchards, woods, trees with shadows, gates, streams, ponds, roads with destinations. Distinguishes arable land and marks one small piece of common E of Hacton [SO 389417]. Gives plot numbers but no field-names.
 'Picturesque' title cartouche. Compass rose. Ruled border.

SCALE: [1:3520].

SIZE: 950 x 570. Ink and grey wash on paper. Repaired and linen-backed.

SURVEYOR: J. Powell.

REFERENCE: HRO, R34/22.

c.1785, 1786, 1788
LEDBURY

Three versions of a town plan with differing dedicatory titles.

[1] [c.1785]
LEDBURY

'To the Right Hon.ble Lord Som[m]ers This Plan of the Town of Ledbury is humbly Dedicated ... [as above]. Estates Measured and Maped'[1]
 Plan of the town centre. Marks the houses of the principal inhabitants in perspective, church, inns, Corn and Butter Market Houses, Upper Market House, Butchers Row, streets with names, a brook runnning down Bishop [Bye] Street, six turnpike gates, road destinations and mileage.
 The reference table has 21 numbered and 2 lettered landmarks, with the note 'NB J. Miles Esq.r M.rs Bayliss &c &c are not mentioned as the Houses are less remarkable than those already particularized'.
 Compass rose. Scale bar.

SIZE: 412 x 360. Ink and watercolour on paper, cloth-backed.

REFERENCE: Private.

1. Sir Charles Cocks bt. was created Lord Sommers of Evesham in 1784. The plan may have originally been drawn in order to seek his patronage.

[2] 1786

'PLAN of the Town of Ledbury in Herefordshire — J. Lidiard delin.ᵗ in 1786'

Similar map. Framed reference table with key to 23 landmarks.

Simple cartouche. Compass points. Scale bar. Single-ruled border. Neatly drawn and painted.

At foot, dedication 'To the Honourable Rich.ᵈ Cocks Esqʳᵉ this map is humbly Inscribed by his most Obed.ᵗ humble Serv.ᵗ J Lidiard'.

SIZE: 483 x 363. Ink and watercolour on paper.

SURVEYOR: J. Lidiard.

REFERENCE: Private.

[3] 1788

LEDBURY 'To Mic.ˡ Biddulph Esq this Plan of LEDBURY is Dedicated by his most obed.ᵗ serv.ᵗ J: Lidiard' (Plate 60)

Similar map. In a ruled frame is a brief description of the town, its fair and markets, its trade in corn and especially in cider. Recites a local tradition that it was the burial place of St Catherine.

Dedicatory title within a sketch of a waterside ruin. Drawing (? unfinished) of dividers, ruler, protractor, pen. Compass points.

SCALE: [c.1:6034].

SIZE: 318 x 395. Ink and watercolour on paper. Repaired.

SURVEYOR: J. Lidiard.

REFERENCE: HRO, G2/III/55. Reproduced in J.G. Hillaby, *The book of Ledbury*, Buckingham 1982, p. 54.

1786
LEDBURY

'A REDUC'D MAP of Dundridge and Tirrells Frith Estates, belonging to the Master [& Demesne] of S.ᵗ Catherines Hospital, Ledbury: Also of Malms:pool Mill and other Lands belonging to the said Hospital, situate in the Parish of Ledbury'

Dunbridge [SO 718363] to S of Ledbury town, Tyrrells Frith [SO 684381] to the W and Malmspool mill on the river Leadon [see 1720 LEDBURY]. Shows buildings in block plan, scatterd lands in enclosed and open fields, orchards, woods, trees and shadows, streams, roads, gates. Gives names of adjoining owners. Plot numbers marked, but no reference table.

Title cartouche in frame. Compass points. Scale bar. Ruled border.

With a note that it was produced and referred to in an affidavit by Richard Jones sworn before Samuel Rickards, Commissioner, 9 May 1808.[1]

SCALE: 8 chains to 1 inch [1:6336].

SIZE: 295 x 380 [trimmed edges]. Ink and grey wash on paper, linen-backed.

SURVEYOR: Jno. Harris.

REFERENCE: DCA, 3570.

1. Samuel Rickards was presumably a Commissioner of Oaths. The endorsement on the map does not relate to parliamentary inclosure at Ledbury.

[?1786]
STOKE PRIOR, LEOMINSTER, MIDDLETON-ON-THE-HILL, PENCOMBE

4 maps, apparently part of a series relating to the Hampton Court estates of the Arkwright family. All show buildings in perspective, fields in brown (for arable) and green (for pasture), orchards, streams, pools, roads. Names of adjoining owners.

Compass roses. Coloured borders. Scales in chains are written out. The decorative features and scales differ slightly. All are well drawn by the same hand, possibly by John Harris of Wickton, although they are not in his usual style of delicately fine penmanship.

[1] 'Wickden Mr Harris' [in Stoke Prior]

Wickton Court. [SO 523545] (244a.) in S of parish. Marks dovecote, hopyards, landmark stones in open fields, river Lugg. Field-names include Pidgeon House Close, Well Close, Roap Mans Orchard, Redstreak Orchard, and a strip in Ellliott's Field called Long Friday.

Endorsed with part of the reference table, which continues on a separate sheet

SCALE: '8 in an inch' [1:6336].

SIZE: 270 x 365.

[2] 'Birches farm at Stagsbatch [in Leominster] James Bowman Clarke Esq:ʳ'

52a, in scattered closes and strips in open fields between the Leominster-Monkland road and the river Arrow [SO 4657]. Marks Birches farm house by the Leominster-Monkland road.

Field-names include Portman Warden Field, Longall Field, Stagbatch late Ebnal Field.

Endorsed with reference table.

SCALE: '4 in an inch' [1:3168].

SIZE: 275 x 350.

[3] 'Redwood In Middleton' [-on-the-Hill]
Redwood Farm [SO 567654]. 172a. in NE of parish. Shows enclosed fields, including hopyards. Field-names include Merry Oak, Stonefield, Claypit Field.

Reference table.

SCALE: '8 in an inch' [1:6336].

SIZE: 273 x 305.

[4] 'March Court in Pencombe Winston Spencer'
Marsh Court [SO 585520]. 195a. in SW of parish. Field-names include hopyards, Windmill Pitt Field, Little and Great Doles.

SCALE: 8 in an inch' [1;6336].

SIZE: 273 x 302.

SIZE: Various, as above. Ink and watercolour on paper with no dated watermarks.

SURVEYOR: Not named. [Possibly John Harris of Wickton].[1]

REFERENCE: HRO, C69/1-4.

1. Joan Grundy of Ullingswick has drawn my attention to a letter from John Harris of 4 March 1816 [HRO, A63/III/41/5] referring to his survey of Wickton, Ford and other lands of the Hampton Court estate in 1786. These four maps might form part of that survey. But the misspelling of Wickton and the style of draughtsmanship make the attribution questionable.

1787
HEREFORD, ST OWEN'S

'ACCURATE SURVEY of S.t Owens PORT FIELDS. To RICH: JONES Esq.r this PLAN is humbly inscribed by his most obedient servant J: Careless'

52a. immediately outside St Owen's Gate [on N side of St Owen's Street between the Gate and Ledbury Road, SO 515395]. Shows buildings in block plan, terraced houses near the Gate and at Williams's Hospital and St Giles Hospital, field boundaries, trees, ponds, roads, gates. Marks [St Owen's] Gate and nearby barn, Town Ditch, Bye Street Road, Scut Mill Lane [now Ledbury Road] and Mill Pond. See also 1784 HEREFORD ST OWEN'S.

Reference table gives field-names, acreage.

Dedicatory cartouche. Compass rose. Scale bar. Plain border. Intricate penmanship of a writing master.

SCALE: 3.8 chains to 1 inch [1:3010].

SIZE: 230 x 318. Ink and grey and sandy wash on paper, repaired.

SURVEYOR: 'J. Careless, Sur: & Del'.

REFERENCE: HRO, BJ42.

1787-1790
MUCH MARCLE

No title.

Surveys of Hellens estate [SO 661333] with signed valuations of woods on the estate (1 bundle). Other associated papers contain columns of letter and number plot references, apparently relating to a map (not traced).

SURVEYOR: Nathaniel Kent. Some are dated from Malvern, Worcestershire.

REFERENCE: HRO, F35/RC/M III/12-24.

1788
BRIDSTOW

'A Survey of the River Wye and land adjacent in the Parish of Bridstow ... Taken by John Green 1788'(Plate 59)

Area S and E of Wilton Castle [SO 590245]. Shows Wilton Castle and Bennell [Benhall] in perspective, fields outlined in green with field-names and names of owners, river Wye, streams and Black Pool [SO 592240], roads. Marks Wilton bridge with six piers and turnpike gate at Wilton and the former course of the river above Wilton running across the meadows beneath the walls of Wilton Castle. Annotated later with the continuing south-westward erosion of the river opposite Black Pool measured in 1788, 1799 and 1814.

Title cartouche in oval frame. Grid. Compass points. Scale bar with dividers. Ruled yellow border.

SCALE: 88 yards to 1 inch [1:3168].

SIZE: 425 x 275. Ink and watercolour on parchment.

SURVEYOR: John Green.

REFERENCE: HRO, A97/1. Reproduced in P. Hughes and H. Hurley, *The story of Ross-on-Wye,* Almeley [Herefs.] 1999, p. 106.

1788
ULLINGSWICK, LITTLE COWARNE

'A MAP of an Estate in the Parish of Ullingswick ... the property of [blank]'[1]

250a. in scattered open and enclosed fields, including 99a. in Little Cowarne, consisting of Ullingswick Farm, Shortwood Farm and Sheep-cott. Shows buildings in block plan, orchards, roads. Marks church [SO 597499].

Reference table gives some field-names, acreage and land-use.

Rococo cartouche with foliage. Compass rose. Scale bar.

SCALE: 3 chains to 1 inch [1:2376].

SIZE: 1050 x 1050. Ink and watercolour on paper, linen-backed, attached to a roller.

SURVEYOR: Thomas Buckle.

REFERENCE: Private.

1. See J.E. Grundy, 'Ullingswick: a study of open fields and settlement patterns' in D. Whitehead and J. Eisel eds., *A Herefordshire miscellany*, Hereford 2000, pp. 287-300.

1789
KINGTON AND NORTH HEREFORDSHIRE

'A PLAN of the intended CANAL from KINGTON in the COUNTY of HEREFORD to the RIVER SEVERN near STOUR PORT in the COUNTY of WORCESTER'

Plan for a public scheme. Shows proposed route of the canal from Kington following the valleys of the rivers Arrow and Lugg to Leominster and thence by the valley of the river Teme and tunnels at Lindridge and Pensax to the river Severn at Stourport. Marks principal landmarks including the forge at Staunton-on-Arrow, mills and the sites of locks. Shows hills by hachuring.

Table of length of the canal.

Compass points. Scale bar. Plain border.

SCALE: 1 mile to 1 inch [1:63360].

SIZE: 243 x 820. Ink and watercolour on paper.

SURVEYOR: Thomas Dadford, junior.

REFERENCE: HRO, C83/1.

(1789)
MADLEY

'MAP of LULHAM COURT ESTATE and the TODDAGE LANDS situate in the Parish of Madley &c ... the PROPERTY of F. Bumford Eves Esq.' [deleted and replaced by 'G Holloway Esq', also subsequently struck out and replaced in the same later hand by 'copyhold under the Dean & Chapter of Hereford']

134a. in N of parish comprising Lulham Court 123a. [SO 408412] and the Toddage 10a. Shows buildings in block plan, arable and pasture lands including some strips in open fields, hopyards, roads. Gives plot numbers and names of adjoining owners. Marks smith's shop [at Bage Mill]. Field-names include Brick Close at Lulham.

Reference table with field-names and acreage.

Reference table drawn as a piece of curling paper. Compass points. Scale bar. Ruled border.

SCALE: 8 chains to one inch [1:6336].

SIZE: 358 x 440. Ink and watercolour on paper, in a volume, vellum, ½-leather bound. Copied *c*.1820; see 1784 CLEHONGER.

SURVEYOR: Not named [?John Harris].

REFERENCE: DCA, 7004/1, pp. 21-2.

1790
MUCH COWARNE (STOKE LACY)

'A MAP of Hopton Corner Farm in the Parish of Much Cowarne[1] ... belonging to Mes.rs John Freeman, James Yapp, John Yapp, Richard Bennett, [blank] Fisher and William Jenks Measured and Mapped by Thos Buckle February 1790'

Hopton Corner Farm [SO 629500] in the E of Stoke Lacy parish, being enclosed lands lying between Piddle Common Field on the SW and Hedgeley [Edgeley] Common Field on the NE. Small parts of the mapped area appear to have lain outside the ancient parish of Much Cowarne in Avenbury and Stoke Lacy. Shows buildings in block plan, fields with coloured bounds, hopyards, orchards, woods, gates, footpaths, streams, a pit or pond near the farm, roads. Isolated trees drawn with shadows.

Reference table gives plot numbers, field-names, land-use and acreage [but this last right-hand column is missing]. Field-names include Upper and Lower Flaggs, Flaggs Hopyard, Worestone Grove and Coppice [at or about SO 636506], Drakeswell Coppice.

Cartouche with crude floral decoration. Compass rose. Scale bar.

SCALE: 3 chains to 1 inch [1:2376]

SIZE: 650 x 850. Ink and watercolour on paper. In poor condition, torn at folds with the top right-hand part detached and the right edge missing.

SURVEYOR: Thomas Buckle.

REFERENCE: HRO, B9/7. [Not available for research in September 2002 pending repair].

1. Hopton, a northern tongue of the parish of Much Cowarne, was transferred to Stoke Lacy in 1884.

[c.1790]
BLAKEMERE, TYBERTON, PRESTON-ON-WYE

'Plan of the different Roads from PRESTON to TIBBERTON over the Stoney-Field and by Blakemere'

W part of the parishes of Preston-on-Wye and Tyberton and central part of Blakemere [SO 3640]. Shows roads and lanes with gates and swing gates passing by or through lands, some with field-names. [Made in connection with a dispute about the upkeep of the roads and the gates across lanes, most particularly the lane from Tyberton to Holywell].

Reference table identifying gates and including notes about a dirty length of road W of Tyberton village and the 3-foot worn depth of the lane from Tyberton near Holywell in Preston-on-Wye.

Urn-like cartouche. No compass points. No scale bar.

SCALE: [c.1:7040].

SIZE: 245 x 345. Ink and water colour on paper (no watermark).

SURVEYOR: Not named.

REFERENCE: HRO, R34/20.

1791
BLAKEMERE, BROBURY

'PLAN of KINLEY ESTATE situate in the Parishes of BLAKEMERE and MOCCAS … the property of Captain Kyrwood'

99a. in N of parish with small area in adjoining part of Moccas [SO 364421]. Shows buildings in block plan, garden at Kinley Farm, fields, mostly enclosed but some strips, woods, trees, streams, ponds, gates. Gives names of adjoining owners. Does not mark windmill [SO 361420], then rough ground adjoining the road named Wildy-grove Gorst.

Title cartouche in pen and ink incorporating a pastoral scene with windmill and sheep. Compass points. Scale bar [mostly torn off].

SCALE: 4 chains to 1 inch [1:3168].

SIZE: 490 x 480 [? about 700 when complete]. Ink on parchment, much torn and the right-hand part missing.

REFERENCE BOOK: 'Reference to the Plans, and Valuation of the Estates of CAPTAIN KYRWOOD situate in the Parishes of Blakemere, Moccas, & Brobery … by J. Powell 1791'.

Gives field-names, acreage, land-use, valuation and recommendations for improvements. For Brobury [map missing] gives the condition of the farm house and buildings. Field-names include Redstreak Orchard, Cowpasture Hopyard at Blakemere, Castle Common Field at Brobury.

Title cartouche in pen and ink incorporating pastoral scene with windmill and cows.

SIZE: 230 x 150, paper, soft leather-bound with title in gold leaf 'VALUATION OF KINLEY AND BROBERY', 16pp. of writing; at end reversed, memorandum 1792 and receipt 1795.

SURVEYOR: Joseph Powell.

REFERENCE: HRO, J56/III/89.

1791
BOSBURY

'PLAN of several Estates situate in the Parishes of BOSBURY, MUNSLEY and LEDBURY … the Property of Francis, William, Thomas, Brydges Esqr' (Plate 61)

The Farm [SO 693423], The Nelmes and Upleadon Court [SO 665421] estates in SW of Bosbury and small adjoining parts of Ledbury and Munsley, and The Overing [Noverings Farm] in N of Bosbury. Shows buildings in block plan, gardens, fields with reference nunbers but not field-names, mostly enclosed but some strips in open fields, orchards, woods, trees with shadows, river Leadon, Shilo brook, ponds. Marks L-shaped pond [? moat] at Upleadon [but, curiously, not the house of c.1600].

No reference table or book.

Title cartouche with baroque decoration and rustic scene with fisherman and sheep. Compass points. Scale bar with baroque surround, dividers and protractor. Ruled border.

SCALE: 6 chains to 1 inch [1:4752].

SIZE: 1253 x 730. Ink and grey wash (buildings in red) on paper. Repaired.

SURVEYOR: Joseph Powell.

REFERENCE: HRO, AS80/5.

1791
HEREFORDSHIRE

'PLAN of a Propos'd NAVIGABLE CANAL from HEREFORD to GLOCESTER with a Collateral Branch to NEWENT by Jos;h Clowes Engineer & Rich:d Hall Surveyor'[1]

Plan for a public scheme. Shows the route of the proposed canal from its Hereford terminal at Widemarsh [SO 511409] via Shelwick (Holmer), Wergins Bridge (Sutton St Nicholas), Withington Marsh (Withington), a tunnel at Wazzington [Walsopthorne] (Ashperton) and Ledbury and thence to Gloucester. Shows the lands through which it would pass, with plot numbers, the rivers Wye, Lugg, Leadon and Severn, streams, locks, aqueducts, tunnels and mills. Relief is indicated by hachures. Marks parish churches, Ethelbert's Camp [Backbury] (Dormington), Wall Hills (Ledbury), Tidnor Forge (Lugwardine).

Reference table gives plot numbers, names of owners and occupiers and the length of the proposed canal in chains and links crossing their property, a total of 38 miles, 5 furlongs and 5 chains.

The plan pre-dates the standing orders of the House of Commons 1792, which ordered the deposit of plans of proposed public schemes with the clerk of the peace of the relevant counties. It was instead signed by the Speaker, Henry Addington, on 30 May 1791 as one of the maps or plans of references required to be certified for the Act of Parliament, 31 George III, c.89 (1790-91) already passed for making the canal.

The route of the Gloucester to Ledbury section of the canal was amended in 1792 and completed to Ledbury in 1798 under the direction of Robert Whitworth, following Clowes's death in 1795. The Ledbury to Hereford section was constructed 1839-45 under Stephen Ballard of Colwall (1804-91), who made minor variations to Clowes's line.

SCALE: 4 inches to 1 mile [1:15840].

SIZE: 1280 x 2400. Ink on paper. Some damage to the edges with a few pieces missing. Repaired by Gloucestershire RO.

SURVEYORS: Josiah Clowes, engineer, and Richard Hall, surveyor.

REFERENCE: Gloucestershire RO, Q/Rum 1.

1. There is a 'Sketch map of the intended Herefordshire & Gloucestershire Canal with the River Severn and the several canals with which it communicates', 1791, at a scale of 1:633600, in the Worcestershire RO, 705:174/4464/56/ii.

(1791)
KINGSTONE

'PLAN of several ESTATES AT KINGSTON lately purchased by Sir George Cornewall Baronet. Taken by Joseph Powell 1791 & copied by Hall & Son 1809'

Most of centre and S of parish [SO 4235]. Shows buildings in block plan, enclosed fields and strips in open fields outlined in colour. Marks church, Henley [Hanley] Court, Bigstye. Gives names of some adjoining owners.

Scale bar. Ruled border.

SCALE: 4 chains to 1 inch [1:3168].

SIZE: 1130 x 640. It appears to be the left-hand part of a larger map with about one-third or one-half missing. Ink and watercolour on paper, cloth-backed.

SURVEYOR: Joseph Powell 1791. Copied 1809 by Hall & Son in the style of an early 19th-century parish or tithe map.

REFERENCE: HRO, A6/1.

1791
MUCH MARCLE

'Map of Chandois and other Grounds'[1]

Chandos [SO 644345] with scattered lands, mostly in open fields to its E and W. Shows buildings in block plan, closes and strips in open fields, outlined in colour with plot numbers, field-names and land-use, orchards, roads. Marks Common Meadow. Field-names include The Redstroke Orchard.

Oval ornamental title cartouche. Compass rose. Scale bar with dividers. Engraved ornamental strip pasted on as a border.

SCALE: 3 chains to 1 inch [1:2376].

SIZE: 635 x 1193. Ink and watercolour on parchment. Torn at bottom edge without loss of evidence.

SURVEYOR: Not named.

REFERENCE: Wiltshire & Swindon RO, 1720/588H.

1. I am grateful to Andrew Crookston of Wiltshire & Swindon RO for his help in compiling this entry.

1791
OCLE PYCHARD

'PLAN of HILLHAMPTON ESTATE in the Parish of Oacle-pitchard…the property of the Rv:d M:r Tomkyns'

    129a. in N of parish, mostly in open fields [SO 590474]. Shows buildings in block plan. Gives field-names and land-use, including orchards and hopyards. Indicates the numbers of ridges cumulated or amalgamated in the several open fields, though the number of ridges do not correspond to their acreage. For example 40 ridges are given as measuring 1a. 0r. 35p., 28 ridges measuring 1a. 1r. 1p. and 20 ridges 2a. 0r. 31p. Gives names of adjoining owners.

    Compass points. Scale bar. Ruled border.

SCALE: 4.5 chains to 1 inch [1:3564].

SIZE: 548 x 615. Ink on parchment, in a crude style.

SURVEYOR: Thomas Davis.

REFERENCE: HRO, R53/1.

[1792]
AYMESTREY

'Shirley Farm in the Parish of Aimstrey …'

    135a. in W of parish [SO 384653]. Shows field boundaries, roads. Gives names of adjoining landowners. Marks Darvels [Deerfold] bridge.

    Aligned with S at top.

SCALE: [c.1:5280].

SIZE: 185 x 240. Ink on tracing paper, frail, with photocopy.
    Found loose, folded inside a conveyance by lease and release dated 26-27 March 1792.

SURVEYOR: Not named.

REFERENCE: HRO, T78/21.

1792
WHITBOURNE

'MAP and Survey of an ESTATE in the Parish of Whitbourn belonging to Mr Jas. and Han. Davis Survey'd January 1792 by T. Allen'

    187a.[Poplars Farm] in scattered closes in centre and N of parish [SO 7257]. Shows buildings in block plan, enclosed fields, hopyards, hedges, woods, trees with shadows. Marks river Teme, the principal house at Meadows Green and others at Rosemore and Smeeths. Field-names include Brick Ground, Omenlay.

    Reference table gives field-names and acreage.

    Reference table with floral border. Compass points. Ruled border. Neat draughtsmanship.

SCALE: [c.1:2816].

SIZE: 575 x 710. Ink on parchment.

SURVEYOR: T. Allen.

REFERENCE: HRO, B92/1.

1793
EYE (MORETON)

[1]
No title or date

    37a. comprising four fields called Upper and Lower Bower Walk, Road Field (part) and Yesall, lying apparently between Ridgemoor Brook and Berrington park in the S of the parish [SO 5062]. Outline drawing, giving field-names and acreage.

SCALE: [c.1:3168].

SIZE: 135 x 240. Ink on paper.

SURVEYOR: [Thomas Matthews].

[2]
Valuation: Associated with the sketch plan is 'Mr Mathews Valuation of Moreton Estates and Lands at the Bower', relating to an exchange of lands belonging to Thomas Harley [of Berrington Hall] (37a.) and the Corporation of Ludlow (29a.). Signed by Thomas Matthews, 1793.

[3]
'Plan of an Estate at Moreton in the Parish of Eye … belonging to the Corporation of Ludlow'. Endorsed Map of Corporation Estate at Moreton exchanged with Mr Harley for Lands near the Bower', undated.

    29a. of scattered plots [SO 4965] including lands in Orleton Moors, a coppice (3a.) near the Ludlow to Leominster road and strips in open fields [crossed by the Leominster to Stourport canal, on which construction began in 1793]. Outline drawing, giving names of adjoining owners, including Thomas Harley.

    Reference table gives plot numbers, field-names, acreage. Field-names include Common Shuttocks, Common Lurky, Field adjoinning Dog Kennel.

    Cardinal points.

SCALE: [*c.*1:3168].

SIZE: 330 x 530. Ink on paper.

SURVEYOR: [?Thomas Matthews].

REFERENCE: Shropshire RO, [1] LB/4/4/84/1; [2] LB/84/4/4/84/2; [3] LB/4/4/92.

1793
MONKLAND

'PLAN of ESTATES at Monkland near Leominster ... held on Lease by Mr Edward Coleman and Mr Powell under the Dean and Canons of Windsor'

228a. throughout the parish [SO 4557] in six separate groups of lands of 210a. leased to Edward Coleman and 19a. leased to Mr Powell. Shows buildings in block plan, field boundaries with abuttals and names of adjoining owners, streams, roads, bridge. Field-names include Ox's Nest, Broomy Hurst and Beast Pieces, Woosen Field, Moor Gobbett, Leadalls, Moors Hill. Fields numbered, with additional later pencil numbering.
Reference table gives field-names, land-use, acreage.
Compass points. Scale bar.

SCALE: *c.*5 chains to 1 inch [*c.*1:3960].

SIZE: 590 x 510. Ink, pencil and watercolour on parchment.

SURVEYOR: J. Lucas.

REFERENCE: Dean and Canons of Windsor Archives, CC11210.[1]

1. I am grateful to Dr E.P. Scarff, Librarian and Archivist, and to her assistant Jude Dicken for this catalogue entry.

[1793]
TEDSTONE DELAMERE

No title. 2a., including Mill Meadow [SO 6958]. Shows fields outlined in colour.
Reference table gives field-names, acreage, land-use, valuation.
Compass rose. Scale bar. Ruled border.

SCALE: 3 chains to 1 inch [1:2376].

SIZE: 340 x 530. Ink and watercolour on parchment.

SURVEYOR: Thomas Buckle, surveyor.

REFERENCE: HRO, A100/296. Associated with a deed of exchange of land between Revd P.G. Tomkyns of Buckenhill, Norton in Bromyard, rector of Tedstone Delamere, and R. Moore of Great Shelsley, Worcestershire, esq., 2 May 1793 [HRO, A100/295]. See also *c.*1793 TEDSTONE DELAMERE.

[*c.*1793]
TEDSTONE DELAMERE

No title. 22a. in scattered closes and some strips in open fields [SO 6958]. Shows fields outlined in colour. Field-names include Parsons Mill Rough.
Reference table give field-names, acreage, land-use, valuation. Similar to 1793 TEDSTONE DELAMERE.
Compass rose. Scale bar. Ruled border.

SCALE: 3 chains to 1 inch [1:2376].

SIZE: 390 x 495. Ink and watercolour on parchment.

SURVEYOR: Thomas Buckle, surveyor.

REFERENCE: HRO, A100/297.

1794
CLODOCK [LONGTOWN]

'A Survey of An Estate Situate in the Parish of Clodock ... commonly called NEUADDLWYD Belonging to JOHN LEWIS Esqr' (Plate 64)[1]

32a. [Neuadd-Lwd] on boundary of Longtown and Newton parishes [SO 330325], formerly in the parish of Clodock. Shows buildings in block plan, fields outlined in colour, woods, gates, road from Hay to Longtown.
Reference table gives land-use, acreage.
Title cartouche decorated with flowers. Reference table in ruled frame. Compass rose. Colourful but crude [perhaps in imitation of Benjamin Fallowes; see 1718 EWYAS HAROLD].

SCALE: 2.25 chains to 1 inch [1:1782].

SIZE: 515 x 360. Ink and watercolour on parchment.

SURVEYOR: Edward Penry.

REFERENCE: HRO, N44/39.

1. Longtown, an ancient chapelry of Clodock, was created a separate civil parish in 1866.

1794
STRETTON SUGWAS

'A Map of Stretton Court Estate, in the Parishes of STRETTON SUGWAS. KENCHESTER. CREDENHILL and BURGHILL … the Property of the President and Governors of Guy's Hospital' (Plate 63)

651a. N of the Hereford-Brecon road [SO 466429]. Shows buildings in block plan, except for the church in perspective view, enclosed fields, orchards, woods, trees with shadows, streams, roads. Marks a formal garden (almost a maze) S of the church at Stretton Court, Sugwas Pool. Gives names of adjoining owners.

Reference table gives acreage and field-names, including Tumpy Orchard [SO 474432].

Pen and ink decorative title cartouche. Compass rose. Scale bar. Plain border.

SCALE: 4 chains to 1 inch [1:3168].

SIZE: 1020 x 800. Ink and watercolour on parchment.

SURVEYOR: James Cranston.

REFERENCE: HRO, C99/III/218.

[1794]
DORSTONE, CLIFFORD

Survey book, lacking front cover and title page, of the estates of Jesus College, Oxford.[1] It includes:

'XVIII Sidecomb Farm in the Parish of Dorston'

Sidcombe Farm [SO 298445] (94a.) in E of Clifford and NE of Dorstone parishes. Shows buildings in block plan, enclosed fields in outline, coppices, road. Adjoining lands or owners include Merbach Hill Common at the N end of Merbach Hill.

Reference table gives plot numbers, field-names, acreage. Field-names include Mill Close [SO 297435].

Compass points. Ruled border. Neatly drawn, as if engraved.

SIZE: 270 x 380, double page.

'XIX Cross Way Farm in the parish of Dorston'

[Lower] Crossway Farm [SO 318421] (86a.) in Dorstone village with lands running up to Arthur's Stone. Shows buildings in block plan, enclosed fields in outline, orchards, coppice, roads. Adjoining lands or owners include Common [on Dorstone Hill]. Marks 'The Golden Brook', the road from Hereford to Hay [Spoon Lane].

Reference table as above. Field-names include Lower Chappell [SO 319424], Arthur Stone [Field] [SO 319431] with later pencil note 'Arthur's Stone cromlech'.

Compass points. Ruled border. Note 'NB. A scale of Eight Chains to One Inch'.

SCALE: 8 chains to 1 inch [1:6336].

SIZE: 305 x 210. Volume, 1/2-vellum bound with marbled back cover. Front cover and first pages missing; the first gathering loose.

SURVEYOR: Not named.

REFERENCE: Jesus College, Oxford, archives.

1. Date of watermark.

[1794][1]
LITTLE BIRCH

No title. Endorsed 'Ailstons Wood in Little Birch & Aconbury', 'Ailston's Wood The Property of His Grace the Duke of Norfolk'

Outline boundary of Athelstans Wood [SO 520310], comprising 208a. statute measure, 128a. wood measure. Marks a N-S track through the wood.

SCALE: [c.1:3168].

SIZE: 570 x 380. Ink on paper, damaged.

SURVEYOR: Not named.

REFERENCE: BL, Add. MS. 36307, G12.

1. Date of watermark.

n.d.[c.1794][1]
MICHAELCHURCH ESCLEY

'A Survey of an Estate situate in the Parish of Michaelchurch-Escley … Commonly Called Lower Cavalla Belonging to John Lewis Esq.r'

74a. on W side of the road to Longtown. Shows buildings in block plan [?at Llanbaddon, SO 316332], enclosed fields with reference numbers.

Reference table gives acreage but not field-names.

Cartouche in frame. Compass rose. Four decorative flowers. Ruled border in red.

SCALE: [*c.*1:1584].

SIZE: 540 x 355. Ink and watercolour on parchment.

SURVEYOR: Not named. [Edward. Penry].

REFERNECE: BL, Egerton MS 3021/E.

1. The date of 'before 1803' is given in the British Library *Catalogue of Manuscripts*. The distinctive style of pale green and orange colouring of features in the map with the drawings of flowers suggest that the unnamed surveyor was Edward Penry. See 1794 CLODOCK (LONGTOWN).

1795
EASTNOR, LEDBURY

Four maps of the Eastnor estate '… the property of the Right Honorable Lord Sommers' by John Harris.

The maps are all similar, showing most buildings in block plan, fields in outline with land-use indicated by delicate representation of arable and pasture, orchards, trees, gates, roads. Give names of adjoining owners.

No title cartouches. Compass points. Scale bars painted yellow within a frame. Simple ruled borders.

SCALE: 4 chains to 1 inch [1:3168].

SIZE: Various, see below. Ink and grey wash on parchment, with a little other watercolour.

[1] 'THE FARM. in the Parish of Eastnor'
[Eastnor Farm, SO 730371]. Marks Eastnor church in block plan, tree at the road junction to its S. Field-names include Squirrel Hill W of the church.

SIZE: 580 x 840.

[2] 'THE WHITEHOUSE in the Parish of Eastnor'
White House Farm, [SO 727384] mostly enclosed but indication of former strips in open fields. Marks tree at road junction [SO 727383].

SIZE: 479 x 569.

[3] 'LOW'S HURST. <and Lower Mitchell> [added later in pencil] in the Parish of Eastnor'
[Lower Mitchell Farm, SO 725388]. Mostly enclosed lands E of White House Farm but indication of former strips in open fields. Marks the turnpike road from Ledbury to Malvern with destinations shown by pointing hands.

SIZE: 522 x 615.

[4] 'UPPER MITCHELL. in the Parish of Ledbury'
Upper Mitchell Farm [SO 723393]. Mostly enclosed lands, woods.

SIZE: 483 x 694.

SURVEYOR: John Harris signed [1] and [3]. The other two maps, though unsigned, are clearly in the same hand.

REFERENCE: Private.

1795
LEDBURY

'PLAN of the ARGUS FARM &c belonging to JOHN BIDDULPH Esq: in the Parishes of Ledbury and Donnington …'

165a. [SO 710353] in S of Ledbury and centre of Donnington. Shows buildings in block plan, enclosed fields and a few strips in open fields, orchards, ponds, roads. Marks Donnington church, Dinchill [Dinchall]. Gives names of adjoining owners.

Reference table gives acreage and field-names, including Pye Corner (W of Argus Farm), Hill Field [SO 717353].

Title written over an oval grey wash drawing of a rustic scene with a farm house and half-timbered barn. Reference table as a scroll. Compass point. Scale bar. Ruled border.

SCALE: 3 chains to 1 inch [1:2376].

SIZE: 833 x 533. Ink and grey wash and red watercolour on parchment.

SURVEYOR: George Young, surveyor.

REFERENCE: HRO, G2/III/37.

(1795)
WEOBLEY

'Fenington Farm in the Parish of Weobley…the property of John Peploe Birch Esq.' (photocopy)

217a. in W of parish [Fenhampton, SO 391504]. Shows buildings in block plan, arable and pasture land, mostly enclosed, hopyard, orchards, plantation or woodland with widely spaced trees, roads and tracks. Trees drawn with shadows. Gives names of adjoining owners. Field-names include The Parks.

Reference table, with field-names, land-use, acreage. Title cartouche incorporating a pedestal with urn flanked by trees, and a drawing of an imaginary house on a river bank. Floral compass rose. Scale bar with flowers, measure, dividers and protractor. Ruled border.

SCALE: 4 chains to 1 inch [1:3168] (of photocopy).

SIZE: 625 x 404 (of photocopy). [Ink and watercolour on parchment].

SURVEYOR: David Pain.

REFERENCE: HRO, BN72/2.

n.d. [c.1795]
AYMESTREY

No title. Lands at Lower Lye in N of parish [SO 406669]. Basic sketch giving outline of fields with names of fields or landholders. Field-names include Pound Close and Piece Next Door.

SCALE: [c.1:7040].

SIZE: 200 x 250. Ink on paper, watermarked 1794.

SURVEYOR: Not named.

REFERENCE: HRO, AD17.

[c.1795]
HEREFORD, ST MARTIN'S & ST JOHN'S

'A MAP of the MANOR of Winaston Situate in Blackmarston near the City of HEREFORD: Being an undivided Estate; Part held by Lease (under St Ethelbert's Hospital and the College of Hereford, and the Residue a freehold, the Property of James Donnithorne Esqr.'[1]

144a, of which 77a. belonged to St Ethelbert's Hospital, a charitable institution associated with Hereford cathedral, 53a. to the College of the Vicars Choral of the cathedral and 14a. to James Donnithorne. The lands were mostly in the open fields on both sides of the road from Hereford to Abergavenny SW of the Wye Bridge in the Hunderton and Newton Farm area of the city. Shows buildings in block plan and distinguishes arable and meadow by yellow and pale green, including a small piece of detached meadow in a loop of the river Lugg near Worging [Wergin] Swath [in Sutton St Nicholas, about SO 532446]. Marks Causeway House [SO 506392], Upper House [SO 504390], Hunderton Tump [SO 493389], roads, river Wye. Field-names include Wye Common Field, Newtons Common Field [SO 498380] E of Great Newton Farm, Little and Great Gallows Fields [SO 500388 and SO 500384], Hunderton Croft, Brick Plock, Vineyard Croft [SO 505385].

Reference table gives plot numbers, field-names, land-use, and acreage. Note that computed acreage was 205a.

Cardinal points. Scale bar. Ruled border.

SCALE: 4 chains to 1 inch [1:3168]

SIZE: 470 x 584. Ink and watercolour on paper.

SURVEYOR: Not named.

REFERENCE: HRO, N32/60

1. The map may have been made some ten years earlier (see (1784) HEREFORD, ST MARTIN'S & ST JOHN'S), soon after James Donnithorne (1744-c1820) inherited the property of his father, Revd Isaac Donnithorpe, one of the vicars choral of Hereford Cathedral, in 1782. But the extent of the estate is noted identically on the map and in a lease book of the College of the Vicars Choral, apparently in connection with a lease of 1795. The paper of the map has no watermark. [DCA, 7003/3/1 and 3877/1-4; F.C. Morgan 'Causeway Farm, Hereford; its surroundings and history', *Trans WNFC*, vol.37, (1961), p. 29].

1796
MADLEY

'A PLAN of [blank] –HOUSE [& Land in the] PARISHES OF MADLEY and KINGSTON … Belonging to Thomas Maddy Gen^t.'

155a. N of Madley village [Town House Farm, SO 421388]. Buildings shown in block plan, enclosed fields and some strips in open fields, orchards, roads.

Reference table with acreage and field-names including Widgatts [Woodyatts] Cross, Dragon Field, Bady Pits.

Oval title cartouche. Compass rose. Scale bar. Ruled border.

SCALE: 4 chains to 1 inch [1:3168].

SIZE: 630 x 730. Ink, faint, on parchment, torn.

SURVEYOR: J. Maddy.

REFERENCE: HRO, R52/1.

1796
LEDBURY, MUNSLEY, BOSBURY

'PLAN of an Alteration of the HEREFORDSHIRE and GLOCESTERSHIRE CANAL from the Lane leading from Asperton to Bosbury to BYE STREET ROAD AT LEDBURY'

Plan of a public scheme. Shows the route proposed in the Act for the Herefordshire & Gloucestershire Navigation 1791 and the alternative route west of Ledbury, 12 September 1796. Shows property boundaries in outline, rivers, roads with destinations. Relief indicated by hachuring. Marks Ledbury, Bosbury and Munsley churches in perspective and Ledbury town by solid shading. Reference table gives plot numbers, names of proprietors and occupiers of land, and length of the canal crossing their property.

Scale bar. Compass points.

SCALE: [c.1:15000].

SIZE: 514 x 920. Ink and watercolour on paper.

SURVEYOR: Robert Whitworth and Robert Wright Hall, engineers.

REFERENCE: HRO, Q/RW/C4.

(1797)
KING'S PYON

'Meer Place Farm, Wallhouse Farm & the Hill Farm in the Parishes of Kingspion and Weobley…the Property of J.P. Birch Esq.ʳ' (photocopy)

145a., 32a. and 133a. respectively in N and W of King's Pyon [SO 4250] and at Hill Farm in Weobley [SO 414494]. Shows buildings in block plan, lands, distinguishing arable and meadow, mostly enclosed but a few strips in open fields, hopyards, orchard, woods, ponds, roads and tracks. Trees are drawn with shadows. Gives names of adjoining owners. Marks Wallhouse Farm [SO 419506] and a large pond at Hill Farm. Field-names include Lidgemoor, Gabbots, Crabstock Orchard, lands of Poor of Pembridge.

Reference table, with field-names, land-use, acreage.

Title cartouche incorporating a pedestal, urn and landscape. Compass rose with ribbon. Scale bar with flowers and ribbon. Ruled border.

SCALE: 4 chains to 1 inch [1:3168] (of photocopy).

SIZE: c.545 x c.660 (of photocopy). [Ink and watercolour on parchment].

SURVEYOR: D. Pain.

REFERENCE: HRO, BN72/1.

1797
MUCH MARCLE

No title.

Parliamentary inclosure maps (10 sheets) of corn rent schedules sealed with the award under the Act of Parliament 1795 for the inclosure of the whole parish comprising 1,150a. in the townships of Marcle [SO 6533], Wolton [SO 6433] and Kynaston [SO 6435]. Show buildings in block plan, roads, inclosure allotments, woods, ponds. Marks church and neighbouring motte and bailey to its N, turnpike [SO 645327].

A private plan and survey of the township of Kynaston, not traced, was also produced to the Commissioners by William Freeman Esq. Each map has table(s) arranged by landholders, giving land-use, acreage, field-names, value of small tithes, quantity of wheat equal to small tithes in bushels. No compass points. No scale bars. The scale and alignment of the maps vary.

SCALE: [c.1:7040, except for map no. 1, which is c.1:4874].

SIZE: 520 x 690-700. Ink and watercolour on parchment. Duplicate maps of nos. 8-10 (loose) with some other papers.

INCLOSURE COMMISSIONERS: John Stone of Pull Court, [Queenhill], Worcestershire, Francis Webb of Salisbury, John Harris of Wickton, Leominster.

SURVEYOR: Henry Clark of Shipston-on-Stour, Worcestershire [now Warwickshire].

REFERENCE: HRO, Q/RI/36 (Commissioners' copy) and A78/1.

(1798)
CLIFFORD

'Broadmeadow Farm', [a tracing made 1868 from a map by E. Penry 1798]

Upper Broadmeadow Farm, 25a. [SO 264419], and Lower House Farm, Westbrook, 31a. in S of parish. Shows buildings in block plan, enclosed fields, roads with destinations. Field-names at Lower House Farm include The Llan.

Reference tables (2) with field-names and acreage.

N point.

SCALE: [c.1:3960].

SIZE: 555 x 430 (of tracing). Ink and watercolour on thin tracing paper, torn and fragile.

SURVEYOR: E. Penry.

REFERENCE: HRO, C97/6.

1798
COLLINGTON

'A MAP of Estates in the Parish of COLLINGTON … the property of Jonathan Pytts Esq:[r]'

[Felton Court, SO 649599] and lands in the centre of the parish. Shows buildings in block plan, enclosed fields with letter and number references and distinguishing land-use, rough grazing, orchards, woods, streams with direction of flow, relief indicated by hachure and grey wash, roads with destinations given by a pointing hand, gates across roads and lanes. Gives names of adjoining owners.

Reference book missing.

Cartouche, shield-shaped within a frame of leafy stalks. Cardinal points with an E cross, the N-S pointer being a wavy eel-like line, the E-W pointer ruled. Scale bar. Ruled border.

SCALE: 6 chains to 1 inch [1:4752].

SIZE: 640 x 610. Ink and grey wash on paper, linen-backed.

SURVEYOR. T. Davis.

REFERENCE: Worcestershire RO, (BA8683) 899:396.

1799
CLIFFORD

'An Exact Survey of GLANN Y COED an Estate in the Parish of Clifford … belonging to JOHN SMITH GENT. Survey'd A.D. 1799 by E. Penry'

132a. [Llanerch-y-coed Farm, SO 275425] in S of parish adjoining Little Mountain common. Shows buildings in block plan, enclosed fields, hedges, gates, roads with directional signs. Field-names include Limekiln Field, Cae Tumpen.

Reference table with field-names, acreage, land-use.

Compass rose. Scale bar with dividers. Ornamental border. Crude penmanship.

SCALE: 2.7 chains to 1 inch [1:2138].

SIZE: 535 x 875. Ink and watercolour on parchment.

SURVEYOR: E. Penry.

REFERENCE: HRO, J12/III/1.

1799
CLODOCK (LONGTOWN)[1]

'Caue Tack Belonging to Bryn farm'. Endorsed 'Middle Bryn farm Clodock'.

56a. [Cae Tack, SO 341284], E of Longtown. Shows fields outlined in colour. Marks road and well.

Table of areas. Ruled border.

SCALE: 'Same as Cumberlog [Cwmbologue] Scale'. [c.1:7920].

SIZE: 250 x 170, left-hand part cut off a larger map to exclude Ty Canol farm, which was part of the same estate according to a map of 1856 in the same bundle of deeds. Ink and watercolour on paper.

SURVEYOR: D[avi]d Davies.

REFERENCE: HRO, T24/2.

[1] Longtown, an ancient chapelry of Clodock, was created a separate civil parish in 1866.

(1799)
LEOMINSTER

'A MAP OF WHARTON ESTATE situate in the Parish of LEOMINSTER …' (photocopy)

Large area in S of parish[SO 5055], consisting chiefly of lands in open fields. Shows buildings in block plan, river Lugg, roads. Gives names of adjoining landowners.

Rococo cartouche. Compass points.

SCALE: [c.1:4526] (of photocopy).

SIZE: 260 x 250 (of photocopy). Part of map only, photocopy reduced in size from original. Probably ink and watercolour on parchment.

SURVEYOR: Edward Yeld.

REFERENCE: HRO, X89/4.

1799
WEOBLEY

'MAP OF LAND AT GANSON in the Parish of WEOBLEY … the Property of IOHN PEPLOE BIRCH ESQ.[R] and the POOR of the Parish of PEMBRIDGE taken July 6.th 1799 by William Galliers'

Exchange of 14a. of small parcels of land in E of parish [SO 4050]. Includes strips in open field, one reverse S-shaped. Marks arable fields, wood, trees, gates, roads. Gives names of adjoining owners. Field-names include Bellrope Meadow.

Two lists of the lands exchanged.

Compass points. Scale bar. Ruled border.

SCALE: 4 chains to 1 inch [1:3168].

SIZE: 425 x 540. Ink on paper, torn.

SURVEYOR: William Galliers.

REFERENCE: HRO, L57/77.

n.d.[c.1799][1]
HOLME LACY

No title. Endorsed 'Cuneagar Farm.— thrown with Hom Lacy part' [?*recte* park]

Oval area (58a.) of enclosed fields bisected by a track, adjoining Holme Lacy park [SO 5534], wood and hopyard. Shows outline boundaries.

Reference table gives plot numbers and acreage of the Cunniger (52a.) and small adjoining plots.

SCALE: [c.1:2376].

SIZE: 383 x 466. Ink on paper.

SURVEYOR: Not named.

REFERENCE: BL, Add. MS. 36307, G38.

1. Date of watermark.

n.d. [late 18th century]
BIRLEY

'A Plan of BIRLEY COURT, LYE COURT, BUCKNELL and the HILL FARM with TENEMENTS in the Parishes of BIRLEY, SARNESFIELD and KINGS PYON Belonging to RICHARD GORGES Esqr'.

906a. in centre and W of Birley [SO 4553], mostly E of the Roman road, with small parts in Sarnesfield and King's Pyon.. Shows buildings in block plan, fields, mostly enclosed but some strips in open fields, outlined in colour, orchards, woods, roads. Marks church in perspective, Roman road. Distinguishes arable. Gives names of adjoining owners.

Reference table gives acreage, land-use, field-names, including Oldhouse Croft, Cross Field, Brickyard, Thorney Marsh, The Hundred Crofts, all in Birley, Tinkers House in Hill.

Compass rose. Scale bar. Ruled border.

SCALE: 8 chains to 1 inch [1:6336].

SIZE: 563 x 675. Ink and watercolour on parchment, stained, damaged at edges.

SURVEYOR: Not named.

REFERENCE: HRO, F49/46.

n.d. [late 18th century]
HEREFORD

'Plan of the City of Hereford for the watch & ward'

City centre. Shows buildings in block plan, streets (named) and roads with destinations, city walls and gates. Marks [Bishop's] Palace, Deanery, College [of the Vicars Choral], Castle Hill and Castle Green.

Notes that 'The numbers 1, 2, 3, &c denote the stations of the different watchmen & the colours [of] their respective districts. Those streets which are left white are not included in any station'. The stations were at High Town, the Deanery, Widemarsh, Eign, St Owen's and Bysters Gates, outer Bye Street, the corner of Bridge Street and King Street, and St Martin's Street.

N point. Ruled border.

SCALE: [c. 1:6336].

SIZE: 214 x 287. Ink and watercolour on paper. The paper is browned, ink faded and the colouring lost. Repaired and rebacked, obliterating any watermark.

SURVEYOR: Not named.

REFERENCE: HCL, L.C. 912/4244.

n.d. [late 18th century]
HEREFORDSHIRE

'A Map of HEREFORDSHIRE Describing the situation of the several Woods & Farms belonging to the President and Governors of Guys Hospital'.

Whole county. Names the towns and parishes, indicated by conventional drawings of churches with towers or spires. Marks river Wye, Guy's Hospital property with reference numbers, main roads.

Reference table gives names of properties and tenants. A column for acreage is left blank.

SCALE: [*c.*1:63360].

SIZE: 1120 x 1120. Ink on paper, linen-backed, torn and dirty.

SURVEYOR: Not named.

REFERENCE: HRO, C99/III/215.

n.d. [late 18th century]
HOLME LACY ESTATES

Undated and unsigned maps of the Holme Lacy estate forming part of the archive of maps presented to the British Library by Mr D.H. Barry in 1900. See also 1695 HOLME LACY ESTATES and 1771-1775 HOLME LACY ESTATES.

[1] LITTLE DEWCHURCH
'Prothether in the Parish of Little Dewch[h]. belonging to Mr Styant'
  Prothither [SO 536300] (162a.) in the S of the parish. Shows buildings in block plan, field boundaries in outline with plot numbers and acreage, streams, roads. Gives names of adjoining owners, including the Duke of Norfolk.
  Reference table gives field-names, land-use, acreage. Field-names include Vides.

SCALE: [*c.*1:2376].

SIZE: 585 x *c.*375. Ink on paper, roughly drawn. In poor condition, repaired.

REFERENCE: BL, Add. MS. 36307, G19.

[2] HOLME LACY, BOLSTONE
No title. Unfinished or draft.
  Most of the parishes of Holme Lacy [SO 5535] and Bolstone. Shows buildings in block plan, field boundaries in outline, with the Holme Lacy demesne left blank.

SCALE: [*c.*1:7680].

SIZE: 774 x 760. Ink on paper.

REFERENCE: BL, Add. MS. 36307, G25.

[3] HOLME LACY
No title. Endorsed 'Part of Bogmarsh'.
  Area in the N of the parish [SO 5335]. Shows field boundaries in outline with plot numbers and field-names, mostly enclosed but some strips in meadows, stream. Field-names include Holm Bridge, Much Moor, Oozle Gate, Portugal Piece, Limekiln Field, Fish Pool Meadow.
  Unfinished, with pencilled bearings and boundaries.

SCALE: [*c.*1:3520].

SIZE: 723 x 702. Ink on paper, repaired.

REFERENCE: BL, Add. MS. 36307, G34.

[4] HOLME LACY
No title. Endorsed 'Plan of Road & Lands adjoining from Bower farm to Hom Lacy House previous to alteration of Entrance to Park'.
  Shows buildings in block plan and outline boundaries [SO 552355].

SCALE: 2 chains to 1 inch [1:1584].

SIZE: 933 x 454. Ink on paper. No watermark.

REFERENCE: BL, Add. MS. 36307, G37.

[5] HOLME LACY
'Fido's Meadow'
  13a. including the hedges [SO 5535]. Shows the meadow and its divisions, hedgerow trees, road adjoining the Park, footpath, gates.
  Scale bar, Ruled border.

SCALE: 4 chains to 1 inch [1: 3168].

SIZE: 150 x 145. Ink on paper.

REFERENCE: BL, Add. MS. 36307, G40.

[6] HOLME LACY
'A Plan of Ganner Farm in the Parish of Hom-Lacy ... The property of the Duke of Norfolk'
  [Gannah Farm, SO 547334] in the S of the parish (123a.). Shows buildings in block plan, trees with shading in bold penmanship and watercolour.
  Reference table gives plot numbers, field-names, land-use, acreage. Field-names include Redyer [Red Deer] Pool Piece, Old Park, Quarry Piece, Old Hopyard.
  Scale bar. N point. Ruled border.

SCALE: 2 chains to 1 inch [1:1584].

SIZE: 880 x 653. Ink and watercolour on paper, repaired. Watermark 'J Whatman'.[1]

REFERENCE: BL, Add. MS. 36307, G43.

1. James Whatman began making paper at Turkey Mill, Maidstone in 1739.

CATALOGUE OF MANUSCRIPT MAPS

[7] HOLME LACY
No title. Endorsed 'Intermixed Lands in Hom Lacy — Duke of Norfolk, Mr Bodenham and Mr Guy'

59a. in strips adjoining a stream [?in the W of the parish, SO 5534]. Field boundaries shown in outline.

Reference table gives plot numbers, field-names, land-use, acreage. Field-names include Fish Pool Meadow, Common Field, Goose Plunge, Fox-holes.

Scale bar.

SCALE: 3½ chains to 1 inch [1:2772].

SIZE: 382 x 540. Ink on paper.

REFERENCE: BL, Add. MS. 36307, G26

[8] ST WEONARDS
No title. Endorsed 'Troughland farm. Folio 58 in Map book. X[d] with New Survey & found correct'. 'St Weonards & Garway Parishes'

Trothland Farm [SO 479223] in the SW of St Weonards parish. Shows one house in perspective, enclosed fields, with outline boundaries, field-names and acreage, roads. Field-names include Great and Little Capendar.

SCALE: [c.1:6336].

SIZE: 321 x 387. Ink on paper.

REFERENCE: BL, Add. MS. 36307, G54.

n.d. [late 18th century]
HOLME LACY, ACONBURY, BOLSTONE

Endorsed 'Road from Hom Court to Coldecott'[1]

Outline sketch of the roadway [Green Drive] running through the Deer Park from Holme Lacy House to its junction with the Hereford-Ross road near Caldicott [at SO 524327]. Marks adjacent farms, including 'Hom Lodge' [about SO 545337]. From Newtown Grove [SO 532338] to Caldecott the drive, or proposed drive, ran parallel, on the E, to the public road.

No scale bar or N point.

SCALE: 8 chains to 1 inch [1:6336].

SIZE: 139 x 783. Ink on paper.

SURVEYOR: Not named.

REFERENCE: PRO, Kew, C115/103/7889.

1. Undated, no watermark. The plan is among Chancery Masters' Exhibits relating to the estates of Frances Duchess of Norfolk. The archive of her Scudamore family estates entered Chancery in 1816 when she was deemed lunatic. It would seem more likely to be associated with the changes in the Deer Park suggested by the previous map of Holme Lacy 1799 than Frizell's maps of 1771-5.

n.d. [late 18th century]
[KENTCHURCH]

No title. Endorsed 'Leeks Farm'.

Location unidentified, but in NE of parish adjoining Park Farm and Old House Farm [SO 4227].[1] Shows fields with field-names and a road.

SCALE: [c.1:4356].

SIZE: 190 x 235. Ink on paper.

SURVEYOR: Not named.

REFERENCE: HRO, M26/6/13 (Kentchurch Court estate papers).

1. Both place-names are recorded on the Kentchurch tithe map 1846.

n.d. [late 18th century]
CLODOCK [LONGTOWN]

A series of five maps of Longtown in Clodock by the same country land-measurer.

[1] 'Map of Gridal occupied by Benj. Titley'
246a. [Greidol Farm, SO 349273] in E of parish. Shows buildings in crude block plan, field boundaries with field-names, acreage and land-use, woods, ponds, quarries, some gates, names of some adjoining owners. Field-names include Try Kenol [Ty Canol], The Tumpin Mead [SO 353275].

Aligned with E at top.

[2] 'Trelandan'
161a. in E of parish [SO 347267]. Similar features. Shows a small plot for potatoes. Field-names include Spout Mead near farm house.

Aligned with W at top.

[3] 'Lower Keven'
64a. in E of parish [Lower Cefn, SO 367304]. Similar features.

[4] 'Upper Keven'
85a. in E of parish [Upper Cefn, SO 360310]. Similar features. Marks parish boundaries.

183

[5] 'Longtown Farm' (Plate 62)
79a. at S end of Longtown village [SO 326285]. Similar features. Marks the Crown inn, 'Pont Hendry' house and motte [SO 326282], roads including 'road to Lanthony' from Longtown [not a right of way in 2001]. Field-names include Pidgeonhouse Meadow, Mill Croft, Pont Annis [Pont yr Ynys], The Park, Blacksmiths Field.

Reference table, on a separate sheet, to lands of Gridal, Trelandon, Longtown, Lower Keven and Upper Keven farms. Gives plot numbers, acreage and tenants' names. Distinguishes woods.

None of the maps has N points.

SCALE: [c.1:25200].

SIZE: [1] 500 x 383; [2] 367 x 275; [3] 335 x 270; [4] 255 x 386; [5] 252 x 310. Ink on paper pasted on scraps of parchment deeds, of which [1] is dated 22 August 1691. All are crudely drawn in the same hand. [2]—[5] have been heavily repaired.

SURVEYOR: Not named.

REFERENCE: [1] HRO, B75/1; [2]—[5] HRO, R57/1-4.

n.d. [c.1800]
CANON FROME

'Plan of Radland Farm situate in the several Parishes of Canon Froome and Stretton Grandsome…in the Occupation of M.r William Lane'

137a. [SO 6544], mostly in N of Canon Frome between the road to Castle Frome and the river Frome, with small part in E of Stretton Grandison. Shows buildings in block plan, fields mostly enclosed, river Frome, Filly Brook, Mear Brook on parish boundary with Castle Frome. Marks Gains Cottage [SO 655443].

Reference table gives acreage, land-use, field-names, including Fish Pool, Stapla Meadow.

Scale bar. Ruled border.

SCALE: 3 chains to 1 inch [1:2376].

SIZE: 442 x 713. Ink and watercolour on parchment.

SURVEYOR: Not named.

REFERENCE: HRO, E3/457

n.d. [c.1800]
MADLEY

'PLAN of SHENMORE ESTATE'

c.104a [Bower Farm, SO 395381] and lands at Lower Shenmore in W of parish. Marks buildings in block plan, roads and lands in outline. Distinguishes land-use.

No scale. No N point.

SCALE: c.8 inches to 1 mile [1:6336].

SIZE: 270 x 365. Ink and watercolour on paper with no watermark.

SURVEYOR: Not named.

REFERENCE: HRO, AW 56/41.

Herefordshire Map-Makers before 1800

Herefordshire Map-Makers

THE BIOGRAPHY

Much of the following listing of surveyors active in Herefordshire before 1800 is derived from Susan Bendall, *Dictionary of land surveyors and local map-makers of Great Britain and Ireland 1530-1850*, London 1997, 2 vols. Just as almost every modern map in the kingdom is ultimately based upon the Ordnance Survey's work, so any study of British map-makers and their lives is now necessarily built upon Bendall's *Dictionary*. The debt is gratefully acknowledged. In borrowing so much from another scholar it is some small comfort to a former archivist that Bendall's definitive work was itself begun two generations earlier by Francis Steer, county archivist of West Sussex 1953-69, and owed much to the initial information supplied by the staff, including myself, in county record offices.

In the entries below, the biographical material taken from the *Dictionary* has been given first with Bendall's reference, for example, William Adams, (*Bendall* A031). Bendall's massive infrastructure of sources can therefore be easily followed up from the *Dictionary*. Among these R.V. and P.J.Wallis, *Biobibliography of British mathematics and its applications, part ii 1701-60*, Letchworth 1986, replaced by the enlarged and updated R.V. and P.J. Wallis, *Index of British mathematicians 1700-1800, part iii*, (Project for Historical Biobibliography) Newcastle 1993, hereafter *Wallis,* includes a number of the surveyors noted here. I found very little additional material not noted by Bendall apart from speculative dates of birth and death, which I too have not included. A new source, unavailable both to Bendall and myself, is the *Oxford Dictionary of National Biography*, which will replace the familiar *DNB* in September 2004. Locally, David Whitehead of Hereford has supplied me with invaluable biographical information taken from sources in the Herefordshire and Worcestershire record offices and the *Hereford Journal*, for which I am most grateful. It is acknowledged here as *ex inf.* Whitehead.

My own contributions come last, after the cross-references to *Bendall*. Further biographical information about those few surveyors who were based in the county has been obtained from the International Genealogical Index (IGI) compiled by the Mormons and the personal names indexes available in the Herefordshire Record Office. These led to evidence in Herefordshire parish registers and estate papers. In some cases a common name defied identification, in others no relevant evidence could be found. Some twenty surveyors not known to Bendall have been discovered.

Surveyors and maps recorded in this Catalogue are indicated in **bold.** The surveyors' place of residence and dates of birth and death, where known, are given first, followed by the dates and areas of their activity and concluding with other biographical information. A number of surveyors active in Herefordshire early in the nineteenth century have been included although no relevant maps by them before 1800 have been found.

ADAMS, William. Apparently of Shropshire. 1671-1702, Cheshire, Herefordshire and Shropshire. Worked with Christopher Adams, Mathias Asley and Walter Rose in Shropshire 1702. (*Bendall A031*). Brother of John Adams, a more significant figure who was educated at. Shrewsbury, barrister, instrument-maker and Savilian professor of astronomy, Oxford, d.1738. (*Bendall A019.9*).

ADDAMS, Samuel.
1758 Colwall.

AIRD, J, ?1772, Worcestershire. (*Bendall A048.5*). Of Worcester, surveyor to Worcester corporation 1763-84, schoolmaster and teacher of land surveying 1777, (*ex inf.* Whitehead),writing master 1789 (*Wallis*). Style reminiscent of the Dougharty family of Worcester 1716-55.
1775 Wacton; 1780 Mathon.

ALLEN, Elias. Of Spelsbury, Oxfordshire, 1616, gentleman. From 1601, Gloucestershire, Herefordshire, Worcestershire. Employed by the Crown. Died before 1637, (*Bendall, A069*).

ALLEN, Thomas. 1792-(?1830), Herefordshire, ?Worcestershire. (*Bendall A084*). Payments to him by Worcester corporation 1817-30 included map of Worcester 1823 (*ex inf.* Whitehead).
1792 Whitbourne.

BACH, John. Of Hereford, schoolmaster 1766 (*Bendall B004.5*). He advertised as a surveyor from 1770 in the *Hereford Times*, which reported his death on 3 August 1777 (*ex inf.* Whitehead). Paul Foley commissioned him to design the gardens at Stoke Edith in 1766 (*WNFC* 1980) and at Newport (Nieuport), Almeley in 1767 (*Whitehead 2001*). The Bach family occurs in north Herefordshire, especially in Leominster and Eyton from the 17th century. Among them was John Bach, bailiff of Leominster in 1733 and 1749, and Robert Bach of Lucton, described as a gentleman, who held a mortgage on the Plume of Feathers, Shobdon, in 1738; the property was sold to John, 2nd Viscount Bateman, in 1748 when Bach was described as a writing master. John Bach, the Hereford schoolmaster, may be the John Bach of St. Nicholas, Hereford, husband of Elizabeth, whose sons were baptised, John in 1765, William in 1772 and Robert in 1777. He was a competent surveyor, but drew somewhat stiff and boldly coloured maps in the heavy style of a school atlas. A distinctive feature was his practice of overlaying his maps with a grid.
1766 Stoke Edith gardens; **1767 Almeley**; **1768 Much Birch**; **1771 Monnington-on-Wye**; **1774 Widemarsh, Hereford.**

BRIDGEMAN, Charles. Of London, b.?1685, d.1738. Landscape gardener, Master-Gardener to George I and George II. Worked throughout south England (*Bendall B565*)
1722 Brampton Bryan.

BROOME, John. Of Worcester, Stanford on Teme, Astley. 1760-80, Staffordshire, Worcestershire (*Bendall B595*). Succeeded John Doharty, (*Bendall D270*) as surveyor of Worcester cathedral fabric 1755 (*ex inf.* Whitehead)
1758 Ledbury.

BROWN, Lancelot, 'Capability'. Of London, b.1715, d.1783. Landscape gardener, designing gardens and parks throughout south England and parts of north-east Wales (*Bendall B628*). Brown worked at Moccas and Monnington-on-Wye in 1778 and visited Eywood in 1775 and Berrington in 1780 (*Whitehead 2001*).

BUCKLE, Thomas. Of Kempsey, Worcestershire. 1794-5, Warwickshire (*Bendall B678*). Earliest commissions were in Herefordshire 1783-93.
1783 Donnington; **1788 Ullingswick**; **1790 Stoke Lacy**; **1793 Tedstone Delamere (2).**

BUDGEN, Charles. 1794-1822. From a family of surveyors. He was a military surveyor employed by the Ordnance Survey, working throughout south England and south Wales, including Herefordshire (*Bendall B685*).

BUDGEN, Thomas. 1788-1818. Apparently not related to Charles Budgen, but he too was employed by the Ordnance Survey, working throughout south England and south Wales, including Herefordshire. (*Bendall B688*).

BUISHOP, John. 1648-9, Herefordshire (*Bendall B689.7*). Received payment of 60s. 'for Surveying Alterennis [Alltyrynys, Walterstone] and the plotts thereof' for William Earl of Salisbury 1649 (Hatfield House, Receiver General's account book, 157/50). No map has been traced.

BURNS, A. n.d. Tyrone. Perhaps the same as Arthur Burns of Tarporley, Cheshire, school and writing master and instrument maker, 1771-6 (*Bendall B731*).
1774 Shobdon.

CAMPBELL, Colin or **Colen.** Born 1796, died in London 1729. Graduated at Edinburgh 1695. Lawyer, architect and garden designer (*Bendall C031*). Author of *Vitruvius Britannicus or, The British Architect*, 1717-25.
1731 Hope under Dinmore.

CARELESS, J. Joseph Careless of St Owen Street, Hereford, advertised 1776-79 as a carpenter and joiner, timber dealer and auctioneer. His son Joseph, timber dealer, went bankrupt 1797 (*ex inf.* Whitehead).
1787 Hereford.

CARPENTER, Robert. Of Ross, surveyor. Occurs as a subscriber to an unnamed mathematical publication 1783 (*Wallis*).

CLARK(E), Henry. Of Drayton, Pebworth and Shipston-on-Stour, Warwickshire. 1778-c.1825, Midlands including Herefordshire and once in co. Monaghan. Inclosure commissioner, estate agent. (*Bendall C220*).
1797 Much Marcle inclosure.

CLEER, Thomas. 1685-1706, east and south-east England, employed by the Crown 1705 (*Bendall C264*). A 'Mr Cleare' is reported to have surveyed Mynde Park, Much Dewchurch (see p. 37)
1700 Stoke Prior.

CLOWES, Josiah. Canal engineer (*Bendall C285.3*). 1737-1795. First engineer for the Thames & Severn canal 1783. Died during construction of the Herefordshire & Gloucestershire canal 1795.
1791 Herefordshire & Gloucestershire canal.

COLLET, Henry.
1721 Ledbury.

CORBET[T], J[ohn]. 1735-69, Northamptonshire, Shropshire, Warwickshire (*Bendall C440*).
1743 Richards Castle.

COVE, Morgan (Revd.). b.?1753, d.1830. Only recorded as a surveyor in Cornwall (*Bendall C482*). Of Sithney, Cornwall 1782-99, Eaton Bishop 1799-1830. LL.B Cambridge 1776, DCL Oxford 1810, prebendary of Gorwall and Overbury, Hereford 1801, chancellor of Hereford cathedral 1828.

CRANSTON, James (I). 1748-1835. Of King's Acre, Hereford, nurseryman and landscape gardener. Came from James Lee of Hammersmith, nurseryman, to be gardener to Uvedale Price at Foxley 1771. He supplied trees for Tyberton Court 1778-89 and worked later at other Herefordshire estates, including assisting Humphry Repton at Garnons. Married Elizabeth Barrett of Yazor 17 February 1785. Bought property at King's Acre 1785, where he established his nursery and lived at Archenfield House. In partnership with James Douglas of Ludlow, nurseryman 1787-89, with a shop in High Town, Hereford. Uvedale Price recommended him to Sir George Beaumont to lay out grounds of Coleorton, Leicestershire c.1803, when he was described as surveyor and nurseryman (*Whitehead 2001* and *ex inf.* Whitehead). James Cranston (II), of Kings Acre occurs as a parliamentary inclosure and tithe surveyor 1819-42 (*Bendall C517.5*) and also ran the King's Acre nursery. James Cranston (III) was an eminent architect in Birmingham, who also designed buildings in Herefordshire 1852-62 (*ex.inf.* Whitehead).
1794 Stretton Sugwas.

CRAWTER, Henry. Of Cheshunt, Bucks. 1796-1849, home counties, Gloucestershire, Herefordshire. Member of a dynasty of late 18th and 19th-century surveyors. (*Bendall C527*).

CROFT, Thomas.
1698 Garway.

CROOME, James. Of Breadstone in Berkeley and North Nibley, Gloucestershire. 1793-184(?6), Gloucestershire, Herefordshire. (*Bendall C561*).

DADFORD, Thomas (II). Of Cardiff 1794, Welshpool 1795. 1789-d.1806. Canal engineer, south Wales, Herefordshire. (*Bendall D010.1*).
1789 Kington canal.

DARLY, N. (*Bendall D039*). Probably of Worcester.
1763 Mathon.

DAVIES, David. Of Llangattock Court, Brecon 1830. 1795-1830, south Wales, Gloucestershire, Durham. Engineer and surveyor, mostly of public schemes, canals, railways (*Bendall D054*). A surveyor D. Davies mapped a farm in Longtown 1799 and in 1805-6 copied at a reduced scale Lord Abergavenny's estate maps of Ewyas Harold, Longtown and Walterstone made by Benjamin Fallowes 1718.
1799 Clodock.

DAVIS, John Lambe. Apparently of a family of estate stewards of Woburn, Bedfordshire. 1760-78, south and east England, Shropshire. Steward to 4th Duke of Bedford in Cambridgeshire 1761-2 (*Bendall D085*).
1772 Moccas; **?*c*.1775 Moccas**.

DAVIS, Thomas. Three surveyors of this name occur in the late 18th century. None had known Herefordshire connections. Thomas Davis of Moreton-in-Marsh, Gloucestershire, who occurs once in 1766, may be ruled out (*Bendall 092.5*). The map of Ocle Pychard 1791 might be attributed to Thomas Davis (I), b.1749, d.1807, of Horningsham, Wiltshire and steward to the 1st and 2nd Marquesses of Bath at Longleat. (*Bendall 094*). The Marquesses of Bath had property at Weobley. But it appears to be the work of an inexperienced country surveyor and might, therefore, be an early venture by Thomas Davis (II), also of Horningsham. He worked in Worcestershire, Monmouthshire and south-west England 1798-1825 (*Bendall D695*) and was probably responsible for the mature survey of Collington 1798.

However, the name is common in the Welsh Marches and the map of Ocle Pychard is more probably by a fourth, locally-based Thomas Davis. At Shobdon, for instance, Thomas son of Thomas and Susan Davis was baptised on 7 October 1764.
1791 Ocle Pychard; **1798 Collington**.

DAWSON, Robert, b.1771, d.1860. Of Devon and London. South-west England, Midlands including Herefordshire, Wales. (*Bendall D111.5*).

DAYNES, Richard. 1628-65, home counties, Gloucestershire, Herefordshire, Worcestershire. Employed by the Crown (*Bendall D124*).

DEELEY, William. ?1684-1701, Warwickshire (*Bendall D135*). Of Birmingham (*Wallis*). Surveyed Stoke Edith 1680 (*WNFC* 1980).

DOHARTY, John. Of Worcester (*Bendall D272*). b.1709, d.1773. Educated at King's School, Worcester, probably trained in London. Worcestershire, Gloucestershire, Warwickshire and other counties in S England. Prolific surveyor 1731-55. Teacher in Worcester with his father, John Dougharty and elder brother, Joseph Dougharty (see following entry). Their pupils included sons of Herefordshhire landowners. All had a similar style in surveying and draughtsmanship, distinguished by fields outlined in colour, ruled yellow borders and cartouches featuring a shell ornament and human face. For family biographies and catalogue of their maps see B.S. Smith, 'The Dougharty family of Worcester, estate surveyors and map-makers, 1700-60', *Miscellany 2*, Worcs. Hist. Soc., New series, vol. 5 (1967), pp. 138-80; B.S. Smith, 'The Dougharty family, eighteenth-century map-makers', *Transactions Worcs. Arch. Soc.*, 3rd series, vol. 15(1996), pp. 245-82; *Oxford DNB* (forthcoming 2004).
?*c*.1750 Pixley.

DOUGHARTY, Joseph. Of Worcester (*Bendall D272*). b.1699, d.1737. Educated at King's School, Worcester, trained in London. 1723-36, Herefordshire, Shropshire, Worcestershire. Topographical draughtsman, teacher with his father and younger brother, John Doharty. (see previous entry).
1732 Pencombe.

EDWARDS, John. ?1780-1829, Gloucestershire (*Bendall E047.8*). Possibly the surveyor at Yarkhill for the Dean and Chapter of Gloucester *c*.1771.
***c*.1771 Yarkhill**.

FALLOWES, Benjamin (I). Of Essex, living at Danbury 1714, Maldon 1715-1720 when he was described as a widower intending to remarry, and Purleigh, from at least 1726. He died before 1731 when the burial of Ann Fallowes, daughter of Benjamin Fallowes, deceased, was recorded in the Purleigh parish register. He appears to have been a Quaker. His surveys are dated within the period 1714-26, mostly in Essex, but also in Cambridgeshire, Herefordshire, Monmouthshire, Worcestershire. His clients included London-based owners, from one of whom he leased his farm in Purleigh (*Bendall F021*; A.S. Mason, *Essex on the map. The 18th century land surveyors of Essex*, Chelmsford 1990).

His links with London owners presumably led to the commission in 1720 to survey Broome Farm, Eardisland for William Bateman, the son of Sir James Bateman, alderman and lord mayor of London. William Bateman succeeded his father at Shobdon in 1718, was M.P. for Leominster 1721-22 and 1727-34 and was created Viscount Bateman in 1725. Dr Mason has suggested that Fallowes might have had London origins (*private communication*). A Benjamin Fellowes (*sic*), son of Benjamin and Mary Fellowes, was baptised at St Andrew's, Holborn on 18 April 1682; he must have died in infancy for a second Benjamin, born to the same parents, was baptised at St Andrew's on 2 November 1684.

Quaker registers in London have not been searched. Fallowes's style includes precise dating of his maps by day, month and year and decoration incorporating colourful flowers and birds.
1718 Clodock, Ewyas Harold; 1720 Eardisland.

The connection with the Batemans of Shobdon was maintained by Benjamin Fallowes (II), who was the estate steward to William 1st Viscount Bateman at Shobdon (1695-1744). Lord Bateman, his brother Richard, who supervised the management of the estate, and his son John, 2nd Viscount Bateman (1721-1802), maintained a voluminous correspondence with Fallowes 1739-63. This includes a letter referring to a map of The Marshes [Shobdon Marsh], 8 February 1753. Fallowes also acted as the Batemans' political agent locally.

With Lord Bateman's support he secured the appointment of under-sheriff of the county for his son, Benjamin Fallowes (III) 'the younger' of Leominster, in 1753, and the posts of clerk of the peace of Herefordshire 1752-97 and bailiff of the borough of Leominster 1762 and 1776. He may be identified as the Benjamin Fallowes who married Thomasin Watson at Leominster on 15 December 1757.

Their son Benjamin (IV), baptised at Leominster in 1758, was in turn described as Benjamin Fallowes, junior, in 1787. He married Jane Bullock at All Saints, Hereford on 29 April 1792 and died before 1835, having served as clerk of the peace of Herefordshire 1804-17 (E. Stephens, *The clerks of the counties 1360-1960*, 1961, p.100) and as steward of Wormelow manor in the early 19th century.

FORSTER, Richard (Revd.). Of Shefford, Bedfordshire. and Oxford, b.*c.*1704, d.1766. ?Essex, Herefordshire. Fellow 1732-49, vice-president 1744-45 and bursar 1746-48 of Brasenose College, Oxford 1744-48(*Bendall F147.5*).
1747 Leominster.

FOWLER, William. Of Pendleford, Staffordshire, b.*c.*1610, d.1664. Worked throughout the west. Midlands. (*Bendall F175*). Featured drawings of surveyors and instruments on his maps.
1662 Downton.

FRIZELL, Charles, (II). son of Charles Frizell (I) of co. Wexford, farmer. Worked in Ireland with his brother Richard 1758-68 (*Bendall F224*). This map, also with Richard Frizell, is his only recorded visit to England.
1771 Holme Lacy.

FRIZELL, Richard. son of Charles Frizell (I). Of cos. Wexford and Dublin. 1750-97, extensively in Ireland with his father and brother 1758-68 and in England. Agent to the 1st Marquess of Ely 1778 (*Bendall F225*). Surveyed Herefordshire estates of Charles Howard, 11th Duke of Norfolk.
1771-5 Holme Lacy estates; 1774-5 Holme Lacy estates; 1780 Holme Lacy estates.

GALLIERS, William. Of Presteigne, Radnorshire 1807-2(?4) and Leominster 1824-35. Herefordshire and Shropshire (*Bendall G015*). Member of north-west Herefordshire family, occurring especially in Leintwardine and associated with early breeding of Hereford cattle; dynasty of 19th-century surveyors of Leominster. This is his earliest known map. Died October 1839.
1799 Weobley.

GEORGE, John.
1726 Eastnor.

GETHIN, John. Of Brick House, Kingsland (*Wallis*), b.1757, d.1831. Mason, surveyor of bridges for Herefordshire (*Bendall G083*).

GILBERT, Job. (*Bendall G119*). The name is not common in Herefordshire and he has not been traced in parish registers. A family of Gilbert were tenants of the Scudamores of Kentchurch in Kentchurch and Llancillo 1752-54 (HRO, M26/8/132).
1709 Weston under Penyard.

GREEN, J. (I). There are three surveyors named J[ohn] Green associated with Bridstow but their relationship has not been established. John Green I occurs with reference to an untraced map made by him. Possibly father of J. Green II. (HRO, C99/III/256).

GREEN, J. (II). 1727-9 Herefordshire, Shropshire. (*Bendall G259*).
1727 Willersley;

GREEN, J. (III). 1755 Herefordshire (*Bendall 259.2*), with the query that he might be the same as J. Green (II). The gap of twenty-five years between the maps of 1727-9 and 1755 makes this unlikely.

In addition to the maps noted here John Green also surveyed, without maps, the outlying estates of Guy's Hospital

in west Herefordshire in 1766 (HRO, C99/III/244-5). A John Green of Bridstow, gentleman, occurs in a deed of 1769 (HRO, G87/8/29). About the same year a John Green is said to have been buried at Ross (IGI) but no entry has been found in the Ross parish registers. Probably the father of John Green IV.
1755 Bridstow; 1756 Bridstow; 1757 Little Marcle.

GREEN, John (IV). Stylistic differences confirm that he is not the same as J. Green (III) though it would seem likely that they were related. The name occurs in the parish registers of Ross and Brampton Abbots, but not Bridstow or Peterstow. The likeliest of these entries is that of the baptism of John, son of John and Mary Green of Ross, on 12 March 1754.
1788 Bridstow.

GREEN, Jona[h].
1729 Leominster.

GREEN, Walter. 1760-79 Herefordshire, Monmouthshire. (*Bendall G265*)
1761 Goodrich.

HALL, Richard (I). Of Cirencester and Gloucester, 1759-93, founder of the dynasty of Cirencester surveyors. Canal, estate and town (*Bendall H047*).
1791 Herefordshire & Gloucestershire canal.

HALL, Robert Wright. d.1815. Of Newent, Cirencester and Gloucester, son of Richard Hall (I), Estate and public works including canals in Glos., Wilts and S. Wales. (*Bendall H049*)
1796 Herefordshire and Gloucestershire canal.

HARDWICK, Thomas. of Brentford, Middlesex, surveyor (*Bendall H121.3*).
1784-5 Breinton.

HARRIS, John (III). The Harris family was long settled in Stoke Prior. John Harris (I), probably the surveyor's grandfather, was a churchwarden there in the 1730s. John Harris (II) (1723-84), the steward of the Hampton Court estate, was living at Wickton Court in Stoke Prior in 1761, though on his death in 1784 he was described as of Hampton Court.

John Harris III (*Bendall H144*), the son of John Harris II and his wife Sarah (d.1782), was baptised at Stoke Prior on 26 July 1753. He was active 1772-1811 in Herefordshire and Shropshire. He signed himself as of Newton, south of Leominster 1772-81, of Weston, Herefordshire 1784 and, following his father's death, of Wickton Court in Stoke Prior 1784-1813. He styled himself landmeasurer (1783) and gentleman. On 14 July 1786 he married Jane Eaton at Leominster, perhaps hurriedly arranged as their first-born, Samuel Joseph, was baptised at Stoke Prior on 12 November that year. Three further sons were born and baptised at Stoke Prior, John in 1789, William, who died in infancy in 1791, and William in 1794.

John Harris was employed by the dean and chapter of Gloucester in 1783-4, by the dean and chapter of Hereford in 1793 and 1798, by a range of private owners and by inclosure commissioners. He was one of the original shareholders in the Kington and Leominster Canal, 1791. In 1809 he was appointed to partition an estate at Belmont, Hereford between the four successful joint bidders at auction (HRO, C38/16/1, fols. 37-38). He worked on Bodenham inclosure with James Cranston, 1811-13 (HRO, R8/1/1). Later he was an inclosure commissioner for Wigmore 1828, when he was described as late of Wickton and now of Heath, presumably The Heath near Wickton, and Shobdon, Lingen and Aymestrey 1829. He was buried at Stoke Prior on 14 October 1829. Other members of the family continued as Herefordshire surveyors until the mid 19th century.
1772 Stoke Prior; 1773 Wickton; 1774 Westhide; 1777 Bredenbury; 1778 Hope under Dinmore; 1778 Ashperton; ?1780 Croft; 1781 Leintwardine; 1783 Pembridge inclosure; 1783 Ullingswick; 1784 Clehonger; 1784 Pipe and Lyde; 1784 Staunton-on-Wye inclosure; 1785 Holme Lacy; 1786 Ledbury; 1786 Stoke Prior etc; 1789 Madley, 1795 Eastnor; 1797 Much Marcle inclosure.

HARRIS, Joseph. Wiltshire 1780 (*Bendall H149*). ?Same as Joseph Harris of Stanford-on-Teme, Worcestershire. Assistant to Francis Webb of Stow-on-the-Wold, Gloucestershire. 1779-82. Surveyed timber for dean and chapter Hereford 1794, Upton Bishop demesne 1798 and The Mynde, Much Dewchurch with Cranston. Lets house at Wickton 1792 (*ex inf.* Whitehead).

HARRIS, Thomas. (*Bendall H155*). First cousin of John Harris (III) of Wickton. Moved from Worcester to Chertsey, Surrey 1783. Employed by the dean and chapter of Gloucester.
1783 Ullingswick.

HAVERFIELD, John. 1790 Oxfordshire (*Bendall H204.5*). Of Kew, Surrey 1784-5.
1784-5 Breinton.

HAYWOOD, John. Occurs in Hampshire and Northamptonshire 1776 (*Bendall H251*). Of London 1783. A John Haywood of Long Acre, London, architect, occurs in 1793.
1783 Breinton.

HILL, James. Barrister and antiquary. Of Hereford d.1727-8. (J. Cooper, 'Herefordshire' in C.R.J. Currie and C.P. Lewis, eds., *English county histories: a guide*, Stroud 1994, pp. 178-9).
1677 Hereford.

HILL, W. Probably William Hill 1676-?1719, throughout England, including Gloucestershire, Shropshire (*Bendall H377*).
1686 Welsh Newton.

HODGKINSON, John. Of Newport, Monmouth, ?of London 1809. b.1773, d.1861. Canal, road and railway engineer. West Midlands including Herefordshire and south Wales (*Bendall H413*).

HOPE, John. Of Rhayader, Radnorshire 1778 (*Bendall H490.1*).
1774 Brimfield.

HULL, W. Of Kempsey, Worcestershire. Designed court-room at Worcester guildhall 1820 (*ex inf.* Whitehead).
1775 Bosbury.

JONES, Meredith. Of Brecon 1756-7. 1743-66 south Wales, Herefordshire, Shropshire (*Bendall J175*). No biographical information known in 2001 by the NLW, Glamorgan, Gwent and Powys ROs, Brecon Library, Brecon Museum.
1754 Lyonshall; 1757 Aconbury; 1757 Stretton Sugwas; 1759 Hereford (2); 1760 Coddington; 1763 Brilley.

JONES, William. Possibly the Revd William Jones of Clungunford and Clun, Shropshire, schoolmaster, active 1757-87 (*Bendall J153.4*). Other less likely namesakes are recorded in *Bendall* and the name is, of course, common in the Welsh Marches.
1784 Hereford.

KENT, Nathaniel. Of Fulham, Middlesex and London, b.1737, d.1810. Worked throughout south England. Agriculturist and land valuer (*Bendall K088*). First-class surveyor and draughtsman. He also surveyed and valued the Hampton Court estate for Lord Maldon 1786- 87 (*Trans. WNFC*, vol. 37 (1961), p. 43, n. 63).
1772-74 Collington, Kyre Park estate; 1774 Yazor; 1787-90 Much Marcle.

KING, James. Of various addresses in Northamptonshire, d.1781. 17(?60)-1781 throughout Midlands (*Bendall K133*). But this map was perhaps by the James King whose family lived at Staunton Park, Staunton-on-Arrow, and who was probably responsible for the enclosure of the park *c.*1770-5 (*Whitehead 2001*).
1774 Almeley.

LAURENCE or **LAWRENCE, Edward.** Of Stamford, Lincolnshire. but of various addresses elsewhere especially in London, baptised 1674, d.1739. Mostly east Midlands and Yorkshire, but including Gloucestershire and Herefordshire (Goodrich and Credenhill) estates of the Duke of Kent 1716-19. Agriculturist and land valuer (*Bendall L071*). Published *The young surveyor's guide*, 1716, 1717, 1736, and *The duty of a steward to his lord*, 1727, reissued and revised as *The duty and office of a land steward*, 1731, 1743.
1717 Goodrich.

LEGGETT, Thomas. Of Ellesmere, Shropshire 1770. Nurseryman. 1769-70 Herefordshire landscape project (*Bendall L134*). Drew up a plan, not carried out, for Bartholomew Richard Barneby at Brockhampton Park, Bromyard 1769 in the style of William Emes (1730-1803), garden designer, with whom he worked elsewhere. He possibly worked for John Freeman II, Barneby's brother-in-law, at Gaines in Whitbourne *c.*1760s (*Whitehead 2001*).
1769 Brockhampton (Bromyard).

LIDIARD, J. Married Sophia Pool at Ledbury 17 June 1790.
***c.*1785, 1786, 1788 Ledbury.**

LUCAS, John. Of Clungunford, Shropshire 1803. 1792-1803 Herefordshire, Shropshire (*Bendall L282*). Probably the J. Lucas, surveyor at Monkland 1793.
1793 Monkland.

MADDY, J. Possibly a member of a Dorstone family. John, son of Joseph and Elizabeth Maddy, was baptised at Dorstone 24 March 1766.
1796 Madley.

MATTHEWS, Thomas. ?Of Ludlow, valuer. 1793. Unlikely to be Thomas Mathews, who occurs as a town surveyor in Somerset 1794 (*Bendall M243.2*)
1793 Eye.

MEREDITH, J. Of Shrewsbury 1739-45. 1728-35 Shropshire (*Bendall M305, M305.8*)
1733 Weobley.

MERRETT, John. Of Gloucester, 1768-85 Gloucestershire. Canal, estate, parish (*Bendall M311*).
1785 Lea.

MOORE, Ed[ward].
1721 Brockhampton (Ross); 1724 Much Dewchurch; 1726 Eastnor.

MYLNE, Robert. Of Edinburgh and London, b.1744, d.1811. Worked throughout England and Scotland including Herefordshire. Canal and harbour engineer (*Bendall M544*).

NORDEN, John. Of Middlesex, b.15(?48), d 1626. Worked throughout England and Wales, including Herefordshire. Employed by the Crown (*Bendall N113*).

PAIN, David. Of Lugwardine. 1800-(?19) Herefordshire and Shropshire. Worked with Joseph Powell of Sutton 18(?19) (*Bendall P023.7*). Surveyed Pipe and Lyde and Derndale for the dean and chapter of Hereford 1797 (maps not traced); inclosure commissioner for Yarkhill 1799 and Stoke Edith 1804; plan of Stoke Edith estate 1802 (*ex inf.* Whitehead). No reference in Lugwardine parish registers.
1795 Weobley; 1797 King's Pyon.

PENRY, Edward. A Longtown family of this name is recorded in the Clodock parish registers, but not an Edward Penry. The style of his colouring and floral decoration is imitative of Benjamin Fallowes of Maldon, Essex (see above).
1783 Clifford; 1794 Clodock (Longtown); *c.*1794 **Michaelchurch Escley; 1798 Clifford; 1799 Clifford.**

PERKINS, John. 1737. A man of the same name of (?Highmoor) Oxfordshire was an inclosure commissioner in Buckinghamshire 1742/3 (*Bendall P173*)
1737 Brockhampton (Bromyard).

POWELL, Joseph. 1765-1801 Gloucestershire, Herefordshire, Leicestershire, Shropshire, Worcestershire. Of Bridgnorth 1785-95 and Highley, Shropshire 1795 (*Bendall P306*). Signed himself Joseph Powel 1765 and J. Powell of Bridgnorth 1782.

He worked with his son, not otherwise identified, 1789-(?95) (*Bendall P299.3*), who might have been Joseph Powell of Sutton, Herefordshire, who occurs 1819 (*Bendall P306.5*).
1765, 1772 Thornbury; 1775 Little Hereford; ?1777 Moreton-on-Lugg; 1782 Eyton; *c.***1785 Preston-on-Wye; 1791 Blakemere; 1791 Bosbury; 1791 Kingstone.**

PRICE, Charles. Of London, b.1680, d.1733. Worked throughout England, including Worcestershire. Instrument maker, engraver, publisher. (*Bendall P324*), hydrographer to William III (*Wallis*).
1720, 1730 Ledbury.

PRICE, John. Of Hereford 1743, teacher, writing master, drainer (*Bendall P328.7*).

PRICE, Robert Parry. 1744-69 Wales, Herefordshire. Opened school at Hay, Brecon 1765. (*Bendall P330*).
1769 Leominster; 1769 Titley.

PYE, John. 1695-1704 Gloucestershire, Herefordshire, Wiltshire. Employed by the 11th Earl of Kent 1700 (*Bendall P382*). John Pye of Kilpeck, a lawyer and probably an estate steward, was connected with the Pye family which owned The Mynde in Much Dewchurch from the 16th century until its sale to James Brydges first Duke of Chandos in 1723. The Duke sold it on to Richard Symons of London in 1727. The previous year John Pye had written to Richard Symons to explain the measurement of customary and statutory acres in Herefordshire. Apparently an associate of Geoge Smyth, surveyor. Owned property in Much Dewchurch, Treville and St Devereux. His will was proved 11 September 1731 (*Robinson*, pp. 93-6; *Whitehead 2001*, pp. 362-4; Mynde Park papers, 964, 1466).
1695 Holme Lacy; *c.***1695 Garway; 1704 Goodrich.**

RICHARDSON, Richard. Of Darlington, co. Durham 1770-8, Cirencester, Gloucestershire 1778, Wiltshire 1778-93, Bath 1786-1801, London 1800-26. Worked all over England and Wales including. Herefordshire with various partners including his son (*Bendall R147*).

SANKEY, R. 1775 Herefordshire (*Bendall S040.5*).
1775 Elton.

SHERRIFF, James. Of Birmingham 1793-1812, Liverpool c.1816-30. 1774-1830 west Midlands and south England. Artist, land agent, landscape gardener. Possibly father and son of the same name. (*Bendall S198*). The surname occurs earlier in north Herefordshire. An accomplished surveyor and draughtsman, he also surveyed the Foley family's Stoke Edith estate in a fine series of maps (*Annual report of the Friends of the National Libraries*, 2002, p.43).
1778 Leinthall Earls; **1780 Downton**.

SILVESTER, John. Hereford, where he proposed redevelopment of Hereford Castle site.
1677 Hereford.

SMITH, John. None of the other mid 18th-century John Smiths worked in Herefordshire (*Bendall*).
c.1750 Pembridge.

SMITH, Thomas. Two Thomas Smiths were active at this time, one in Kent and Sussex 1756-98 (*Bendall* S356), the other of Shrivenham, Berkshire, who worked more widely 1768-78, mostly in south England including Gloucestershire (*Bendall S357*). This map for the dean and chapter of Hereford may be attributed to the latter.
1776 Breinton.

SMYTH, George. There appears to be more than one surveyor or land agent of this name (*Bendall S372.8*).

George Smyth surveyed woods at Holme Lacy in 1723 for Lady Scudamore. The map has a printed border pasted on in the style of John Pye of Kilpeck, who referred to him in a letter in 1726 as a surveyor of woods at The Mynde, Much Dewchurch (NLW, Mynde Park papers, 1466). He continued surveying in Herefordshire until 1742. A map of Uffculme, Devon by George Smyth, dated c.1750, also has a printed border pasted on (M.R. Ravenhill and M.M. Rowe, eds., *Devon maps and map-makers: manuscript maps before 1840*, Devon and Cornwall Record Society, new ser., vol. 45 (2002), p. 364), making it probable that this is by the same surveyor. The same or another George Smyth surveyed two Norfolk estates in 1732 and 1734 (*ex inf.* Norfolk RO).

It is possible that the Herefordshire surveyor, was the same as, or related to, George Smyth, agent to John Scudamore of Kentchurch in 1752-54, when he was a tenant of Old House Farm, Rowlestone and Upper House Farm, Llancillo and Llancillo Forge (HRO, M26/8/132). He has not been identified in Herefordshire parish registers, but in view of the early connection with John Pye it may be noted that a George Smith or Smyth, perhaps a relative, was baptised at Kilpeck in 1662.
1723 Holme Lacy; **1728 Goodrich**; **1733 Llangarron**; **1741 Much Marcle**; **1742 Bromyard**; **1742 Madley**.

STONE, John. Of Severn Stoke 1768, Pull Court, Queenhill 1778-93, Chambers Court, Longdon 1800-13, all Worcestershire. Active as estate surveyor and inclosure commissioner before 1768 to 1813 in Herefordshire and Worcestershire. A man of the same name occurs in Leicestershire 1765-80 and Shropshire 1778 (*Bendall S352*)

TAYLOR, Isaac. Of Ross 1754-88, b.c.1720, d.1788. Worked throughout S England 1750-77, especially on printed county maps. (*Bendall T049*). Has been confused in the past with Isaac Taylor of London and Essex, engraver and watercolourist, b.Worcester 13 December 1730, the son of William, a brassfounder (IGI; A.S. Mason, *Essex on the map*, Chelmsford 2000; H. Mallalieu, *The dictionary of British watercolourists up to 1920,* Woodbridge 2002, vol. 2, p. 210), d. Edmonton, Middlesex 1807 (*Bendall T048.8*).

Isaac Taylor of Ross was probably born c.1720, but his place of birth and training have not been traced. Isaac, son of Thomas and Martha Taylor was baptised at St. Nicholas, Hereford on 14 October 1720 but the same names can be found in other church registers at that period. For instance, in London, where he might have trained, as he was already a competent surveyor and draughtsman when he published his first town plans in 1750, Isaac son of Isaac Taylor of Hand Alley was baptised at New Broad Street Presbyterian chapel, 27 March 1718 and Isaac son of John and Ann Taylor was baptised at St. Botolph's, Bishopsgate, 19 August 1718. (IGI indexes of parish registers in other counties where he had early connections have been scanned, namely Dorset, Hampshire, Oxfordshire, Staffordshire and Worcestershire, without finding a relevant baptism). An Isaac Taylor drew a maritime chart of Portsmouth and the Isle of Wight as a navigation student's exercise in 1739, but he was not a student at the Royal Academy of Portsmouth. Nothing is known to connect him with Isaac Taylor of Ross (*ex inf.* Gill Rushton, Hampshire RO).

Taylor's first plans of the towns of Oxford, surveyed in 1750 and engraved rather crudely by G. Anderton 1751 for publication by W. Jackson of Oxford 29 October 1751 (BL, Maps 4735/10) and Wolverhampton, surveyed 1750 and engraved by Thomas Jefferys 1750 (Wolverhampton Archives) were skilful and mature works.[1] His first county map, that of Herefordshire 1754, shows that he was subsequently influenced by Henry Beighton's map of Warwickshire 1728, re-issued 1750.

From the date of his printed maps of Herefordshire 1754 and Hereford city 1757 he frequently signed himself as 'of Ross', and sometimes wrote his name as J. Taylor. The cartouche of his map of Hereford shows a map-maker at work, perhaps intended as a self-portrait (Plate 29). He married Eleanor Newman by licence at Ross 23 December 1759. Their two daughters died in infancy, Mary Newman Taylor, baptised 24 February and buried 13 April 1765, and Elizabeth, baptised 1 March 1766 (when Isaac Taylor was described as 'geographer') and buried 7 November 1768, all at Ross. From 1757 Taylor spent much of the following twenty years working away from home. By 1776 he and his wife were living in some comfort at 54-55 High Street, Ross, where part of the property was empty (HRO, L78/3, Ross churchwardens' accounts). The dwelling house was insured with the Sun Life insurance company in 1780 for £200 and the remainder, comprising a stable, brewhouse with an adjoining and still untenanted house, for a further £200. Their household and other goods were insured for £100, including a harpsichord for £25 (*ex inf.* Heather Hurley, from Sun Life archives, Guildhall Library, London). Beneath its floor boards the present owner of the house discovered a pair of heavy iron dividers (Plates 26, 27). Taylor was buried at Ross on 17 June 1788. A grave inscription has not been found. His will was not proved at Hereford. The death of his widow, Elizabeth, has not been traced. (P. Hughes and H. Hurley, *The story of Ross-on-Wye*, Almeley 1999; A.S. Mason, *Essex on the map: the 18th century land surveyors of Essex*, Chelmsford 1990, pp, 25, 89; Ross parish registers).

Taylor was both an estate surveyor and a printed map engraver and publisher, describing himself as 'Author of the MAP of Herefordshire' on his undated estate map of Canon Pyon and advertising on his printed maps his ability and willingness to survey estates. His printed maps of Herefordshire 1754, reissued 1786 and 1757, were followed by Hampshire 1759, reissued 1795-96, Dorset 1765, Worcestershire 1772, reissued 1800, and Gloucestershire 1777, reissued 1800. None of his maps was awarded the premiums offered by the Royal Society of Arts from 1759, critics attributing this failure to his draughtsmanship as an estate surveyor. (T. Chubb, 'A descriptive catalogue of the printed maps of Gloucestershire, 1577-1911', *Transactions of the Bristol & Gloucestershire Archaeological Society for 1912*; P. Laxton ed., *Two hundred and fifty years of map-making in the county of Hampshire*, Lympne 1976; Delano-Smith & Kain, pp. 89-91, 96-7, 264).

His earliest known estate maps were in Hampshire 1757 (Forton in Wherwell and Lower Clatford, Hants. RO, 33M49/1) and undated 1754-59 (Laverstoke, Hants. RO, photocopy 701). Between 1765 and 1777 he spent much time in Devon and Dorset where he compiled atlases containing maps of the estates of the Sturt family of Crichel in Dorset (39), Devon (14), Wiltshire (3) Cambridgeshire (3), Essex (1) and Middlesex (1) 1765-70 and of the Erle-Drax family of Charborough in Dorset (19) 1773-77. These maps remain in private hands but there are copies in the British Library. (*Catalogue of photographs of old cadastral and other plans of Great Britain*, Ordnance Survey, Southampton 1935 (BL, Maps.A.5.(15); S. Bendall's data sheet for entries in the *Dictionary of land surveyors*; A.S. Bendall, *Maps, land and society. A history, with a carto-bibliography of Cambridgeshire estate maps*, c.*1600-1836*, Cambridge 1992, pp. 93, 180, 231-2, 337; M.R. Ravenhill and M.M. Rowe, eds.*Devon maps and map-makers. Manuscript maps before 1840*, Devon and Cornwall Record Society, vols. 43, 45 (2002); P. Laxton, ed., *Two hundred years of map-making in the county of Hampshire*, Lympne 1976, n. 81 and *ex inf.* by letter 2 Dec. 2003). Before and after these absences Taylor made a few surveys in Herefordshire and in 1780 advertised as agent for the sale of property and timber at Wilton and Peterstow (*ex inf.* Whitehead).

c.1755-60 Canon Pyon; 1763 R. Wye; 1764 Preston-on-Wye; c.1780 Madley.

1. The town plans of Oxford and Wolverhampton are signed merely 'Isaac Taylor'. Their lavish dedications and the inclusion of framed views of buildings in the style of Isaac Taylor of Ross strongly indicate them to be his work rather than that of Isaac Taylor of Brentwood, Essex, who would have been only twenty at the time. But in the absence of further proof the attribution is arguable. Isaac Taylor of Brentwood did not leave his birthplace of Worcester, relatively close to both Oxford and Wolverhampton, until *c.*1752 and after heading for London he was then employed by Josiah Jefferys, a cutler of Brentwood, whose daughter Sarah he married in 1754. Josiah was brother of Thomas Jefferys, the London engraver and Geographer to the Prince of Wales, subsequently George III, who engraved the map of Wolverhampton.

THOMAS, E[dward]. Of Margam. 1772-97 south Wales, Herefordshire (*Bendall T110*).

WARD, J. A writing master and teacher of mathematics and navigation called John Ward occurs at Leominster 1747 (*Wallis*).
1759 Monkland.

WARD, Thomas. A mathematician, Thomas Ward of Leominster, is recorded in the periodical *Gentleman's diary* 1760 (*Wallis*).
1767 Eyton.

WATKINS, Benjamin. 1791 Brecon and Herefordshire, possibly still active in 1833 (*Bendall W144*)

WHITTELL, William. Of Bodenham 1705-8 (*Bendall W317*) A Whittell family had owned property at Wharton in the late seventeenth century, which they sold to Thoms Lord Coningsby of Hampton Court (HRO, A63/II, p. 10). William Whittell, probably the surveyor's grandfather, was vicar of Aymestrey in 1669 when he performed a marriage service at Orleton. Thomas Whittell was vicar of Bodenham from 1687, where he was buried 1710. His widow Margaret was buried there 1722.

Their two sons, Thomas and William were baptised at Orleton in 1678 and on 2 July 1684. The whereabouts of William Whittell, the surveyor, has not been traced after 1709. His brother Thomas was instituted as vicar of Yarpole in 1718 (Parish registers of Bodenham, Orleton and Yarpole).
1705 Ocle Pychard; **1708 Aymestrey**; **1709 Kingsland**; **1709 Wigmore**.

WHIT[T]LESEY, Robert. 1724-(?56) throughout south England including Gloucestershire, Herefordshire, Ireland (*Bendall W321*).
1729 Holme Lacy.

WHITWORTH, Robert. Of London, b.1734, d.1799. Canal engineer throughout Britain (*Bendall W324*).
1779 Wye navigation; **1796 Herefordshire and Gloucestershire canal**

WILLIAMS, Daniel. (*Bendall W365*).
1758 Goodrich; **1758 Whitchurch**.

WILLIAMS, John. Of Worcester. The owner of the property rather than a surveyor.
1768 Much Birch (2).

WILLIAMS, Mathew. Of Llangadog and Rhosmaen, Carmarthen. b.?1732, d.?1819. Brecon, Carmarthen., Gloucestershire (*Bendall W381*) and Llandeilo, Carmarthen (*Wallis*).
1785 Welsh Newton.

YELD, Edward.
1799 Leominster.

YOUNG, George. Of Mealcheapen Street, Worcester, 1798. b.1750, d.1820. Throughout west Midlands, mostly public schemes. Member of the Society of Civil Engineers (*Bendall Y019*). A teacher of mathematics (*Wallis*). Of Silver Street, 1790-92 and 52 Broad Street, Worcester 1802. Worcester maps and Worcester bridge surveys 1769-81(*ex inf.* Whitehead).
1795 Ledbury.

INDEX

Figures in **bold** indicate the principal entries referring to the authors of the printed maps and the place-names of manuscript maps listed in Part Two. The biographies of the surveyors, arranged alphabetically by personal names on pp. 188-97, have not been indexed again here. With a few exceptions the names of engravers and publishers of printed maps, of landowners who commissioned estate maps, and of places outside Herefordshire have not been indexed.

Aconbury 12, 18, 48, 97, **122**, Plate 49
Acton Beauchamp 18
Addams, Samuel 36, 53
agriculture 7, 8, 10-12, 141-4, 146-7, 151, 158-60, 183
Aikin, J. 90
Aird, John 37, 47, 149, 160
Allen, T. 174
Almeley 10, **127-8**, **144**
Alltyrynys, *see* Walterstone
Amberley **136**, **148**, **160**
archaeological sites 26, 81; *Ariconium*, Kenchester 22, 65, 66, 68, 82, 90; Arthur's Stone 12, 27, 140, 142, 176; iron age forts 88, 90, 109, 111, 114, 115, 119, 142, 156, 162, 166, 173; King Arthur's Cave 123; Roman roads 81, 161; tumps 82, 128, 167. *See also* boundaries
Ashperton **152**
Aston Ingham 167
Aston, Pipe 94, 156
Aubrey, Richard 33
Avenbury 171
Aymestrey 35, **103-4**, **130**, **145**, **156**, **174**, **178**, Plate 31

Bach, John 47-8, 128, 133-4, 145, Plate 51
Badeslade, Thomas **77-8**
Baker, Benjamin **90**
Bakewell, Thomas 74
Ballingham **99-100**, **134**, **136**, **137**, 138, **148**, **158-9**
Baskerville family 7, 9, 19
Bassett, Thomas 66, 68
Bateman family 84. *See also* Shobdon
Bateman, Sir James 33, 44
Bateman, William, 1st Viscount Bateman 46
Beighton, Henry 26, 49
Benning, Richard 48-9, 84
Bickham, George 5, **80**
Bill, John 23, **69**
Birch, Little **135**, **176**
Birch, Much 47, **128-9**, Plate 51
Birley **181**. *See also* Hope-under-Dinmore

Bishopstone **131-2**, **147**
Black Mountains (Hatterrall Hill) 5, 6, 18, 32, 83
Blaeu, Joannes 23-4, **69-70**, 71, Plate 13
Blakemere 10, 52, **112**, **172**
Blome, Richard 24, 25, **71-2**, **74**, Plate 15
Bockleton 142, **143**
Bodenham 52, 53, **105**, 153, **159-60**
Bolstone **100**, **111**, **134**, **136**, **182-3**
Bosbury 47, 52, **149**, **172-3**, Plate 61
Boswell, Henry (alias Alexander Hogg) 83, **87**, Plate 18
boundaries (ditch and hedge, merestones) 36, 94, 105, 108, 113, 114, 132, 140, 152, 167, 169. *See also* Wergins Stone
Bowen, Emanuel 27, 28, **76**, 77, **80**, **83-4**, **85-6**, **86-7**, Plate 7
Bowen, Thomas 80, **86-7**
Bowes, William 23, 65, 69
Bowles, Carrington 73, 79, 84, 85, 86, 87
Bowles, John 79, 83-4
Bowles, Thomas 83-4
Bowyer, W. 75
Brampton Abbots **159**
Brampton Bryan 12, 20, 32, **110**
Bredenbury 52, **151**
Bredwardine 12, 13, **139-40**, **149-50**, Plate 42
Breinton 33, **133**, 150, **162**, **166-7**
Bridge Sollers 5
Bridgeman, Charles 110
bridges 13, 20, 22, 24, 28, 36, 42, 72-3, 82, 88, 89, 90, 96, 103, 105, 108, 110, 111, 116, 120, 121, 124, 126, 140, 149, 150, 153, 157, 160, 164, 170, 175
Bridstow 37, **120**, **121**, 125, **136**, **170**, Plates 48, 59. *See also* Wilton
Brilley 32, 33, **94-5**, **125-6**
Brimfield **144-5**
Bringewood and Mocktree (Downton) 7, 9, 20, 32, 33-4, 37, 39, 44, 89, **94**, **97**, **155**, Plates 30, 37, 53
Brinsop 132
Brobury 172
Brockhampton (Bromyard) 39, **116**, **129**
Brockhampton (Ross) 9, **99**, **109**, **159**

199

Bromyard 6, 7, 8, 10, 12, 18, 43. *See also* Brockhampton, Linton, Norton
Broome, John 123
Buckle, Thomas 162, 171, 171-2, 175
buildings 7-8, 12-13, 15, 27, 36-7, 43, 44, 47, 48, 53, 81-2, 84, 95, 96, 98, 103, 104, 108, 112-3, 115, 121, 122, 149; as conventional signs 94-184 *passim*. *See also* castles, monasteries
Burghill 122, 132, 176
Burns, A. 130, 145-6
Burrington **94**, **104-5**
Byford **132**, **147**
Byton **120**, **127**

Camden, William 3, 21-2, 23, 24, 28, 65, 66, 69, 70, 75, 89
canals 14, 27, 53, 82, 88, 89, 157, **171**, **173**, 174, **178-9**
Caple, How 109
Caple, King's **99-100**, **138-9**, **159**
Careless, J. 170
Carpenter, George, 2nd Baron 33, 78, 116
Cary, John 28, 29, **88**, 89, **89-90**, Plates 8, 9
castles 7, 8, 20, 22, 46, 48, 64, 67, 78, 87, 94, 96, 98, 99, 102, 105, 106, 107, 110, 113, 118, 120, 121, 132, 140, 170; moated sites 111, 116, 117, 126, 132, 140, 147, 172, 181; mottes 103, 129, 138, 144, 161, 164, 179, 184
Cecil, William, 1st Baron Burghley 18-21, 28, 32, 64
Chaplin, Robert 33, 43
Chiswell, Richard 66, 68
cider 10, 14, 81, 83, 158. *See also* orchards
Clark, Henry 179
Clark, John 10, 11, 12, 13
Cleer, Thomas 38, 43, 102
Clehonger 52, **164-5**
Clifford 6, 7, !8, 54, 132, **140**, **151**, **163-4**, **176**, **179**, **180**
Clodock, *see* Craswall, Longtown
Clowes, Josiah 53, 173
Coddington 38, 122, **124-5**, **149**
Collett, Henry 110
Collington **142-3**, **180**
Colwall 36, 53, **113**, **122**, **123**, **160**
Coningsby family 10, 12, 41, 47, 51. *See also* Hampton Court (Hope-under-Dinmore), Pencombe
Coningsby, Thomas, 1st Lord 43, 74, 111, 112
conventional signs 20, 22, 26, 35, 36-7, 43, 49, 64, 80, 81
Corbet, John 35, 39, 117-8, Plate 35
Cornewall, Sir George, bt. 50
Cowarne, Little **171**
Cowarne, Much, *see* Stoke Lacy

Cowley, John 24, **78**
Cradley 18, **160**
Cranston, James 34, 52, 176, Plate 63
Craswall **139-41**
Craven, William, 1st Baron Craven 97
Credenhill 44-5, 122, 176
Croft **118-9**, **154**
Croft, Thomas 44, 101
Cusop 12, **139-40**
Cwmyoy (Gwent) 6, 95

Dadford, Thomas 53, 171
Darley, N. 126
Davies, David 107-8, 180
Davies, John 130
Davis, John Lambe 37, 50-1, 139-41, 149-50, Plates 41, 42
Davis, Thomas 174, 180
Dean, forest of 18, 32, 94, 105
decoration 21, 23-4, 38-40, 46, 48, 51-2, 107-8, 109, 111, 114, 116, 121, 129, 153, 154-7, 161, 164, 166, 172, 175
Devereux, Price, Viscount Hereford 77, 79
Dewchurch, Little **100**, **134**, **137**, **139**, **148**, **159**, **182**
Dewchurch, Much **111**. *See also* Mynde
Dicey, Cluer 66, 71
Dilwyn **116**, **121**, 132, **157**, **160**
Doharty, John 119
Donnington **109**, 133, **162**, **177**
Dore, Abbey 9, 18, 46, **159**
Dorstone 12, 20, **139-40**, **149-50**, **176**
Dougharty family 35, 40, 47
Dougharty, Joseph 12, 36, 47, 115, Plate 47
dovecotes (pigeon houses) 102, 110, 111, 113, 115, 117, 122, 123, 131, 132, 144, 156, 161, 169, 184
Downton 9, 13, 33, 40, 51, **97**, **154-7**, Plates 52, 53. *See also* Bringewood
Drayton, Michael **68**
Dulas 20

Eardisland 46, **108-9**, 116, Plate 38
Eardisley **114**, **144**, **157**
Eastnor 12, **112-3**, **113**, 123, **135**, **177**
Edwyn Ralph **142-4**
Ellis, Joseph 28, **86**
Elton 94, **149**, **156**
enclosure 11-12; parliamentary inclosure awards **145**, **162-3**, **166**, **179**
Ewyas 18
Ewyas Harold 12, 46, **107-8**, Plate 32

Eye **150**, **174-5**
Eyton **128**, **151-2**, **162**

Faden, William 82
fairs and markets 15, 25, 79, 83, 86, 98
Fallowes, Benjamin (I) 34, 40, 45-6, 67, 107-9, 168, Plates 32, 38, 40
Fallowes, Benjamin (II and III) 46
Fawley 37
ferries (boat) 13, 14, 28, 82, 88, 89, 105-6, 114, 132, 134, 137-8, 150, 153, 158, 159, 164
Fiennes, Celia 13
Foley family 12, 33, 41
Ford, *see* Stoke Prior
Forster, Richard 118
Fownhope 6, **161**. *See also* Brockhampton (Ross)
Fowler, William 37, 39, 44, Plate 30
Fownhope 99
Foxley 13, 42, 50, **130-3**, **146-7**
Foy **159**
Frizell, Richard 33, 50-1, 134-9, 148, 158-60
Frome, Bishop's **118**
Frome, Canon **184**

Galliers family 12, 53
Galliers, William 180-1
gallows (hangman) 94, 106, 165, 178
Ganarew 105
gardens 96, 98, 103, 104, 107, 110, 111, 114, 116, 121, 123, 127, 129, 131, 134, 150, 164
Garway 40, 43, 44, **101**, **102**, Plate 44
geology 6
George, John 112-3
Gibson, John **85**
Gilbert, Job 44, 104
Gloucester, dean and chapter 34
Golden Valley 5, 7, 8, 20, 27, 34
Goodrich 10, 12, 20, 37, 43, 44-5, 94, **101**, **102**, **105-6**, **114**, **123**, 125, Plate 45. *See also* Symond's Yat
Green, John (II) 35
Green, John (III) 48, 120, 121, 122, Plate 48
Green, John (IV) **170**, Plate 59
Green, Walter 125
Grendon Warren 12, 36, 47, **115**, Plate 47
Grey, Henry, Duke of Kent 10
Guy's Hospital 11, 32, 33, 34, 41, 48, 52, 122, 181

Hall & Son 173
Hall, Richard 173
Hall, Robert Wright 179
Hampton Charles 142
Hampton Court 12, 43, 50, 51, **111**
Hardwick, Thomas 166-7
Harley family 10, 12, 32. *See also* Brampton Bryan
Harley, Edward, 3rd Earl of Oxford 84
Harris, John 51-2, 53, 141, 151, 152, 154, 161-2, 163, 164, 166, 167, 169, 169-70, 171, 177, Plates 55-8
Hatterrall Hill, *see* Black Mountains
Haverfield, John 166-7
'Haversham' 44, 106-7
Hay-on-Wye (Powys) 13, 18
Haywood, J. **89**, 162
Hentland **100**, **125**, **137**, **148**, **159**
heraldry 19, 23, 25, 26, 35, 64, 67, 68, 70, 72, 74, 77, 78, 79, 81, 83, 84, 103-4, 105, 109, 111, 116, 150, 158
Hereford 5, 6, 7-8, 10, 15, 17, 18, 34, 79, 94, 105, 110; castle 7, 22, 67, 84, 90, 95-6, **98**, **105**, 110-1; city plans 8, 14-15, 22-3, 27, 42, 48, **67**, **70**, **81**, **84-5**, **90**, **95-6**, **181**, Plates 10, 11, 28; communications 13-15; dean and chapter 31, 33, 49, 51, 52, 53, 127, 133, 161, 162, 164-5, 165, 166, 171; economy 7-8, 15; estate maps **110** (2), **124**, **165** (2); friaries 22, 67, 84, 95-6; population 15, 84
Hereford, earls of 23, 67
Hereford, Little **149**
Hickes (Hyckes), Richard 22, 65
Hill, James 105
Hill, W. 98, Plate 43
Hoarwithy, *see* Hentland
Hogenberg, Remigius 21, 64
Hogg, Alexander, *see* Boswell, Henry
Hole, William 21, **66-7**, 68
Holme Lacy 9, 10, 12, 32, 33, 43, 47, 50-1, **97**, **100**, **111**, **114**, **134**, **135-9**, **158-60**, **167**, **181**, **182-3**
Holmer **133**, **148**, **159**, 166
Hondius, Jodocus 22, 67, 70
Hope, John 144-5
Hope-under-Dinmore 52, **105**, **152**, **156**.
   See also Hampton Court
hops (bines) and hopyards 10-11, 14, 81, 100-82 *passim*.
   See also vineyards
Howard, Charles, 11th Duke of Norfolk 33, 50, 52.
   See also Scudamore, Frances
Hull, William 47, 149

Humber **160**
Humble, George 22, 23, 66, 67, 71
hundreds 22, 28, 65, 67, 70, 72, 74, 75, 77, 78, 80, 81, 83, 86-7, 88, 89, 102
Huntington **132**
Hutchinson, Thomas 24, **79**

industries: brick-making, clay pits 9, 36, 117, 119, 121, 122, 130, 131, 140, 146, 148, 155, 156, 158, 162, 170, 171, 178, 181; cloth industry (fulling or tuck mills) 8, 102; coal pits (?charcoal) 109, 121; forges, furnaces 26, 36, 81-2, 89, 96-7, 99, 106, 123, 148, 153-4, 155, 159, 162, 171, 173; glass-making 9; gloving 8; lime kilns and pits 9, 36, 53, 98, 102, 109, 116, 118, 119, 123-4, 126, 140, 148, 153, 155, 156, 160, 180, 182; mining 81, 106, 124; paper-making 82, 97; potteries 9; oven 121; smithies (blacksmith's shops) 36, 106, 107, 124, 126, 131, 132, 143, 152, 155, 161, 171, 184; tanning, tanhouses 9, 84, 111, 121. *See also* mills, quarries
inns 24, 72, 82, 88, 113, 119, 121, 167, 184
instruments (drawing, surveying) 23, 24, 35, 38-9, 39, 42, 44, 81, 90, 97, 155, 162
Ivington, *see* Leominster

Jansson Jan 23-4, **70-1**, Plate 4
Jefferys, Thomas 25, 26, 48, 65, **79**, Plate 17
Jenner, Thomas 23, 69, Plate 12
Jones, Meredith 27, 33, 34, 38, 48, 120, 122, 124, 125, 126, 127-8, Plates 36, 49
Jones, William 165

Keere, Pieter van den 23, **66**, 69
Kenchester 22, 176
Kent, Nathaniel 13, 34, 42, 50-1, 131, 141-4, 146-7, 170
Kentchurch **183**
King, James 144
Kingsland **104**, **156-7**
Kingstone **173**
Kington 8, 13, 14, 171
Kinnersley **144**
Kitchin, Thomas 25, 27, 28, 48, 73, **79**, 80, 81, **83**, 85, **86**, **87**, Plate 17
Knight, Richard Payne 13, 15, 33, 51, 154-7
Kyre Park (Worcs.) 34, 50-1, **141-4**
Kyrle, John 5, 80, 81

Langeren, Jacob van 23, **69**, Plate 12
latitude and longitude, *see* meridian

Laurence, Edward 35, 37, 44-5, 105-7, Plates 45, 46
The Lea **167-8**
Ledbury 8, 10, 14, 18, 33, 47, 52, 53, **109**, **110**, **113**, **123**, **133**, **168-9**, 172, 173, **177**, 178, Plates 33, 60. *See also* Wellington Heath
Leggett, Thomas 39, 129
Leinthall Earls **116**, **152-3**
Leinthall Starkes **104-5**, **156**
Leintwardine 10, 18, **156**, **161-2**, Plate 54
Leominster 8, 14, 18, 34, 37, **115**, **118**, **129**, **142**, 157, **169-70**, 180
Leybourne, William 26, 35
Lhuyd, Humphrey 18, 20
Lidiard, J. 14, 168-9, Plate 60
Lingen **146**
Linton (Bromyard) **117**
Linton (Glos.) 9
Livers Ocle, *see* Ocle Pychard
Llangarron 43, 52, **100**, 105, **115**, **138**, **148**, 159
Llanwarne **100-1**, **136**, 159
Lodge, John 89
Longtown 7, 37, 40, 46, 54, **107**, **175**, **180**, **183-4**, Plates 40, 62, 64
Ludford (Shropshire) 94
Lugwardine **133**
Lyonshall 7, **120**, **127-8**, **130**, 144

Maddy, J. 178
Madley 10, 43, 48, **117**, **161**, **171**, **178**, 184
Malvern Hills 5, 7, 18
Mansel Lacy 10, 11, **131-2**, 147
Mansell Gamage **132**, 146
mappa mundi 3, 17
Marcle, Little **122**
Marcle, Much 6, 10, 11, 12, 20, 22, 32, 34, 43, 50, 83, **117**, **168**, 170, **173**, **179**. *See also* Yatton
Marden 9, 10, 11, 32, 35, **94**, **111-2**, 150-1, **160**. *See also* Amberley
Marshall, William 5, 10, 11, 13, 15
Marstow **125**
Mathon 18, 37, 47, **126**, **160**
Matthews, Thomas 174-5
measures and scales xv; area (statutory, customary) 37-8, 93, 104, 122, 125, 126, 128, 133, 152, 166, 174, 176; distance 23, 24, 26, 37, 42, 43, 63, 72, 73, 74, 77, 81, 83; scales 31, 37, 42, 63, 72, 75, 76, 81
members of Parliament 8, 25, 76, 77, 79, 83, 86
Mercator, Gerard 18, 66

Meredith, J. 35, 47, Plates 34, 39
meridian (longitude) 23, 25, 69, 73, 74, 75, 76, 77, 78, 79, 81, 88, 89
Merrett, John 166-7
Michaelchurch Escley 20, 54, 83, 86, **176-7**
Middleton-on-the-Hill **170**
milestones 122
mills, water 8, 26, 36, 67, 81-2, 84, 88, 95, 96, 97, 99, 100-84 *passim*
mills, wind 81, 82, 109, 113, 116, 138, 159, 170, 172
Moccas 6, 34, 37, 50-1, **139-40**, **149-50**, 172, Plate 41
Mocktree Forest, *see* Bringewood
Moll, Herman 24, 25, 76, **77**
monasteries and friaries 7, 22, 67, 84, 95-6, 102, 106, 130, 141, 161
Monkland **124**, **156**, 175
Monnington-on-Wye 40, 48, **133-4**
Moore, Edward 37, 109, 111, 113
Morden, Robert 24, 25, 37, 74, **75**, **76**, 77, 80, 83, 85, Plate 16
Mordiford 18
Moreton-on-Lugg **152**
Munsley 172
The Mynde 34, 43

Neville, George, 11th Baron Bergavenny 40, 45-6
Newland (Glos.) 105, 167
Newton (Leominster), *see* Croft
Newton, Welsh 7, 20, **98**, **101**, **136**, **159**, **168**, Plate 43
Noble, E, **89**, Plate 9
Norton (Bromyard) **161**
Nowell, Laurence 19

Ocle Pychard 32, 44, **102**, **174**
Ogilby, John 11, 13, 24-5, 28, 42, **72-3**, 76, Plate 5
open fields and meadows 11, 26, 32, 33, 34, 36, 44, 81-2, 102-82 *passim*
orchards 10, 36, 98, 102-80 *passim. See also* cider
Ordnance Survey 4, 25, 26, 28-9, 37, 80, 82, 88
Ortelius, Abraham 18, 22
Overton, John and Henry 24, 68, **71**, 83, 85
Owen, John **76**

Pain, David 52, 177-8, 179
Paris, Matthew 17
parks 7, 13, 20, 64, 67, 81-2, 94, 97, 107, 113, 115, 116, 117, 118, 127, 134, 137, 140, 142, 143, 144, 145, 149, 150, 153, 155, 158, 164, 178, 182-3, 183, 184

Parry family 7, 19
Pembridge 8, 9, **96-7**, **198**, **162-3**
Pencombe **115**, **119**, **163**, **170**
Penry, Edward 37, 40, 54, 163-4, 175, 176, 179, 180, Plate 64
Perkins, John 116
Peterchurch 9, **139-40**, **150**
Peterstow 50, **120**, **121**
Pipe and Lyde 52, **166**, Plates 55-8
Pixley **119**
playing cards 23, 25, **65**, **73-4**
Pontrilas 9
population 7, 8, 15, 84
Powell, Joseph 50, 52, 127, 149, 152, 162, 168, 172, 173, Plate 61
Preston-on-Wye 49, 52, **127**, **168**, **172**
Price, Charles 33, 47, 109, Plate 33
Price, Henry 4, 9, 14, 26, 28-9, 80, 82, 152
Price, John **90**
Price, Robert Parry 129-30
Price, Uvedale 13, 15, 50, 130-3, 146-7, 152-3
prisons 96, 106
Pye, John 34, 37-8, 40, 42-3, 99-101, 102, Plate 44
Pyon, Canon 9, 12, 27, 49, **121**
Pyon, King's 52, 132, **179**, 181
Pytts, Edmund 50

quarries (quarell) 82, 99, 101, 107, 109, 118, 122, 124, 125, 127, 139, 140, 144, 148, 151, 156, 159, 164, 182, 183

Rea, Roger 66, 68
Read, Thomas **78**
relief 20, 24, 26, 28, 37, 44, 53, 64, 65, 78, 79, 81, 83, 88, 89, 90, 105-6, 123, 146, 150, 151, 152, 153, 162, 173, 179, 180
Richards Castle 11, 12, 32, 35, 39, **94**, **98**, **117**, Plate 35
rivers 6, 7, 8, 14, 18, 20, 28, 36, 94-181 *passim. See also* Wye
roads 13, 17, 20, 22, 24-5, 26-7, 28, 36, 42, 49, 72-3, 74, 76, 77, 81-2, 83, 88, 89, 90, 94-184 *passim*; road books 23, 24-5, 69, 72-3, 76-7, 90
Rocque, John 3, **78**, 79
Ross-on-Wye 5, 6, 8, 13, 15, 18, 26, 28, 48-50, 80, 81, 94, 105, 120, 121
Rowlstone 37, **151**
Rudd, John 19

203

St Weonards 9, **101**, **139**, **148**, **159**, **160**, 183
Sankey, R. 149
Sarnesfield 181
Saxton, Christopher 3, 5, 9, 13, 18-22, 25, 27, 41, 42, **64-6**, 75, Plates 1, 2
Sayer, Robert 65, 73, 79, 83-4, 85, 86
scales, *see* measures
schools 96, 111, 153
Scudamore family 7, 9, 10, 12, 19, 32, 41, 43, 99. *See also* Holme Lacy
Scudamore, Frances (Howard, Duchess of Norfolk) 33, 50, 135
Scudamore, John, 1st Viscount 9, 10, 12, 74
Seckford, Thomas 19, 64
Sellack **129**
Seller, John 24, 25, 47, **74-5**, Plate 6
Senex, John 28, 73
Sheldon, Ralph 22
Sherriff, James 13, 33, 51, 152-3, 154-7, Plates 52-4
ships (trows) 14, 81, 84, 110, 120, 123
Shobdon 33, 34, 35, 43, 46, **103**, **145-6**
Shucknall 6
signposts 81, 82
Silvester, John 98
Simpson, Samuel **79**
Smith, C. 28, 89
Smith, John 119
Smith, Thomas 150
Smyth, George 38, 43, 111, 117
Society for the Arts 25, 49
Speed, John 14, 22-3, 24, 27, 42, 66, **67-8**, 70, 71, 74, 75, 76, **96**, Plates 10, 11
Staunton-on-Wye **166**
Stoke Bliss **142**
Stoke Edith x, 12, 33
Stoke Lacy 132, **171-2**
Stoke Prior 10, 43, **101-2**, **141**, **169-70**
Stretford **156**
Stretton Grandison 13, **119-20**, 184
Stretton Sugwas 11, 27, 34, 52-3, **122**, 166, **176**, Plates 36, 63
Sudbury, John 22, 66, 67
Sutton St Michael and Sutton St Nicholas **111-2**. *See also* Wergins
Symond's Yat 7, 9, 14, 37, 43, 44, 53, 106, 154
Symons, Richard 37-8, 43, 53

Tarrington **102**
Taylor, Isaac, 3, 13, 14, 48-50, 80, 121, 126, 127, 161, Plates 26, 27; county map 9, 11, 25-7, 28, 29, 31, 48, 63, **80-2**, 83, 85, 88, 89. Plates 19-25; Hereford city map 14, 15, 48-9, **84-5**, Plates 28, 29; self-portrait 49, 84, Plate 29
Taylor, Thomas 74
Taynton (Glos.) 113
Tedstone Delamere **175**
Thomas, Edward 37, 151
Thornbury 52, **127**, **141**, **143**
Titley **129-30**
Tomkins family 8, 12
Toms, William Henry 77-8
Tonson, Jacob 33, 47
Tretire **101**, **148**, **159**
Tyberton 172

Ullingswick 11, **163**, **171**
Upton Bishop 52, **159**

Vaughan, Rowland 7, 8, 9, 34
vineyards 82, 90, 100, 109, 156, 165, 178. *See also* hopyards

Wacton 47, **142-3**, 149
Wales (Welsh) 5, 6, 7, 11, 18, 20, 32, 64, 67, 83, 144
Walford 105, **159**
Wallwyn, Edward 11
Walterstone (Alltyrynys) 9, 18, 19, 32, 34, **95**, **108**, Plate 3
Ward, John 47, 124
Ward, Thomas 47, 128
Wathen, James 14, 15
Web, William 23, 65
Wellington 111, 121
Wellington Heath **115**
Weobley 8, 18, 33, 34, 35, 47, 52, **116**, **120**, **177-8**, **179**, **180-1**, Plates 34, 39
Wergins (Sutton St Nicholas) 165, 178; Wergins Stone 111
Westhide **147**
Weston-under-Penyard 22, 44, **104**, 167
Whitbourne 10, **174**
Whitchurch 9, 105, **123-4**, Plate 50. *See also* Symond's Yat
Whitney-on-Wye 6, 13, 54, **163-4**
Whittell, William 33, 35, 38, 39, 43, 102-4, 105, Plate 31
Whittlesey, Robert 47, 114
Whitworth, Robert 14, 53, 153-4, 179

Wigmore 6, 7, 8, 10, 18, 20, 37, **97**, **104-5**, **114**
Wildey, George 65
Willersley 35, **114**
Williams, Daniel 37, 53, 123, Plate 50
Williams, John, 128-9
Williams, Mathew 168
Wilton 13, 14, 18, 48, 50, 105, **120**, **121**, **170**
Withington 111
woodland and forests 7, 9, 11, 18, 20, 22, 33-4, 36, 89, 94, 97, 103, 104-5, 111, 114, 122, 123-4, 127, 141, 158-9, 164, 176; timber 9, 14, 107-8, 133, 158, 167. *See also* Dean, forest of

Woolhope 6, 7, 20
Wye, river 5-7, 13-14, 15, 17, 18, 25, 49, 53, 82, 89, 90; river features, 106, 120, 121, 123, **126**, 127, 133, 140, 150, **153-4**, 163-4, 166, 170. *See also* bridges, ferries.

Yarkhill **134**
Yarpole **112**
Yatton **159**
Yazor **131**, **146**
Yeld, Edward 180
Young, George 177